PROGRESS IN CLINICAL AND BIOLOGICAL RESEARCH

Series Editors
Nathan Back
George J. Brewer

Vincent P. Eijsvoogel
Robert Grover
Kurt Hirschhorn

Seymour S. Kety
Sidney Udenfriend
Jonathan W. Uhr

Vol 1: **Erythrocyte Structure and Function,** George J. Brewer, *Editor*
Vol 2: **Preventability of Perinatal Injury,** Karlis Adamsons and Howard A. Fox, *Editors*
Vol 3: **Infections of the Fetus and the Newborn Infant,** Saul Krugman and Anne A. Gershon, *Editors*
Vol 4: **Conflicts in Childhood Cancer,** Lucius F. Sinks and John O. Godden, *Editors*
Vol 5: **Trace Components of Plasma: Isolation and Clinical Significance,** G.A. Jamieson and T.J. Greenwalt, *Editors*
Vol 6: **Prostatic Disease,** H. Marberger, H. Haschek, H.K.A. Schirmer, J.A.C. Colston, and E. Witkin, *Editors*
Vol 7: **Blood Pressure, Edema and Proteinuria in Pregnancy,** Emanuel A. Friedman, *Editor*
Vol 8: **Cell Surface Receptors,** Garth L. Nicolson, Michael A. Raftery, Martin Rodbell, and C. Fred Fox, *Editors*
Vol 9: **Membranes and Neoplasia: New Approaches and Strategies,** Vincent T. Marchesi, *Editor*
Vol 10: **Diabetes and Other Endocrine Disorders During Pregnancy and in the Newborn,** Maria I. New and Robert H. Fiser, *Editors*
Vol 11: **Clinical Uses of Frozen-Thawed Red Blood Cells,** John A. Griep, *Editor*
Vol 12: **Breast Cancer,** Albert C.W. Montague, Geary L. Stonesifer, Jr., and Edward F. Lewison, *Editors*
Vol 13: **The Granulocyte: Function and Clinical Utilization,** Tibor J. Greenwalt and G.A. Jamieson, *Editors*
Vol 14: **Zinc Metabolism: Current Aspects in Health and Disease,** George J. Brewer and Ananda S. Prasad, *Editors*
Vol 15: **Cellular Neurobiology,** Zach Hall, Regis Kelly, and C. Fred Fox, *Editors*
Vol 16: **HLA and Malignancy,** Gerald P. Murphy, *Editor*
Vol 17: **Cell Shape and Surface Architecture,** Jean Paul Revel, Ulf Henning, and C. Fred Fox, *Editors*
Vol 18: **Tay-Sachs Disease: Screening and Prevention,** Michael M. Kaback, *Editor*
Vol 19: **Blood Substitutes and Plasma Expanders,** G.A. Jamieson and T.J. Greenwalt, *Editors*
Vol 20: **Erythrocyte Membranes: Recent Clinical and Experimental Advances,** Walter C. Kruckeberg, John W. Eaton, and George J. Brewer, *Editors*
Vol 21: **The Red Cell,** George J. Brewer, *Editor*
Vol 22: **Molecular Aspects of Membrane Transport,** Dale Oxender and C. Fred Fox, *Editors*
Vol 23: **Cell Surface Carbohydrates and Biological Recognition,** Vincent T. Marchesi, Victor Ginsburg, Phillips W. Robbins, and C. Fred Fox, *Editors*

Vol 24: **Twin Research,** Proceedings of 2nd International Congress on Twin Studies
Walter E. Nance, *Editor*
Published in 3 Volumes:
- Part A: Psychology and Methodology
- Part B: Biology and Epidemiology
- Part C: Clinical Studies

Vol 25: **Recent Advances in Clinical Oncology,** Tapan A. Hazra and Michael C. Beachley, *Editor*

Vol 26: **Origin and Natural History of Cell Lines,** Claudio Barigozzi, *Editor*

Vol 27: **Membrane Mechanism of Drugs of Abuse,** Charles W. Sharp and Leo G. Abood, *Editors*

Vol 28: **The Blood Platelet in Transfusion Therapy,** G.A. Jamieson and Tibor J. Greenwalt, *Editors*

Vol 29: **Biomedical Applications of the Horseshoe Crab (Limulidae),** Elias Cohen, *Editor-in-Chief*

Vol 30: **Normal and Abnormal Red Cell Membranes,** Samuel E. Lux, Vincent T. Marchesi, and C. Fred Fox, *Editors*

Vol 31: **Transmembrane Signaling,** Mark Bitensky, R. John Collier, Donald F. Steiner, and C. Fred Fox, *Editors*

Vol 32: **Genetic Analysis of Common Diseases: Applications to Predictive Factors in Coronary Disease,** Charles F. Sing and Mark Skolnick, *Editors*

Vol 33: **Prostate Cancer and Hormone Receptors,** Gerald P. Murphy and Avery A. Sandberg, *Editors*

Vol 34: **The Management of Genetic Disorders,** Constantine J. Papadatos and Christos S. Bartso, *Editors*

Vol 35: **Antibiotics and Hospitals,** Carlo Grassi and Giuseppe Ostino, *Editors*

Vol 36: **Drug and Chemical Risks to the Fetus and Newborn,** Richard H. Schwarz and Sumner J Yaffe, *Editors*

Vol 37: **Models for Prostate Cancer,** Gerald P. Murphy, *Editor*

Vol 38: **Ethics, Humanism, and Medicine,** Marc D. Basson, *Editor*

Vol 39: **Neurochemistry and Clinical Neurology,** Leontino Battistin, George A. Hashim, and Abel Lajtha, *Editors*

Vol 40: **Biological Recognition and Assembly,** David S. Eisenberg, James A. Lake, and C. Fred Fox, *Editors*

Vol 41: **Tumor Cell Surfaces and Malignancy,** Richard O. Hynes and C. Fred Fox, *Editors*

Vol 42: **Membranes, Receptors, and the Immune Response: 80 Years After Ehrlich's Side Chain Theory,** Edward P. Cohen and Heinz Köhler, *Editors*

Vol 43: **Immunobiology of the Erythrocyte,** S. Gerald Sandler, Jacob Nusbacher, and Moses S. Schanfield, *Editors*

Vol 44: **Perinatal Medicine Today,** Bruce K. Young, *Editor*

Vol 45: **A Century of Mammalian Genetics and Cancer, 1929–2029: A View in Midpassage,** Elizabeth S. Russell, *Editor*

Vol 46: **Etiology of Cleft Lip and Cleft Palate,** Michael Melnick, David Bixler, and Edward D. Shields, *Editors*

Vol 47: **New Developments With Human and Veterinary Vaccines,** Avshalom Mizrahi, Israel Hertman, Marcus A. Klingberg, and Alexander Kohn, *Editors*

Vol. 48: **Cloning of Human Tumor Stem Cells,** Sydney E. Salmon, *Editor*

Cloning of Human Tumor Stem Cells

Cloning of Human Tumor Stem Cells

Editor
SYDNEY E. SALMON, MD

Professor of Medicine
Head
Section of Hematology / Oncology
and
Director
University of Arizona Cancer Center
Arizona Health Sciences Center
Tucson, Arizona

Alan R. Liss, Inc. • New York

Address all Inquiries to the Publisher
Alan R. Liss, Inc., 150 Fifth Avenue, New York, NY 10011

Copyright © 1980 Alan R. Liss, Inc.

Printed in the United States of America

Under the conditions stated below the owner of copyright for this book hereby grants permission to users to make photocopy reproductions of any part of or all of its contents for personal or internal organizational use, or for personal or internal use of specific clients. This consent is given on the condition that the copier pay the stated per copy fee through the Copyright Clearance Center, Inc. 21 Congress Street, Salem, MA 01970, as listed in the most current issue of "Permissions to Photocopy" (Publisher's Fee List, distributed by CCC, Inc) for copying beyond that permitted by sections 107 or 108 of the US Copyright Law. This consent does not extend to other kinds of copying, such as copying for general distribution, for advertising or promotional purposes, for creating new collective works, or for resale.

Library of Congress Cataloging in Publication Data
Main entry under title:
Cloning of human tumor stem cells.

(Progress in clinical and biological research; 48)
Includes bibliographical references and index.
1. Cancer cells. 2. Cloning. I. Salmon, Sydney E. II. Series. [DNLM: 1. Clone cells — Congresses. 2. Drug evaluation — Congresses. 3. Neoplasms, Experimental — Congresses. 4. Cell transformation. Neoplastic — Congresses. W1 PR668E v. 48 / QZ206 C644 1989–80]
RC267.C58   616.99'4027   80-19600
ISBN 0-8451-0048-3

Dedication

I dedicate this book to the fine people who have helped me or participated in these efforts to develop the field of human tumor cloning. This includes the outstanding staff of the University of Arizona Cancer Center, the authors of the chapters, who also served as faculty for our workshops, and my wife, Joan, and our five children.

Faculty — First Workshop on Human Tumor Cloning Methods
January 3–5, 1979   Tucson, Arizona

1. Thomas Moon, PhD
2. Jeffrey Trent, PhD
3. Brian Durie, MD
4. David Alberts, MD
5. Ronald Buick, PhD
6. Daniel Von Hoff, MD
7. Ms Dale Curtis
8. Frank L. Meyskens, Jr., MD
9. Anne Hamburger, PhD
10. Ms Laurie Young
11. Ms Barbara Soehnlen
12. Sydney E. Salmon, MD

Faculty — Second Workshop on Human Tumor Cloning Methods

January 3–5, 1980 Tucson, Arizona

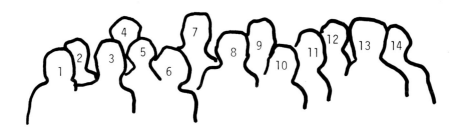

1. Lois Epstein, MD
2. Mr. Roger Morton
3. Mark Rosenblum, MD
4. Robert Ozols, MD
5. Ronald Buick, PhD
6. Anne Hamburger, PhD
7. Thomas Moon, PhD
8. David Alberts, MD
9. Jeffrey Trent, PhD
10. Daniel Von Hoff, MD
11. Thomas Stanisic, MD
12. Mr. Alex Martens
13. Sydney E. Salmon, MD
14. Frank L. Meyskens, Jr., MD

Contents

CONTRIBUTORS xv
PREFACE xix

I. GENERAL PERSPECTIVES

1. Background and Overview
   Sydney E. Salmon 3
2. In Vitro Clonogenicity of Primary Human Tumor Cells: Quantitation and Relationship to Tumor Stem Cells
   Ronald N. Buick 15

II. STUDIES OF VARIOUS TUMOR TYPES

3. Development of a Bioassay for Human Myeloma Colony-Forming Cells
   Anne W. Hamburger and Sydney E. Salmon 23
4. Soft-Agar Cloning of Cells From Patients With Lymphoma
   Anne W. Hamburger, Stephen E. Jones, and Sydney E. Salmon .. 43
5. Blast Colony Formation in Myelogenous Leukemia: Drug Sensitivity and Renewal Capacity of the Leukemic Clone
   Ronald N. Buick 53
6. Development of a Bioassay for Ovarian Carcinoma Colony-Forming Cells
   Anne W. Hamburger, Sydney E. Salmon, and David S. Alberts .. 63
7. Soft Agar-Methylcellulose Assay for Human Bladder Carcinoma
   Thomas H. Stanisic, Ronald N. Buick, and Sydney E. Salmon .. 75
8. Human Melanoma Colony Formation in Soft Agar
   Frank L. Meyskens, Jr. 85
9. Cloning of Human Neuroblastoma Cells in Soft Agar
   James T. Casper, Jeffrey M. Trent, and Daniel D. Von Hoff ... 101
10. Initial Experience With the Human Tumor Stem Cell Assay System: Potential and Problems
    Daniel D. Von Hoff, Gary J. Harris, Gloria Johnson, and Daniel Glaubiger 113

III. SPECIALIZED STUDIES RELEVANT TO IN VITRO TUMOR CLONING OF HUMAN TUMORS

11. Variables in the Demonstration of Human Tumor Clonogenicity: Cell Interactions and Semi-Solid Support
    Ronald N. Buick, and Sydney E. Salmon 127

12. Morphologic Studies of Tumor Colonies
    Sydney E. Salmon 135
13. Cell Kinetic Analysis of Human Tumor Stem Cells
    Brian G.M. Durie and Sydney E. Salmon 153
14. Cytogenetic Analysis of Human Tumor Cells Cloned in Agar
    Jeffrey M. Trent 165
15. Use of an Image Analysis System to Count Colonies in Stem Cell Assays of Human Tumors
    Bernhardt E. Kressner, Roger R.A. Morton, Alexander E. Martens, Sydney E. Salmon, Daniel D. Von Hoff, and Barbara Soehnlen 179

IV. MEASUREMENT OF DRUG SENSITIVITY AND CLINICAL CORRELATIONS

16. In Vitro Drug Assay: Pharmacologic Considerations
    David S. Alberts, H.-S. George Chen, and Sydney E. Salmon ... 197
17. Quantitative and Statistical Analysis of the Association Between In Vitro and In Vivo Studies
    Thomas E. Moon 209
18. Clinical Correlations of In Vitro Drug Sensitivity
    Sydney E. Salmon, David S. Alberts, Frank L. Meyskens, Jr. Brian G.M. Durie, Stephen E. Jones, Barbara Soehnlen, Laurie Young, H.-S. George Chen, and Thomas E. Moon 223
19. Human Ovarian Cancer Colony Formation: Growth From Malignant Washings and Pharmacologic Applications
    Robert F. Ozols, James K.V. Willson, and Robert C. Young 247
20. Chemosensitivity Testing for Human Brain Tumors
    Mark L. Rosenblum 259
21. Further Experience in Testing the Sensitivity of Human Ovarian Carcinoma Cells to Interferon in an In Vitro Semisolid Agar Culture System: Comparison of Solid and Ascitic Forms of the Tumor
    Lois B. Epstein, Jen-Ta Shen, John S. Abele, and Constance C. Reese 277
22. Applications of the Human Tumor Stem Cell Assay to New Drug Evaluation and Screening
    Sydney E. Salmon 291

V. AFTERWORD

23. Perspectives on Future Directions
    Sydney E. Salmon 315

VI. APPENDICES

1. Standard Laboratory Procedures for In Vitro Assay of Human Tumor Stem Cells
   Barbara Soehnlen, Laurie Young, and Rosa Liu 331

2. An Enzymatic Method for the Disaggregation of Human Solid Tumors for Studies of Clonogenicity and Biochemical Determinants of Drug Action
   Harry K. Slocum, Z.P. Pavelic, and Y.M. Rustum 339

3. Protocols of Procedures and Techniques in Chromosome Analysis of Tumor Stem Cell Cultures in Soft Agar
   Jeffrey M. Trent 345

4. Tabular Summary of Pharmacokinetic Parameters Relevant to In Vitro Drug Assay
   David S. Alberts and H.-S. George Chen 351

Index 361

Contributors

John S. Abele [277]
 Moffitt 576, University of California at San Francisco, San Francisco, CA 94143

David S. Alberts [63, 197, 223, 351]
 Associate Professor of Medicine, University of Arizona Cancer Center, Arizona Health Sciences Center, 1501 N. Campbell Avenue, Tucson, AZ 85724

Ronald N. Buick [15, 53, 75, 127]
 Ontario Cancer Institute, 500 Sherbourne Street, Toronto, Ontario, Canada M4X 1K9

James T. Casper [101]
 Midwest Children's Cancer Center, Medical College of Wisconsin, Milwaukee Children's Hospital and Blood Center, Milwaukee, WI 53200

H.-S. George Chen [197, 223, 351]
 Research Associate, University of Arizona Cancer Center, Arizona Health Sciences Center, 1501 N. Campbell Avenue, Tucson, AZ 85724

Brian G. M. Durie [153, 223]
 Associate Professor of Medicine, University of Arizona Cancer Center, Arizona Health Sciences Center, 1501 N. Campbell Avenue, Tucson, AZ 85724

Lois B. Epstein [277]
 Associate Professor, Department of Pediatrics and Research Associate, Cancer Research Institute, University of California at San Francisco, San Francisco, CA 94143

Daniel Glaubiger [113]
 Pediatric/Oncology Branch, Building 10, Room 3B04, National Cancer Institute, 9000 Rockville Pike, Bethesda, MD 20205

Anne W. Hamburger [23, 43, 63]
 Cell Culture Department, American Type Culture Collection, 12301 Parklawn Drive, Rockville, MD 20852

Gary J. Harris [113]
 Department of Medicine, Division of Oncology, University of Texas Health Sciences Center, 7703 Floyd Curl Drive, San Antonio, TX 78284

Stephen E. Jones [43, 223]
 Professor of Medicine, Department of Internal Medicine, Arizona Health Sciences Center, 1501 N. Campbell Avenue, Tucson, AZ 85724

Gloria Johnson [113]
 Pediatric/Oncology Branch, Building 10, Room 3B04, National Cancer Institute, 9000 Rockville Pike, Bethesda, MD 20205

The bold face number in brackets following each contributor's name indicates the opening page number of that author's paper.

xvi / Contributors

Bernhardt E. Kressner [179]
Bausch and Lomb Corporation, 820 Linden Avenue, Rochester, NY 14625

Rosa Liu [331]
Research Associate, Cancer Center Division, University of Arizona Cancer Center, Arizona Health Sciences Center, 1501 N. Campbell Avenue, Tucson, AZ 85724

Alexander E. Martens [179]
Vice President, Image Analysis, Bausch and Lomb Corporation, 820 Linden Avenue, Rochester, NY 14625

Frank L. Meyskens, Jr. [85, 223]
Assistant Professor of Medicine, University of Arizona Cancer Center, Arizona Health Sciences Center, 1501 N. Campbell Avenue, Tucson, AZ 85724

Thomas E. Moon [209, 223]
Head, Biostatistics, Cancer Epidemiology, and Processing of Information Services Unit (BICEPS), Assistant Director, University of Arizona Cancer Center, Arizona Health Sciences Center, 1501 N. Campbell Avenue, Tucson, AZ 85724

Roger R. A. Morton [179]
Bausch and Lomb Corporation, 820 Linden Avenue, Rochester, NY 14625

Robert F. Ozols [247]
Senior Investigator, Medicine Branch, National Cancer Institute, 9000 Rockville Pike, Bethesda, MD 20205

Z. P. Pavelic [339]
Department of Experimental Therapeutics and Grace Cancer Drug Center, Roswell Park Memorial Institute, 666 Elm Street, Buffalo, NY 14263

Constance C. Reese [277]
HSW 763 Brain Tumor Research, University of California at San Francisco, San Francisco, CA 94143

Mark L. Rosenblum [259]
Brain Tumor Research Center, Department of Neurological Surgery, University of California at San Francisco, San Francisco, CA 94143

Y. M. Rustum [339]
Associate Director, Department of Experimental Therapeutics and Grace Cancer Drug Center, Roswell Park Memorial Institute, 666 Elm Street, Buffalo, NY 14263

Sydney E. Salmon [3, 23, 43, 63, 75, 127, 135, 153, 179, 197, 223, 291, 315]
Professor of Medicine, Head, Section of Hematology/Oncology, Director, University of Arizona Cancer Center, Arizona Health Sciences Center, 1501 N. Campbell Avenue, Tucson, AZ 85724

Jen-Ta Shen [277]
Division of Gynecologic/Oncology, Department of Obstetrics and Gynecology, University of Southern California School of Medicine, Los Angeles, CA 90033

Harry K. Slocum [339]
Department of Experimental Therapeutics and Grace Cancer Drug Center, Roswell Park Memorial Institute, 666 Elm Street, Buffalo, NY 14263

Barbara Soehnlen [179, 223, 331]
Research Associate, Arizona Health Sciences Center, 1501 N. Campbell Avenue, Tucson, AZ 85724

Thomas H. Stanisic [75]
Assistant Professor, Department of Surgery, Arizona Health Sciences Center, 1501 N. Campbell Avenue, Tucson, AZ 85724

Jeffrey M. Trent [101, 165, 345]
Assistant Professor of Medicine, University of Arizona Cancer Center Arizona Health Sciences Center, 1501 N. Campbell Avenue, Tucson, AZ 85724

Daniel D. Von Hoff [101, 113, 179]
Professor of Medicine, Division of Oncology, Department of Medicine, University of Texas Health Sciences Center, 7703 Floyd Curl Drive, San Antonio, TX 78284

James K. V. Willson [247]
Johns Hopkins Hospital, Division of Internal Medicine, Harvey 402, Baltimore, MD 21205

Laurie Young [223, 331]
Research Assistant, Department of Internal Medicine, Arizona Health Sciences Center, 1501 N. Campbell Avenue, Tucson, AZ 85724

Robert C. Young [247]
Chief, Medicine Branch, National Cancer Institute, 9000 Rockville Pike, Bethesda, MD 20205

Preface

In 1975, my investigative group at the University of Arizona Cancer Center initiated a research program with the goal of developing and applying techniques to clone human tumor stem cells in vitro. This goal is reflected in the title of this volume. When our early results with in vitro human tumor cloning showed potential for having a major impact on a wide front in cancer research, we opted to disseminate this information as quickly and effectively as possible and held two large-scale "tumor cloning workshops" for basic and clinical investigators. The workshops (held in Tucson, under the auspices of the University of Arizona Cancer Center) were in January, 1979 and 1980, and were attended by almost 400 scientists and clinicians from 15 countries. This book is a direct outgrowth of our 1980 workshop and has been updated to provide the latest methods and results available in the field. Its contents reflect efforts of individuals who have been investigators associated with the University of Arizona Cancer Center or our former "students," who learned much of the basic technology either in our laboratories or at our first workshop. The chapters do not represent workshop presentations per se, but overviews of the investigators' work with appropriate supporting data and appendix material, considered to be of practical value to the reader. As the editor, I have attempted to interweave the chapters so that appropriate cross-references and relationships between various areas can be more readily appreciated. The editor and the authors hope that our efforts will be "cloned by the readers," so that the scientific developments in this field will be more rapid and fruitful, as the opportunities appear to be immense.

I would like to acknowledge that the investigations of human tumor cloning carried out at the University of Arizona Cancer Center were initially supported by a generous donation, and subsequently, by research grants from the National Institutes of Health (CA 21839, CA 17094, and CA 23074). In relation to preparation of this book, a special word of thanks is due to Mrs. Toni Meinke, of the Cancer Center staff, for her outstanding secretarial skills and dedicated assistance.

Sydney E. Salmon, MD
Tucson, Arizona
May, 1980

I. General Perspectives

1
Background and Overview

Sydney E. Salmon

Tissue-culture techniques per se are not new to cancer research, and there has been continued development of new culture techniques and media as well as knowledge regarding the effects of hormones in growth factors on normal and neoplastic cells [1]. General knowledge about various tissue-culture techniques for short-term culture of tumor cells is not the focus of this volume, which is more acutely directed toward assays for tumor colony-forming cells (TCFUs). The text edited by Dendy in 1976 [2] provides a reasonable overview of short-term tissue-culture techniques with human tumors — including developmental efforts at drug-sensitivity assays using various forms of liquid culture for isotope incorporation (although clonogenic assays are virtually not mentioned).

Cloning of established tumor cell lines in agar culture was accomplished as early as 1955 by Puck and co-workers [3] and has been used subsequently for various transplantable tumors and tissue-culture cell lines. Clonal assays for normal pluripotent hematopoetic progenitors were first developed in a transplantable model by Till and McCulloch [4]. A major developmental step thereafter was the development of in vitro clonal assays for normal granulocyte-macrophage progenitors by Bradley and Metcalf [5, 6] and Pluznik and Sachs [7]. In the 1½ decades that have passed since their efforts, colony-forming assays in semisolid media have come to be the major approach to study of normal and abnormal hemopoiesis as well as for analysis of hormones and growth factors affecting clonal proliferation of granulocytic, erythrocytic, megakaryocytic, and lymphocytic progenitors [6]. Similar cultural approaches have recently been used to assess leukemic progenitors (Chapter 5) as well as abnormalities in normal hemopoiesis in the myeloid leukemias.

The concept of using clonal assays for study of tumor stem cells was first delineated by W.R. Bruce and his colleagues [8, 9] and was followed by an impressive series of studies by investigators at the Ontario Cancer Institute [9–11]. They showed that tumor stem cells responsible for population renewal and the colonizing property of a metastatic neoplasm from transplantable lymphoma or myeloma could be assayed in vivo or in vitro with tumor colony-forming assays. Such clonogenic assays have appeared to be more reliable than other measures of cytotoxicity [12, 13]. Measurement of in vitro colony-forming ability after exposure to drugs showed that differential sensitivity was present between different mouse myeloma cell lines, and results of a quantitative in vitro colony assay were predictive of in vivo results [11]. Thus, each individual cell line or transplantable myeloma system was as different in its sensitivity in comparison to the next one as any individual myeloma patient might be from another!

My interest in development of a clonal bioassay for human neoplasms developed from clinical research interests in multiple myeloma. I have studied this form of cancer since the early 1960s because it has many features that make it a "model" human cancer for clinical research. Specifically, it was a disseminated cancer. It was known to be of monoclonal origin [14] and involved the bone marrow and, therefore, was available for serial biopsy for research studies. Additionally, the myeloma cells secreted a marker protein, the monoclonal myeloma immunoglobulin (M component), which could be readily quantitated. Myeloma could also be induced (with mineral oil or pristane) in the BALB/c mouse [15] providing a superb animal model for investigation of carcinogenesis, growth regulation, and drug sensitivity of this neoplasm. It is well-known to most readers that studies of myeloma cells and their secretions also led to the concept of clonal proliferation in the normal immune response [16], definition of the structure of immunoglobulins [17], and the hybridoma technology for producing monoclonal antibodies [18].

In the late 1960s and early 1970s, my group developed relatively simple techniques to quantitate the total body-tumor burden present in myeloma patients using in vitro and in vivo measurements of the rate of synthesis of the M component [19, 20]. Subsequently, this was further simplified with a clinical myeloma staging system to predict the total tumor burden from presenting clinical features [20]. Despite the availability of such detailed clinicopathologic information on our patients, we found that our ability to predict clinical response to the most active drug then available (melphalan) was no better than 50% [20], and thus was no better than flipping a coin! We speculated at that time that it was likely that there was a major variable which we were not measuring, namely the inherent drug sensitivity of the individual patient's tumor to the drugs we used [20]. Lacking a test for drug sensitivity, it seemed reasonable to try to develop an assay for human myeloma stem cells analogous to the technique of Park et al [10] for murine myeloma, as Park's assay was unfortunately not useful for human myeloma cases. In early 1975, I had the good fortune to have Dr. Anne Hamburger contact me about the possibility of a postdoctoral fellowship, and I quickly recruited her. She had prior experience in experimental hematopoiesis and tissue-culture techniques, and it seemed that she would have an ideal background to pursue the problem of developing a clonal assay for human myeloma stem cells. I wrote to her and suggested that we try using cells from BALB/c mice that had been primed with mineral oil to condition medium to induce clonal proliferation of human myeloma cells. The oil-primed mouse provides the microinductive environment for initiation of murine myeloma [15] and facilitates transplantation in that species. We subsequently explored that approach and discovered that adherent spleen cells from such animals did, in fact, condition medium that would then support the growth of human myeloma colonies in soft agar. Dr. Hamburger's own discussion on the development and application of the myeloma stem cell assay appears in Chapter 3 of this text. Morphologic assessment of the adherent spleen-cell population indicated that it was comprised predominantly of macrophages, which is also the predominant population that Potter had noted to be present intraperitoneally after mineral oil injection [21].

We initially thought that our assay was specific for myeloma. However, in control experiments wherein we tested biopsies from various other human cancers, we were pleasantly surprised to find that the culture system facilitated colony growth by a wide variety of human neoplasms. An updated listing of tumors that have grown successfully in this system appears in Table I. Some of these tumors (eg, ovarian cancer, myeloma, melanoma, lymphoma, bladder cancer, and neuroblastoma) have now been studied in greater depth, and characteristic features have emerged with respect to clonal proliferation of each of these types of cancer. Overall, approximately 70% of all tumors plated in the stem cell assay exhibit colony formation, but in over one half of the cases, the cloning efficiency is too low (eg, 0.001%) to permit detailed studies. Dr. Von Hoff learned our techniques from us and reports his findings on neuroblastoma in Chapter 9. Additionally, he has provided a very valuable overview and critique of the tumor stem cell assay in Chapter 10. In the latter chapter, he has also provided additional evidence for the neoplastic nature of cells that give rise to colonies, and of the predictive value of in vitro drug-sensitivity assays. In general, samples from malignant ascites or widespread metastatic disease have had higher cloning efficiencies, and, in some instances, a rise in cloning efficiency in serial samples from an individual patient has heralded that patient's imminent demise. The low clonogenicity that we have observed in some "hard" solid tumors (eg, breast) may, in part, result from difficulties experienced in attempting to mechanically disaggregate solid tumors.

Until recently, Arizona group has used entirely mechanical techniques to dissociate solid tumors as discussed in the various chapters. Techniques incorporating enzymatic steps such as described by Dr. Rosenblum for brain tumors, (Chapter 20) and for various other neoplasms by Dr. Slocum et al (Appendix 2), are worthy of careful testing, as there is great need for improvement in techniques for solid tumor disaggregation. In our studies of the gross colony morphology of various tumor types, we found that they differed significantly from tumor type to tumor type in a manner analogous to the differences in bacterial colony formation that are observed among various species.

TABLE I. Human Tumor Types Successfully Cultured With In Vitro Clonogenic Assay at the University of Arizona*

(More than 1,000 biopsy samples tested including primary tumors and metastases)

Carcinomas	Sarcomas
Adrenal	Acute myeloid leukemia
Bladder	Chronic lymphocytic leukemia
Breast	Diffuse lymphoma
Colon	Ewing's tumor
Kidney	Fibrosarcoma
Liver	Glioblastoma
Lung	Liposarcoma
Ovary	Macroglobulinemia
Pancreas	Melanoma (melanotic and amelanotic)
Prostate	Mesothelioma
Stomach	Multiple myeloma
Testis	Nephroblastoma (Wilms' tumor)
Thyroid	Neuroblastoma
Upper airways (head and neck)	Nodular lymphoma
Uterus (corpus and cervix)	Osteosarcoma
Unknown primary	Rhabdomyosarcoma

*The assay of Hamburger and Salmon [22] was used for all tumor types (with or without conditioned medium) except for acute myeloid leukemia, wherein the technique of Buick (Chapter 5) was utilized. In each instance, more than 10 tumor colonies arose per 500,000 cells plated. Among the carcinomas, squamous, adenocarcinoma, undifferentiated, and other histologic subgroups have grown in specific tumor categories.

Such differences in gross morphology are well reflected in Figure 1, which was published in our first report on the tumor stem cell assay that appeared in *Science* in 1977 [22]. In that first paper we also predicted the potential broad applications such an assay could have for important studies of cancer biology and treatment. During the course of subsequent studies, we found that the mouse-spleen-conditioned medium was not required for nonhematologic neoplasms, perhaps because the presence of endogenous macrophages and/or other host cells within the tumor cell population might be providing conditioning factors directly. In fact, in our ovarian cancer studies (Chapter 6), Dr. Hamburger and I found that the mere removal of phagocytic macrophages from ascites markedly reduced the cloning efficiency for the TCFUs. This led us to hypothesize that the macrophage might be a "two-edged sword" regulating and promoting the growth of both immune cells and clonogenic tumor cells via soluble factors (Chapter 23). Subsequently, Dr. Ronald Buick has markedly extended the analysis of the "feeder layer" effect of host cells as discussed in Chapter 11. Cell-separation procedures play an increasingly important role in studies of cell interaction in the clonogenic assays.

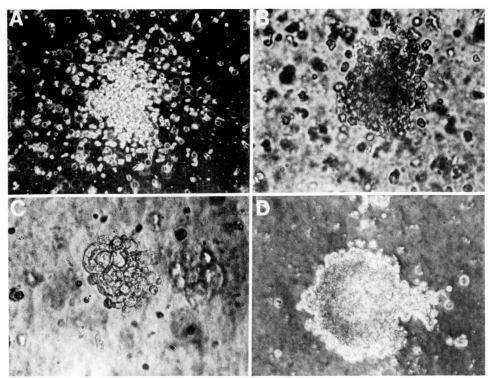

Fig. 1. Morphological characteristics of human tumor stem cell colonies (inverted microscopy of viable cultures). A) Sunburst appearance of non-Hodgkin's lymphoma colony at 13 days of culture ($\times$ 100). B) Heaped-up myeloma colony at 13 days ($\times$ 200). C) Fourteen-day-old ovarian carcinoma colony ($\times$ 200). The morphology is consistent with a mucin-secreting adenocarcinoma. D) Spherical neuroblastoma colony at 21 days of culture ($\times$ 200). Reproduced for [22] with permission of the publishers.

It is largely through careful analytic application of separation methodology that the significance of heterogeneity of the cell population in a tumor sample can be adequately appreciated, allowing the interactions that affect in vitro clonogenicity to be potentially defined. It now appears likely that a variety of hormones (eg, insulin) and growth factors (eg, epidermal growth factor) modulate clonogenicity of tumor stem cells.

One of the concerns that some readers might have about the tumor stem cell assay is the low plating efficiency for the TCFUs. With our current technology, less than 0.1% of tumor cells are usually clonogenic. While the cloning efficiency will likely be improved somewhat with better techniques for cell disaggregation, adjustment in medium components, or addition of appropriate growth factors and hormones, I am intellectually prepared to accept a low cloning efficiency (eg, less than 1.0%) as an inherent feature of spontaneous human tumors that have not been subjected to serial in vitro passages or animal transplantation. It is, in fact, these latter serial selection procedures which appear to lead to selective enrichment of the clonogenic compartment in long-term cell lines. Dr. Rosenblum's observations on brain tumor clonogenicity (Chapter 20) are clearly in support of this concept. Thus, the 10%–100% cloning efficiency of some such lines may, in fact, be an artifact resulting from the technique of their development. As such, I believe that some tumor-cell biologists may have been misled by such high cloning efficiencies in the model systems and accordingly assumed that spontaneous human tumors would have similar cloning efficiencies. It was, in fact, because we held the hypothesis that tumor stem cells might be present in situ in low frequency (in numbers similar to those of normal bone marrow stem cells) that we plated various tumors at sufficiently high density so that we could identify TCFUs at a frequency of $1:10^3 - 10^5$. Fortuitously, our efforts were also facilitated by the fact that normal fibroblasts (which often can overgrow in liquid culture) are inhibited from proliferation by agar [24].

In our studies of various tumors, it has been clear that it is necessary to look for a series of features which would permit us to identify the colonies growing in the semisolid matrix as being of neoplastic origin. These features are summarized in Table II. In all instances, morphology of the colonies on dried slides, standard histologic sections, or electron microscopy has been consistent with the cell types present in the tumor's origin. Examples of morphology of some of the various tumor types appear in Figures 2–6. In many tumors, we have been able to capital-

TABLE II. Identifiable Features of Human Tumor Colony-Forming Cells Consistent With In Situ Tumor Stem Cell Origin

1. Expression of Reproducible Markers of Clonality by Tumor Colonies.
   a) Histology: eg, melanin or mucin production; characteristic electron microscopic features.
   b) Cytogenetics: marker chromosomes, aneuploidy, and other characteristic chromosomal abnormalities.
   c) Production of tumor-associated biomarkers: eg, monoclonal immunoglobulin, carcinoembrionic antigen, hormones.
2. Uniformity of Cell Type in Developing Colonies.
   (Colonies composed of neoplastic cell types and not mixed with normal cells.)
3. TCFUs are Physically Separable from Non-Neoplastic Elements in the Biopsy. (Centrifugal gradients or elutriation, "staput" analysis, phagocytic or rosette depletion, surface adherence, etc.)
4. TCFUs Manifest Self-Renewal Properties.
5. TCFUs Reflect in vivo Tumor Behavior.
   a) Serial increases in relative cloning efficiency are associated with increasing malignancy in vivo.
   b) Strong correlation between in vitro drug sensitivity or resistance and clinical response.

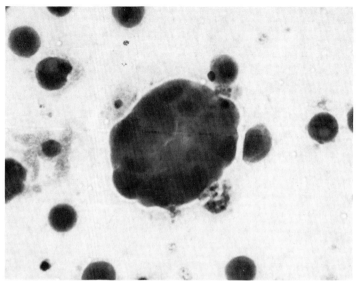

Fig. 2. Breast cancer colony at seven days of growth after soft-agar plating of cells obtained from a malignant effusion (Papanicolaou stain, × 400). This morphological preparation (as well as Figs. 3–6) was prepared with the dried-slide technique detailed in Chapter 12.

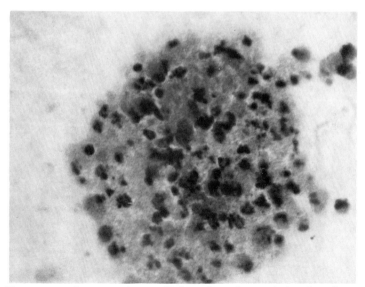

Fig. 3. Human prostatic cancer colony at 13 days after soft-agar plating of a cell suspension prepared from tumor chips from a transurethral resection of a prostate cancer (Papanicolaou stain, × 400).

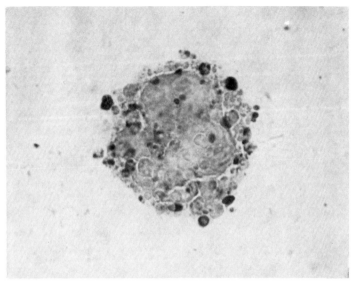

Fig. 4. Small cell carcinoma (oat cell) colony at 12 days after plating a cell suspension prepared from a tumor biopsy. A large central zone of necrobiosis is present, which gives the homogeneous gray appearance (Papanicolaou stain, × 400).

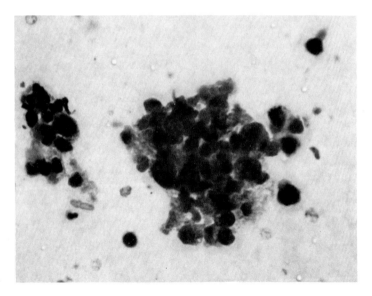

Fig. 5. Developing tumor colony at the 30-cell stage some nine days after plating cells from a soft tissue tumor nodule from a patient with liposarcoma (Papanicolaou stain, × 400).

Background and Overview / 11

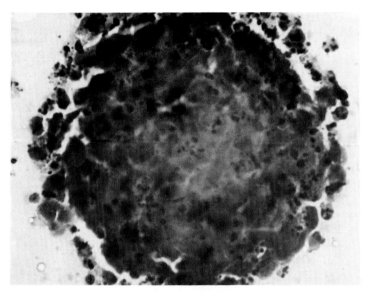

Fig. 6. Large (250-cell stage), rapidly growing tumor colony at seven days after plating single cells from a pleural effusion from a terminally ill child with Ewing's sarcoma. The plating efficiency was 0.1% (much higher than average), and the in vitro doubling time of this patient's clonogenic tumor cells was in the range of 24 hr (Papanicolaou stain, × 400).

ize on additional markers (eg, marker protein, or hormone secretion, or cytogenetic) to further confirm the neoplastic origin of the colonies. As detailed by Dr. Trent (Chapter 14), our development of techniques to analyze cytogenetics in tumor clusters and colonies has, in fact, substantially amplified and broadened the applications of karyology to human solid tumors. Investigations of drug sensitivity began in our laboratories shortly after we had our first successes at growing TCFUs from biopsies. The phenotypic expression of drug sensitivity and resistance of the various tumor types has also been characteristic of cells of a neoplastic origin. When we have studied drugs active in cancer chemotherapy, we have limited ourselves to assessment of pharmacologically achievable dosages (and have avoided higher drug concentrations. Of interest, prior studies of solid tumors using other assays, usually with dye exclusion or thymidine incorporation into whole tumor cell populations, have generally employed substantially higher dosages of cytotoxic agents than are clinically achievable. We believe that our success in predicting clinical drug sensitivity or resistance (Chapter 18) has resulted from a combination of using a relevant assay (of just the clonogenic cells), relevant drug concentrations, and a precise method of quantitating in vitro sensitivity (as outlined by Drs. Alberts and Moon, Chapters 16 and 17). Applications of the agar-culture system for studies of mechanisms of drug resistance may also prove important.

Some new problems have arisen as a result of application of the tumor stem cell assay to clinical measurements of drug sensitivity and new drug screening. One problem is that of having too many samples to count reliably by conventional means! This limitation has, in turn, resulted in efforts to develop an automated system for enumeration of tumor colonies. Advances in this area of necessary technology development are now quite encouraging (Chapter 15). In addition to its application to drug assay, colony counting via automated image analysis will provide a valuable new tool with which important quantitative observations on clonal proliferation of normal and neoplastic cells can be made.

This book is organized so that the newcomer to the field can assimilate needed information in stepwise fashion. The first section of the text presents conceptual material on clonogenicity and tumor stem cells, while the second section summarizes biological features on a series of tumors which have been studied in some detail by in vitro clonogenic assay. The third section of the text covers specialized biological studies and new technology, and the fourth addresses drug sensitivity and related clinical and preclinical studies. Clinical investigators will likely have particular interest in the sections on prediction of clinical response to treatment, as it is reasonable to anticipate that these methods will undergo large-scale testing over the next few years. Effects of biological response modifiers (eg, interferon Chapter 21) and the use of the assay for new drug screening (Chapter 22) have also been included to indicate areas of possible increasing importance in the future. The editor's perspective on potential future directions follows in an additional section. Finally, the appendix provides various technical details which investigators will find useful at the laboratory bench. The reader can appreciate, from comments in the preceding paragraphs, that the in vitro study of clonogenic cells from human tumors is now a rapidly evolving field that has been facilitated by a variety of developments in diverse areas of science and technology.

The editor and all the authors recognize that we are involved in an area of cancer research that is now only in its infancy. We hope that by publishing our efforts in this volume, it will serve as a useful wellspring (or "stem cell') for future investigators interested in following these leads (or even "subcloning"!). Close scientific interaction between basic scientists and clinicians has been one of the hallmarks of the work accomplished to date; we believe that such an approach is essential in the conquest of cancer.

REFERENCES

1. Jakoby WB, Pastan IH (eds): "Methods in Enzymology Vol. LVIII. Cell Cuture." New York: Academic Press, 1979.
2. Dendy PP (ed): "Human Tumors in Short-Term Culture." New York: Academic Press, 1976.

3. Puck TT, Marcus PI: A rapid method of viable cell titration and clone production with HeLa cells in tissue culture: use of x-irradiated cells to supply conditioning factors Proc Natl Acad Sci USA 41:432–437, 1955.
4. Till JE, McCulloch EA: A direct measurement of the radiation sensitivity of normal mouse bone marrow. Radiat Res 14:213–222, 1961.
5. Bradley TR, Metcalf D: The growth of mouse bone marrow cells in vitro. Aust J Exp Biol Med Sci 44:287–300, 1966.
6. Metcalf D: "Hemopoietic Colonies." Berlin: Springer-Verlag, 1977.
7. Pluznik DH, Sachs L: The cloning of "mast" cells in tissue culture. J Cell Comp Physiol 66:319–324, 1965.
8. Bruce WR, Valeriote FA: Comparison of the sensitivity of normal hematopoietic and transplanted lymphoma colony-forming cells to chemotherapeutic agents administered in vivo. J Natl Cancer Inst 37:233–245, 1966.
9. Bruce WR, Lin H: An empirical cellular approach to improvement of cancer therapy. Cancer Res 29:2308–2310, 1969.
10. Park CH, Bergsagel DE, McCulloch EA: Mouse myeloma tumor stem cells: a primary cell culture assay. J Natl Cancer Inst 46:411–422, 1971.
11. Ogawa M, Bergsagel DE, McCulloch EA: Chemotherapy of mouse myeloma: quantitative cell culture predictive of response in vivo. Blood 41:7–15, 1973.
12. Roper PR, Drewinko B: Comparison of in vitro methods to determine drug-induced lethality. Cancer Res 36:2182–2188, 1976.
13. Steel GG: "Growth Kinetics of Tumours." Oxford: Oxford Press, 1977.
14. Waldenstrom JG: Studies on conditions associated with disturbed gamma globulin formation (gammopathies). Harvey Lect 56:211, 1961.
15. Potter M, Boyce CR: Induction of plasma-cell neoplasms in strain BALB/c mice with mineral oil and mineral oil adjuvants. Nature 193:1086–1087, 1962.
16. Burnet M: "The Clonal Theory of Acquired Immunity." Cambridge University Press, 1959.
17. Edelman GM: Antibody structure of molecular immunoglobulins. Science 180(4088):830–840, 1973.
18. Kohler G, Milstein C: Continuous cultures of fused cells secreting antibody of predefined specificity. Nature 256:495, 1975.
19. Salmon SE, Smith BA: Immunoglobulin synthesis and total body tumor cell number in IgG multiple myeloma. J Clin Invest 49:1114–1121, 1970.
20. Durie BGM, Salmon SE: A clinical staging system for multiple myeloma: correlation of measured myeloma cell mass with presenting clinical features, response to treatment and survival. Cancer 36:842, 852, 1975.
21. Potter M: Experimental plasma cell tumors and other immunoglobulin-producing lymphoreticular neoplasms.in mice. In Azar HA, Potter M (eds): "Multiple Myeloma and Related Disorders." Hagerstown, Maryland: Harper & Row, 1973, pp 153–194.
22. Hamburger AW, Salmon SE: Primary bioassay of human tumor stem cells. Science 197:461–463, 1977.
23. Salmon SE, Hamburger AW: Immunoproliferation and cancer: a common macrophage-derived promoter substance. Lancet 1:1289–1290, 1978.
24. Bouck N, DiMayorca G: Evaluation of chemical carcinogenicity by in vitro neoplastic transformation (pp 296–297). In Jakoby WB, Pastan IH (eds):"Methods in Enzymology." Vol LVIII (Cell Culture). New York: Academic Press, 1979.

2
In Vitro Clonogenicity of Primary Human Tumor Cells: Quantitation and Relationship to Tumor Stem Cells

Ronald N. Buick

The term "stem cell" was coined to describe a histologically primitive cell in conventional morphology and to provide practicality to the mathematical descriptions of population dynamics of steady-state tissues. In the study of tumor biology, the term has been applied to cells with the capacity to regrow the tumor after subcurative therapy or to give rise to a new tumor after transplantation. Inherent in this definition is the principle that stem cells must be able to give rise to a large number of descendants while at the same time producing at least one cell with the proliferative capacity of itself (self-renewal). The definition of a stem cell must be extended to include the concept of proliferating and nonproliferating (or potential) stem cells [1]. The latter category is included in the group of cells that can act as stem cells under appropriate conditions (eg, under improved nutritional status after tumor reduction by cytotoxic agents).

Normal renewal tissues maintain appropriate cell number and function through an exquisitely controlled balance between stem cell renewal, cell proliferation, and cell differentiation. However, tumor cell populations have apparently lost control of such balance, as evidenced by population expansion and loss of functional capacity of the involved tissue. On a theoretical basis, the probability of cell-renewal events will in great measure govern the most obvious growth characteristics of the tumor. Such concepts have been described in mathematical models by Vogel, Niewisch, and Matioli [2, 3].

These conceptual approaches to tumor stem cells are, however, theoretical, since they depend on mathematical modelling to explain the stem cell population dynamics from gross measurements of total population parameters. The concepts have been rendered tangible by the development of techniques whereby the capacity of a single cell to establish a clone can be estimated under artificial conditions (clonogenicity). A number of such procedures have been described, including the ability of single-cell suspensions of tumor cells to form colonies in vitro and in recipient animals in vivo, or the capacity to repopulate the tumor in situ.

From such tests has derived the main basis for understanding the relationship between clonogenicity under artificial conditions and concepts of tumor stem cells. It has been noted that fractional survival of clonogenic cells after cytotoxic therapy is correlated in some instances with the time of regrowth of the tumor or with cure. When such correlations are seen, there are good reasons to believe that the clonogenic cells under artificial conditions are identical with, or closely related to, the tumor stem cells in situ. The question of stem-cell/clonogenicity relationship is, of course, of ultimate importance. The rest of this paper will investigate the situation where the artificial conditions employed for measuring clonogenicity is a tissue-culture procedure in semisolid conditions, since the characteristics of this procedure form the basis for this text.

IN VITRO CLONOGENICITY

From a practical point of view, the assessment of colony formation in culture has much to be recommended for simplicity and versatility. However, it must be admitted that in vitro culture conditions are further removed from in situ conditions than in vivo colony-forming or tumor-reconstitution studies. The basis of culture-colony formation is as follows: provided adequate sources of nutritional requirements and some procedure to localize descendants of a single clone, the colony-forming capacity of a group of cells can be enumerated. These provisions were first met by Puck and Marcus in a series of classic publications that set criteria for measurement of clonogenicity, fractional radiation survival, and introduced the use of agar to help maintain colony integrity [4, 5].

A further impetus to tumor clonogenicity studies was provided by the work of Pluznick and Sachs [6] and Bradley and Metcalf [7], who demonstrated the use of in vitro semisolid clonogenic assays for hemopoietic progenitor cells. Such procedures have been extensively utilized and have led to the present-day advanced level of understanding of the biology of human hemopoiesis [8]. Semisolid procedures utilize agar or methylcellulose to hold the colonies together, and often a multilayered system is employed with a "feeder layer" underlying the layer containing the bone marrow cells to be grown [9]. Concurrently with investigation of normal human hemopoiesis, a number of investigations have been made of the clonogenicity of leukemic progenitor cells [10–12]. The colony characteristics of leukemic cells in agar has been shown to be related to the status of the disease [10]. A methylcellulose clonogenic assay described by Buick et al ([11] and Chapter 5) has proven amenable to a direct measurement of self-renewal capacity [13]. As well as providing information on the stem cell nature of the leukemic progenitors, this characteristic shows promise of being prognostically significant and of being an important target for developmental therapeutics. With respect to other hematologic malignancies, human myeloma clonogenic cells have been assessed in agar by Hamburger and Salmon ([14] and Chapter 3), and lymphoma clonogenic cells by Jones et al ([15] and Chapter 4). As is the case with leukemia, assays of myeloma and lymphoma clonogenic cells have appeared to have relevance to clinical manifestations of these disorders.

Similar approaches to clonogenic cells in solid human tumors have proven more difficult than for hemopoietic tumors due to the difficulty in obtaining single-cell suspensions. The absolute requirement for a single-cell suspension to initiate clonogenicity experiments has severely hampered advance in this area. One method used to circumvent this is the performance of clonogenic studies with cells derived from xenografts of human tumors growing in immune-deprived mice [16]. Altman et al have taken advantage of the single-cell nature of malignant effusions to measure clonogenicity [17]. Success with human solid tumors has been reported by McAllister and Reed for certain childhood tumors [18], by Hamburger and Salmon for a variety of tumors [19] including ovarian ([20] and Chapter 6), melanoma by Meyskens (Chapter 8), neuroblastoma by Von Hoff et al ([21] and Chapter 9), and for transitional cell carcinoma of the bladder by Buick et al ([22] and Chapter 7). Many of these developments are summarized in this text.

PRACTICAL CONSIDERATIONS OF QUANTITATION

Ideally, the only clonogenic determinant of an in vitro system should be the intrinsic ability of a cell to proliferate to form a colony. This property will be related to the position of the cell on a renewal hierarchy. However, a number of factors have to be considered when considering the significance of clonogenicity measurements in semisolid culture conditions. We will consider cell disaggregation, adequacy of nutritional support, colony size, and kinetic selection.

As mentioned above, difficulty in preparation of single-cell suspensions of human tumors has slowed development in this field. There are at least three important considerations involved in this problem:

a) The requirement for a single cell suspension is an *absolute* requirement. All considerations of clonogenicity are void unless one can be sure that a colony has been derived from a single cell. Apart from direct visualization, one can estimate unicellular origin of colonies by analysis of linearity with respect to cell number plated and by analysis of karyotypic homogeneity within a single colony. This latter procedure, however, implies conservation of genetic material during successive culture mitoses. Recent evidence of Saxe and Meyskens [23] indicates that asymmetric segregation occurs in human melanoma. Thus, lack of genetic homogeneity in melanoma colonies cannot be taken to mean multicellular origin.

b) The physical trauma experienced by the tumor cells during the process of mechanically or enzymatically producing a single-cell suspension may be a significant influence on subsequent clonogenicity.

c) It is probable that the ability of a cell to perform as a tumor stem cell in situ is governed by cell interactions with nontumor cellular elements (eg, stroma or infiltrating lymphoreticular cells). The disruption of such interactions during cell disaggregation may severely limit clonogenicity, particularly if the normal host cells are depleted during preparative stages. Identification of specific growth factors may permit correction of this problem.

Another factor influencing the enumeration of clonogenic tumor cells is the question of adequate nutritional support. Clearly, tumors arising in different tissue are very likely to require different nutritional factors. This area has not been investigated for human solid tumors, although Mather and Sato [24] have defined precise culture requirements for a number of cell lines. The problem may be very great, however, when one considers that cells from different patients with the same histology and even different clones within a single tumor may show marked heterogeneity with respect to nutritional requirements for clonogenicity in culture.

A third consideration is colony size. As is the case with all clonogenic assays, the distinction between clonogenic and nonclonogenic tumor cells is not always clear-cut. The usual situation is that clonal growth occurs to produce a range of colony size. Criteria for counting are therefore extremely difficult to establish. The usual procedure is to set a minimum criterion of 40–50 cells. This problem may be circumvented by the development of refined instrumentation for colony counting (see Chapter 15). It is important to note that the problem of colony size may be even more critical after in vitro cytotoxicity testing, when cells may be damaged sufficiently to cause retarded growth but not sufficiently to eliminate growth entirely.

Of considerable importance, particularly when dealing with fractional drug or radiation survival, is the question of the integrity of the growth system in measuring a kinetically representative fraction of the clonogenic cells. For instance, in the extreme case, a clonogenic assay might only allow quantitation of cells that were in active cell cycle at the time of plating. The high suicide index noted for most tumor clonogenic assays ([14, 20, 25] and Chapter 8) makes this a real possibility. Clearly, this problem is closely related to the question of nutritional adequacy of the medium. The hormonal requirements needed to induce clonogenicity from a G_0 clonogenic cell may be very different from those required to support growth in an already cycling cell.

CONCLUSION

All these considerations lead one to suggest that the actual *number* of cells identified as clonogenic will likely have very little meaning relative to the in situ situation. However, since during in vitro manipulation (eg, drug-sensitivity testing) culture conditions are held constant, change in the measurement of clonogenicity is a valid parameter [26–28]. This conclusion is, of course, dependent on knowing which determinant of clonogenicity is being altered by the in vitro treatment. Likewise, a comparison of colony size between different patients and/or different histologies would not be expected to produce meaningful information.

The question of relationship between clonogenicity and tumor stem cells remains to be established. As described above for animal tumors, a correlation between in vitro drug sensitivity and time of tumor regrowth or cure is ideally required. A correlation between in vitro drug sensitivity and response to treatment is not totally indicative of such an identity, since reduction in tumor size is not solely a function of tumor stem cells [1]. Should a strong positive correlation be derived between in vitro sensitivity and cure or a strong negative correlation between in vitro sensitivity and time of tumor regrowth, such an inference can be made. In the past, such experiments have not been feasible in human tumors; however, correlations between drug sensitivity or resistance in vitro and rate of tumor regrowth are possible in myeloma, ovarian cancer, and selected other neoplasms.

In conclusion, the enumeration of clonogenic tumor cells is an extremely important concept. A number of considerations render the quantitation inexact, but future optimization of tumor disaggregation, culture conditions, and colony-counting procedures should circumvent such problems. The identity of clonogenic cells with, or their relationship to, tumor stem cells remains to be elucidated. However, the enormous importance of these concepts for developmental therapeutics makes this a high-priority task.

REFERENCES

1. Steel GG: "Growth Kinetics of Tumors." Oxford: Clarendon Press, 1977, pp 217–267.
2. Vogel H, Niewisch H, Matioli G: The self-renewal probability of hemopoietic stem cells. J Cell Physiol 72:221–228, 1968.
3. Vogel H, Niewisch H, Matioli G: Stochastic development of stem cells. J Theor Biol 22:249–270, 1969.
4. Puck TT, Marcus PI: A rapid method of viable cell titration and clone production with HeLa cells in tissue culture: The use of x-irradiated cells to supply conditioning factors. Proc Natl Acad Sci USA 41:432–437, 1955.
5. Puck TT, Markus PI: Action of x-rays on mammalian cells. J Exp Med 103:653,666, 1956.
6. Pluznik DH, Sachs L: The cloning of "mast" cells in tissue culture. J Cell Comp Physiol 66:319–324, 1965.
7. Bradley TR, Metcalf D: The growth of mouse bone marrow cells in vitro. Aust J Exp Biol Med Sci 44:287–300, 1966.
8. Metcalf D: "Hemopoietic Colonies." Berlin: Springer-Verlag, 1977.
9. Pike B, Robinson W: Human bone marrow colony growth in vitro. J Cell Physiol 76:77–81, 1970.
10. Moore MAS, Spitzer G, Williams V, Buckley T: Agar culture studies in 127 cases of untreated acute leukemia: the prognostic value of reclassification of leukemia according to in vitro growth characteristics. Blood 44:1–18, 1974.
11. Buick RN, Till JE, McCulloch EA: Colony assay for proliferative blast cells circulating in myeloblastic leukemia. Lancet 1:862–863, 1977.
12. Dicke KA, Spitzer G, Ahearn MJ: Colony formation in vitro by leukemia cells in acute myeloblastic leukemia cells in acute myeloblastic leukemia with phytohemagglutinin as stimulating factor. Nature 259:129–130, 1976.

13. Buick RN, Minden MD, McCulloch EA: Self-renewal in culture of proliferative blast progenitor cells in acute myeloblastic leukemia. Blood 54:95–104, 1979.
14. Hamburger AW, Salmon SE: Primary bioassay of human myeloma stem cells. J Clin Invest 60:846–854, 1977.
15. Jones SE, Hamburger AW, Kim MB, Salmon SE: Development of a bioassay for putative human lymphoma stem cells. Blood 53:294–303, 1977.
16. Courtenay VD, Smith I, Peckham MJ, Steel GG: The in vitro and in vivo radiosensitivity of human tumor cells obtained from a pancreatic carcinoma xenograft. Nature 263:771–772, 1976.
17. Altman A, Chusi F, Rierdan W, Baehner R: Growth of rhabdomyosarcoma colonies from pleural fluid. Cancer Res 35:1809–1812, 1975.
18. McAllister RM, Reed G: Colonial growth in agar of cells derived from neoplastic and non-neoplastic tissues of children. Pediatr Res 2:356–360, 1968.
19. Hamburger AW, Salmon SE: Primary bioassay of human tumor stem cells. Science 197:461–463, 1977.
20. Hamburger AW, Salmon SE, Kim MB, Trent JM, Soehnlen BJ, Alberts DS, Schmidt HJ: Direct cloning of human ovarian carcinoma cells in agar. Cancer Res 38:3438–3444, 1978.
21. Von Hoff DD, Casper J, Bradley E, Trent JM, Reichert E, Altman A: Direct cloning of human neuroblastoma cells in agar: a measure of response and prognosis. Cancer Res (in press).
22. Buick RN, Stanisic TH, Fry SE, Salmon SE, Trent JM, Krasovich P: Development of an agar-methylcellulose clonogenic assay for cells in transitional cell carcinoma of the human bladder. Cancer Res 39:5051–5056, 1979.
23. Saxe DF, Meyskens FL Jr: Abnormal chromosome segregation in human melanoma. In "Proceedings of American Society of Human Genetics." Minneapolis, 1979.
24. Mather JP, Sato GH: The growth of mouse melanoma cells in hormone supplemented, serum-free, medium. Exp Cell Res 120:191–200, 1979.
25. Minden MD, Till JE, McCulloch EA: Proliferative state of blast cell progenitors in acute myeloblastic leukemia. Blood 52:592, 1978.
26. Buick RN, Messner HA, Till JE, McCulloch EA: Cytotoxicity of adriamycin and daunorubicin for normal and leukemic progenitor cells of man. J Natl Cancer Inst 62:249, 1979.
27. Salmon SE, Hamburger AW, Soehnlen BJ, Durie BGM, Alberts DS, Moon TE: Quantitation of differential sensitivity of human tumor stem cells in anticancer drugs. N Engl J Med 298:1321–1327, 1978.
28. Salmon SE, Alberts DS, Durie BGM, Meyskens FL, Soehnlen BJ, Chen H-S G, Moon TE: Clinical correlations of drug sensitivity in the human tumor stem cell assay. "Recent Results in Cancer Research." Berlin: Springer-Verlag, 1980.

II. Studies of Various Tumor Types

3
Development of a Bioassay for Human Myeloma Colony-Forming Cells

Anne W. Hamburger and Sydney E. Salmon

Although primary explants of murine myeloma had been successfully cloned in soft agar [1, 2], the in vitro cultivation of human plasma cells had met with variable success at the time we initiated these studies. Using methods similar to those he devised for in vitro cultivation of mouse myeloma cells, Dr. Chan Park was able to culture human myeloma cells in soft agar [3]. However, he was unable to establish linearity between numbers of cells plated and the number of colonies formed, and thus could not use this technique as an assay system.

This chapter has two major components. First, we will detail steps in the development of our in vitro tissue culture system (which we first applied to cloning human myeloma cells) and present some practical aspects of this technique. Second, we will describe some of the studies directed toward characterizing the myeloma colony-forming cell (M-CFU-c). Purification of the myeloma colony-forming cell would permit biochemical and antigenic analysis of properties of these tumor stem cells and facilitate preparation of a tumor-specific antiserum. In addition, studies seeking to determine factors that control proliferation of myeloma cells will require purified populations of precursor cells.

MATERIALS AND METHODS

Patient Studies

Patients with well-documented multiple myeloma and normal volunteers were selected for study. Clinical and immunological criteria for diagnosis and clinical staging of myeloma were as described [6].

Collection of Cells

Bone marrow cells were obtained from patients or normal volunteers by sternal or iliac puncture after informed consent was obtained. Cells were aspirated into a heparinized syringe, mixed in an equal volume of 3% dextran–saline, and sedimented at room temperature for 45 min. The cells in the supernatant were collected after centrifugation at 150g for 10 min and washed twice in Hank's balanced salt solution (HBSS) with 10% heat-inactivated fetal calf serum (FCS). The viable nucleated cell counts, determined in a hemocytometer using trypan blue, were routinely more than 95%. Bone marrow differential counts were performed on slides prepared with a cytocentrifuge and stained with Wright–Giemsa.

Culture Assay for Myeloma Colony-Forming Cells (M-CFU-cs)

Colonies derived from myeloma cells were grown in the presence of a 1-ml agar feeder layer in 35-mm Falcon Petri dishes. Two types of feeder layers were prepared (Fig. 1). The first was made by incorporating 0.02 ml of washed human-type O+ erythrocytes in modified McCoy's 5A medium consisting of 15% heat in-activated FCS and a variety of nutrients as described by Pike and Robinson [7]. Immediately before use, 10 ml of 3% trypticase soy broth, 0.6 ml asparagine (6.6 mg/ml), and 0.3 ml DEAE-dextran (MW 500,000) (50 mg/ml) were added to 40 ml of the enriched medium.

The second feeder layer utilized 0.25 ml of conditioned medium in 0.5% agar and enriched McCoy's 5A medium. The conditioned medium was prepared from the adherent spleen cells of BALB/c mice obtained from Jackson Labs (Bar Harbor, ME). This substrain of BALB/c produced the most potent conditioned medium. The mice had been primed with 0.2 ml of mineral oil injected intraperitoneally 8–12 weeks previously. This was based on a protocol of Namba and Hanoka [8] for in vitro cultivation of mouse myeloma cells. The adherent spleen cells were obtained as follows: the spleens were teased with 22-gauge needles to form a single-cell suspension, and 5×10^6 cells were placed in 60-mm Falcon tissue-culture dishes in RPMI 1640 medium containing 15% FCS. A 2-hr incubation was used to permit cellular adherence. Subsequently, the dishes were rinsed three times in cold HBSS. Cells were then incubated for three days at $37°C$ in 5 ml RPMI 1640 with 15% FCS, penicillin (100 U/ml), streptomycin (100 μg/ml), and glutamine (2 mM). The plates were two-thirds confluent at the time of harvest. This cell density was essential for the production of the conditioned medium. Plates with fewer cells adhered were discarded. We found that spleen cells failed to adhere in medium supplemented with certain batches of FCS. Therefore, it is advisable to screen batches of FCS prior to use. More rarely, certain batches of tissue culture plates did not allow cells to adhere. In addition, the 2-hr initial adherence step was sometimes lengthened up to four hours to permit more cells to adhere.

The harvested conditioned medium was decanted and centrifuged at 400g for 20 min. The supernatant was passed through 0.45 and 0.22 Naglene filters and stored up to one month at $-20°C$.

Different batches of conditioned medium appeared to differ in activity.

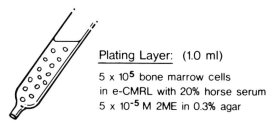

Plating Layer: (1.0 ml)

5×10^5 bone marrow cells in e-CMRL with 20% horse serum 5×10^{-5} M 2ME in 0.3% agar

RBC Feeder Layer: (1.0 ml)
0.5% agar
0.02 ml O+ RBC in McCoy's 5A

Clear Feeder Layer: (1.0 ml)
0.5% agar   McCoy's 5A
0.25 ml RPMI 1640
(conditioned for 3 days
with adherent spleen cells
from oil-primed BALB-C mice)

Fig. 1. Components of the myeloma stem cell assay. The two feeder layers shown have both proven useful, although inverted microscopy is somewhat more difficult with the RBC feeder layer.

Therefore, it is advisable to screen media against cryopreserved myeloma cells with known plating efficiency. If such cells are not available, marrow cells from patients with a plasma cell count of around 30%–40% are used as a standard. As the cells usually clone well, we test the activity of the medium against such a sample.

Bone marrow cells to be tested were suspended in 0.3% agar in CMRL 1066 medium supplemented with 20% horse serum, penicillin (100 U/ml), streptomycin (100 μg/ml), glutamine (2 mM), $CaCl_2$ (4 mM), insulin (3 U/ml), vitamin C (0.3 mM), asparagine (0.6 mg/ml), and DEAE dextran (0.5 mg/ml sterilized by autoclaving). Freshly prepared 2-mercaptoethanol (2-ME) was added at a concentration of 50 μM immediately before plating the cells. One ml of the resultant mixture was pipetted onto the 1-ml feeder layer. Cultures were incubated at 37°C in 7.5% CO_2 in a humidified atmosphere with no additional feeding. Cultures were examined in an inverted phase microscope at 100X and 200X. Final colony counts were made 14–21 days after plating. As cultures containing erythrocytes were opaque, erythrocyte lysis was completed by addition of 0.5 ml 3% acetic acid as necessary before scoring.

The CMRL 1066 medium was purchased from Grand Island Biological (Grand Island, NY) and kept frozen at −20°C for up to three months. Different lots of horse serum have yielded up to a five-fold difference in plating efficiency. Lots of serum should be screened against a positive control as outlined for testing the activity of the conditioned medium. The use of FCS was avoided, as it was often cytotoxic for myeloma cells. Human serum was not used, as it was often an active source of granulocyte colony-stimulating factor (CSF). The level of insulin we used was high, but appeared to be of value in this system. The mercaptoethanol was prepared weekly as a 100X stock in distilled water, aliquoted, stored at 4°C, and discarded after opening tightly capped tubes.

Examination of Cells in Colonies

Individual colonies were removed from the dish using a fine capillary pipette and were suspended in a drop of FCS and air dried 3–4 hr. Cells were stained routinely with Wright–Giemsa, 0.5% orcein in 60% acetic acid, peroxidase, and methyl green pyronin. More recently, we have applied the dried-slide technique [9] as described in Chapter 12 of this text, and have found the Papanicolaou technique quite useful for identifying myeloma colonies and distinguishing them from other types of cells in the agar plates.

For detection of cytoplasmic immunoglobulin, slides of individually plucked colonies were air dried and fixed in −20°C spectrophotometric grade acetone for 20 min. Slides were incubated for 45 min at 20°C with 0.25 ml of a 1:6 dilution of either fluorescein-conjugated rabbit anti IgG, IgA, or IgM (Fab_2 fragments). Cells were washed three times in phosphate-buffered saline at 20°C. Slides were examined in a fluorescent microscope with epillumination. Two hundred cells per slide were counted. Cells with moderate-to-strong fluorescence located only in the cytoplasm were scored as positive.

Determination of the Percentage of Cells in DNA Synthesis by H^3Tdr Killing

The H^3Tdr suicide method of Iscove et al [10] was employed to measure the proportion of M-CFU-cs in the S phase of the cell cycle. Briefly, samples of 2×10^6 cells, suspended in HBSS and 10% FCS, were added to one ml of HBSS containing 40 μCi of H^3Tdr (23 Ci/mM). Control samples were added to HBSS. Cell suspensions were then incubated for 30 min at 37°C and washed twice with 20 ml of cold HBSS containing 100 μg/ml of unlabeled thymidine and 10% FCS. Each suicide and control suspension was cultured in four replicate plates at a concentration of 5×10^5 nucleated cells per plate.

Preparation of Nonadherent and Nonphagocytic Cell Populations

Freshly washed bone marrow cells were separated into adherent and nonadherent populations by the method of Messner et al [11]. Briefly, cells were allowed to adhere for 30 min to 60-mm plastic tissue culture dishes in RPMI 1640 medium and 15% FCS at 37°C. The supernatant was decanted, and the adherence procedure repeated three more times to provide the nonadherent cell populations.

Bone marrow cells were depleted of phagocytic cells by placing 10^7 washed cells in tubes with 40 mg of carbonyl-iron powder and incubating the mixture in a shaking water bath at 37°C. After 45 min, the iron powder and iron-laden cells were attracted to the bottom of the flask with a magnet. The supernatant was carefully poured off, and this step was repeated as often as necessary (generally three or four times) to remove all the iron powder and iron-laden phagocytic cells. Cells were examined in a hemocytometer to ascertain whether all iron-containing cells were depleted. The remaining cells in the supernatant were considered nonphagocytic.

Biophysical Separation Studies

Cell suspensions were separated by velocity sedimentation at unit gravity using the Staput method of Miller and Phillips [12]. This technique separates cells primarily on the basis of size. Briefly, 1×10^8 cells, suspended in phosphate-buffered saline (PBS) with 5% FCS, were sedimented through a 15%–30% FCS gradient in a cylindrical siliconized glass chamber 18 cm in diameter (Johns Scientific Company, Toronto). Sedimentation was carried out at 4°C for 150 min and the chamber was then drained. The first 250 ml, consisting of the fluid in the conical portion of the chamber, was discarded. The remainder of the gradient was collected in 50-ml fractions. Cells were centrifuged at 150g for 10 min and washed. Nucleated cell counts for each fraction were made in a hemocytometer after dilution in 3% acetic acid and cytocentrifuge preparations were made.

In addition, freshly washed bone marrow cells were separated by their ability to adhere to plastic by the method of Messner et al [11]. Cells (5×10^6) were allowed to adhere to 60-mm plastic tissue culture dishes in 5 ml RPMI 1640 and 15% heat-inactivated FCS for 30 min at 37°C. The supernatant was decanted and the adherence procedure repeated three more times to provide the nonadherent cell population. Adherent cells were gently scraped from the bottom of the dishes with a rubber policeman. Cell recovery was about 85% with 90% viability.

Cells were also separated by their ability to adhere to nylon wool columns as described by Julius et al [13]. Freshly washed bone marrow cells (5×10^7) were incubated on a 6-gm nylon wool column packed in a 15-ml plastic syringe for 45 min at 37°C. Cells were washed off with 30 ml of warm RPMI 1640 containing 5% FCS. Cells in the effluent were considered the nonadherent population. The adherent population was recovered by compressing the nylon wool with the syringe plunger until most of the retained medium was expressed [14]. The nylon wool was then teased with a forceps, rinsed with warm medium, and again compressed with a syringe plunger. The procedure was repeated five to six times. Cell recovery was about 80% with 90% viability.

Finally, cells were separated by their adherence to glass beads. Bone marrow cells (5×10^7) in 50% human serum were passed through columns of siliconized glass beads at 37°C according to the active-adherence technique of Shortman et al [15]. The adherent cells were released by first washing the columns with medium containing 0.2% EDTA and then gently disrupting the column bed. Total cell recovery was about 85%.

Immunological Studies

An antiserum generated in rabbits against an established myeloma cell line from a patient with IgG-λ myeloma (RPMI 8226) was prepared and kindly provided by Dr. Robert Krueger (Christ's Hospital, Cincinnati, OH) [16]. The lot of antiserum reacted with RPMI 8226 cells to an endpoint titer of 1:128 as determined by positive immunofluorescence of at least 5% of the cells. The antiserum had an immunofluorescent titer of 1:16 against normal bone marrow plasma cells. Competitive absorption experiments have shown the antigen detected by this antiserum is not HBLA, immunoglobulin, or β-2-microglobulin [16]. One-tenth ml of appropriate dilutions of antiserum was added to the cultures in the experiments presented.

RESULTS AND DISCUSSION

Development and Identification of Colonies

Cell doublings were usually observed within 48 hr of plating, and clusters of 8–40 cells appeared within 5–10 days. Colonies (collections of more than 40 cells) appeared 14–21 days after plating. Cell counts were generally made 21 days after plating. Cell lysis generally occurred 28 days after plating. During the first two weeks of incubation, there was a progressive increase in the number of cells which commenced proliferation. Contaminating granulocyte colonies usually appeared at five days of culture, peaked between 7–10 days, and lysed at about 14 days of culture. They were detected by direct peroxidase staining by the methods of Zucker-Franklin and Grusky [17].

Myeloma colonies consisted of 40 to several hundred large round cells (Fig. 2). They appeared to pile up on one another as opposed to loosely aggregated cells in contaminating granulocyte colonies.

The number of myeloma colonies ranged up to 500 per plate, yielding a maximum cloning efficiency of 0.1%. A linear relationship was obtained between the number of nucleated cells plated and the number of colonies found after 21 days (Fig. 3).

Cells from individual colonies appeared to be plasma cells when examined by light microscopy after staining with Wright–Giemsa (Fig. 4) or methyl green pyronin. They were peroxidase negative and incapable of phagocytosis of latex particles. Immunofluorescent studies demonstrated that 60%–80% of the myeloma cells of individual colonies had the cytoplasmic immunofluorescence specific for the monoclonal immunoglobulin present in the serum of the patient studied (Fig. 5).

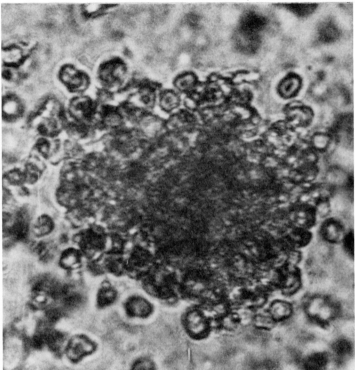

Fig. 2. A typical myeloma colony from a 21-day-old culture grown from the bone marrow of a patient in relapse (×20). Reproduced from J Clin Invest [4] with permission of the publishers.

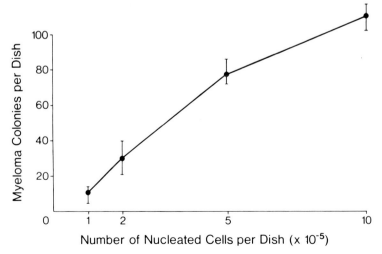

Fig. 3. Linear relationship between colony formation and the number of nucleated bone marrow cells plated. Each point represents the mean of four dishes ± SEM. These are the results of a typical experiment. Three other trials on different patients have yielded similar results. Reproduced from J Clin Invest [4] with permission of the publishers.

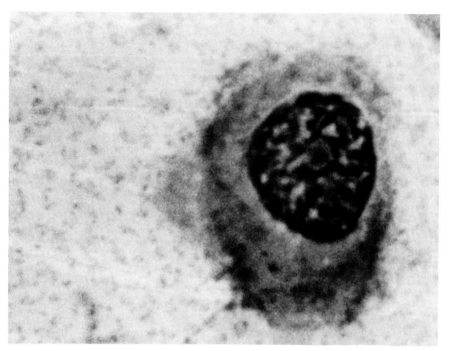

Fig. 4. A plasma cell picked from a 21-day-old culture and stained with Wright–Giemsa (×100). Reproduced from J Clin Invest [4] with permission of the publishers.

Factors Affecting Cell Growth

Substitution of type A, B, or AB erythrocytes usually resulted in decreased numbers of colonies, although differences were not statistically significant. Mouse erythrocytes were cytotoxic. Lysates of type O cells, obtained by hypotonic lysis, also supported growth. Metcalf [2] had found mouse myeloma cells could be cloned in soft agar using allogeneic erythrocytes. However, the activity was not maintained in red-cell lysates, indicating our factors were probably different. Depletion of residual CSF-producing leukocytes from erythrocytes by Ficoll-Hypaque or carbonyl-iron methods did not affect colony growth (Table I).

Media conditioned by spleen cells of oil-primed CD-1, DBA/2, and normal BALB/c mice were not so effective promoters of cell growth as conditioned media from oil-primed BALB/c mice. This might be expected, as BALB/c is the only strain of mice susceptible to induction of myeloma by mineral oil [18]. Conditioned media from W138 cells, MA184 cells (bone marrow fibroblasts), or primary explants of human skin fibroblasts were unable to support cell growth. However, contaminants in the spent media (such as Mycoplasma or endotoxin) may have accounted for the apparent cytotoxic effect of these media. Bacterial

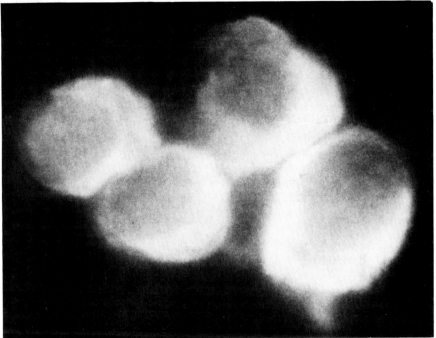

Fig. 5. Cells from a colony grown from the bone marrow of a patient with IgG myeloma. Cells were fixed in acetone and stained directly with fluorescent antihuman IgG. Note the strong specific cytoplasmic fluorescence (×100). Reproduced from J Clin Invest [4] with permission of the publishers.

lipopolysaccharide, mouse peritoneal exudate cells, mouse kidney fibroblasts, or frozen and thawed human leukemic bone marrow cells failed to stimulate myeloma cell growth. Sera from mice injected with endotoxin, bacterial antigens, or mineral oil did not allow colony growth. Autologous serum or urine from myeloma patients had no effect on colony growth (Table I).

We have favored the use of conditioned media from oil-primed BALB/c because it provided a clear feeder layer and made serial observation of colony growth possible. About 80% of myeloma patients had an absolute requirement for the conditioned medium. Figure 6 depicts the effect of different concentrations of BALB/c conditioned medium on colony formation by myeloma cells from the bone marrow of a patient in relapse. Myeloma colony growth was maximally stimulated when conditioned medium was present at a dilution of 1:4 in the underlayer. Fewer colonies grew at a concentration of media of 1:8. Dilutions below 1:4 or above 1:8 did not support colony growth. Three additional experiments with cells from other myeloma patients using different batches of conditioned media have yielded similar results. The shape of the dilution curves sug-

TABLE I. Effect of Substitution of Alternative Feeder Layers on Myeloma-Colony Formation*

	Number of colonies (per 5×10^5 cells)[a]	
	Control (O cell)	Experimental
Erythrocyte feeder layers		
Substituted material (0.02 ml)		
O RBC[b] lysate	158 ± 13.0	134 ± 8.5
O RBC depleted of WBC with Hypaque-Ficoll gradient	40 ± 8.0	49 ± 5.8
O RBC depleted of WBC with Hypaque-Ficoll gradient followed by carbonyl iron	40 ± 8.0	41 ± 4.6
A RBC	98 ± 11.0	74 ± 5.2
B RBC	98 ± 11.0	80 ± 2.5
AB RBC	98 ± 11.0	88 ± 6.1
BALB/c (mouse) RBC	98 ± 11.0	0
	Control (oil-primed BALB/c)[c]	Experimental
Conditioned-medium feeder layers		
Cell source		
Normal BALB/c mice[c]	45 ± 9.6	12 ± 2.3
CD-1 mice (oil-primed)[c]	48 ± 9.6	22 ± 6.5
DBA/2 mice (oil-primed)[c]	45 ± 9.6	8 ± 2.3
W138 (human fibroblasts)	45 ± 9.6	2.0 ± 2
MA194 (human marrow fibroblasts)	45 ± 9.6	1.8 ± 1
Primary skin fibroblasts (human)	48 ± 9.6	0

*Reproduced from J Clin Invest [4] with permission of the publishers.
[a]Mean ± SEM.
[b]RBC, erythrocytes; WBC, leukocytes.
[c]Adherent spleen cells.

gests the presence of an inhibitory factor. However, we have been unable to isolate this factor.

Mercaptoethanol was also essential for growth in 90% of patients tested. However, other sulfhydryl compounds, such as cysteine and dithiothreitol, did not promote proliferation of M-CFU-cs at the concentrations tested. However, monothioglycerol (MTG), at a concentration of 50 μM, proved as effective as 2-ME in supporting myeloma-colony growth. Neither 10 μM 2-ME or 10 μM MTG supported M-CFU-c growth (Table II).

Recently, Izaquirre et al [19] reported on a methylcellulose colony assay for myeloma and related B-cell neoplasms. This assay employs a medium conditioned by phytohemagglutinin (PHA)-stimulated human T lymphocytes. The patient's bone marrow is exhaustively depleted of T cells, which would proliferate in response to PHA. It will be of interest to determine how conditioning factors generated in Izaquirre's system compare to the one that we have described.

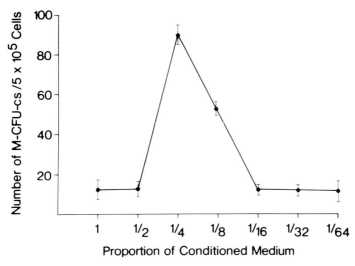

Fig. 6. Effect of different concentrations of conditioned medium on M-CFU-c growth. Bone marrow cells from a myeloma patient in relapse were plated on top of underlayers containing conditioned medium in the proportions indicated. Each point represents the mean of four plates ± SE. Three separate trials on different patients have yielded similar results. Reproduced from J Cell Physiol [5] with permission of the publishers.

TABLE II. Effect of Thiols on M-CFU-c Growth*

Substance	Concentration (μM)	No. M-CFU-c (per 5×10^5 cells)
2-Mercaptoethanol	50	28 ± 5[a]
	10	5 ± 3
Cysteine	50	2 ± 2
	10	8 ± 3
Dithiothreitol	50	3 ± 2
	10	3 ± 3
Monothioglycerol	50	35 ± 6
	10	0 ± 0

*Reproduced from J Cell Physiol [5] with permission of the publishers.
[a]Mean ± SEM, 14 dishes.

Proliferative State of M-CFU-c

H^3Tdr reduced colony survival to as little as 18% of control. Increasing the H^3Tdr concentration to 400 μCi/ml did not further increase the suicide fraction in either of two experiments. The addition of cold thymidine to H^3Tdr completely blocked the suicide effect in three experiments. The relationship of clinical status and various tumor kinetic parameters, including the suicide index, is summarized in Table III. No suicide effect occurred in two patients who had heavy infiltration of the marrow with myeloma cells and low H^3Tdr-labeling indices.

TABLE III. Relation of Clinical and Tumor Kinetic Parameters in Patients With Multiple Myeloma*

Patient	M Component	Status when studied	Clinical stage at diagnosis[a]	Bone marrow myeloma cells (%)	Myeloma cell H^{3}Tdr-labeling index (%)	M-CFU-cs per 10^6 myeloma cells[b]	Plating efficiency (%)	Fraction of M-CFU-cs surviving H^{3}Tdr suicide
1	IgGλ	Untreated	IIA	36	2	1,772 ± 1.78	0.18	0.18
2	IgGκ	Untreated	IIIA	36	14	180 ± 30	0.02	0.33
3	IgAκ	Untreated	IIIA	38	7	640 ± 80	0.06	0.57
4	κ-BJ	Untreated	IIIB	90	1	240 ± 20	0.02	1.25
5	λBJ (amyloid)	Remission	IB	4	1	190 ± 23	0.02	0.25
6	IgGλ	Relapse	IIIA	25	2	560 ± 110	0.06	0.36
7	IgAκ	Relapse	IIIA	73	1	300 ± 50	0.03	1.21

*Modified from [4] with permission of the publishers.
[a]Clinical staging as described in [6].
[b]All cultures were plated at 5 × 10^5 total nucleated marrow cells per plate, and are normalized to 10^6 with correction for plasma cell percentage. Plating efficiency for M-CFU-cs was calculated in relation to the number of myeloma cells plated.

TABLE IV. Effect of Population Depletion on Myeloma-Colony Formation*

Population depleted	Experiment no.	Number of M-CFU-c per 5×10^5 cells plated[a]	
		Before depletion	After depletion
Adherent	1	11 ± 7.2	92 ± 8.3
	2	52 ± 8.7	83 ± 6.8
	3	65 + 8.6	101 ± 11.2
	4	28 ± 5.6	123 ± 9.8
	5	44 ± 4.8	220 ± 28.5
Adherent phagocytic	5	48 ± 8.1	80 ± 22.0
Phagocytic	6	70 ± 14.5	200 ± 37.4

*Reproduced from J Clin Invest [4] with permission of the publishers.
[a]Mean ± SEM of four plates.

Effect of Depletion of Phagocytic or Adherent Cells

Although no exogenous source of colony-stimulating factor (CSF) was supplied in our culture system, adherent phagocytic cells of the bone marrow often produced endogenous CSF that permits granulocyte growth at cell densities of 5×10^5/ml. Therefore, bone-marrow suspensions were depleted of either adherent or phagocytic cells, or both populations, before plating. In every case (Table IV), depletion of the phagocytic or adherent cells increased the number of myeloma colonies seen per 5×10^5 cells. Thus, CSF-producing cells were not required for colony formation.

Effect of Anti-CSF on Myeloma-Colony Formation

A rabbit antiserum to CSF prepared against mouse cell CSF by Shadduck and Metcalf [20] was added to cultures to determine if colony formation was dependent on CSF and to eliminate the possibility that a significant number of granulocyte colonies were growing in these cultures. The top portion of Table V shows the results of a control experiment to determine the potency of the rabbit antiserum to CSF on granulocyte-colony formation. Normal human bone marrow was plated in a standard Pike-Robinson granulocyte assay [7]. The results indicated a 50% inhibition of colony formation at a serum dilution of 1:8. Normal rabbit serum did not depress granulocyte-colony growth more than 10% at this dilution. The results of adding anti-CSF to bone marrow cells from myeloma patients in relapse cultured in our system are presented on the bottom of Table V. Anti-CSF did not inhibit myeloma-colony growth. The studies with CSF antiserum and the cell depletion studies discussed earlier provide clear evidence that CSF for granulocytes does not play a role in the proliferation of myeloma stem cells.

TABLE V. Effect of Anti-CSF on Colony Formation by Granulocyte- and M-CFU-c*

Reciprocal of serum dilution	Number of colonies per 5 × 10⁵ cells	
	Anti-CSF	NRS[a]
Granulocyte-CFU-cs		
No serum	209 ± 21.4	209 ± 21.4
16	182 ± 19.2	198 ± 16.4
8	64 ± 9.3	230 ± 18.1
4	38 ± 3.5	173 ± 32.1
2	55 ± 9.8	192 ± 30.7
M-CFU-cs		
No serum	366 ± 25.7	366 ± 25.7
16	331 ± 29.8	322 ± 28.8
8	345 ± 13.1	320 ± 28.1
4	380 ± 30.5	365 ± 60.0
2	316 ± 40.8	292 ± 15.9

*Reproduced from J Cell Physiol [5] with permission of the publishers.
[a]NRS, normal rabbit serum.
Each point based on four plates ± SEM. One-tenth ml of diluted antiserum was added to each plate.

Biophysical Properties of Myeloma Colony-Forming Cells

The sedimentation velocity of M-CFU-cs was determined using the Staput apparatus. A representative experiment using bone marrow cells from an untreated myeloma patient is depicted in Figure 7. Three additional experiments with cells from other myeloma patients have yielded similar results. M-CFU-cs sedimented as a single broad band with a peak sedimentation velocity of 13 mm/hr. In an additional experiment, marrow cells from a second myeloma patient were separated on a Staput gradient, after which fractions were split, and cells were plated in either a granulocyte culture assay (counting colonies at 14 days) or the myeloma-culture system. The number of granulocyte or myeloma colony-forming cells was evaluated in each fraction. The results are shown in Figure 8. Granulocyte colony-forming cells sedimented more slowly than myeloma colony-forming cells in this marrow.

In addition, 5×10^7 peripheral blood leukocytes obtained by leukophoresis of a patient with IgA myeloma were separated at unit gravity. This patient had a plasma cell count of 9% in his peripheral blood. Peripheral blood M-CFU-cs from this patient sedimented at about 8 mm/hr (Fig. 9, [21]).

Bone marrow cells were also fractionated by three adherence techniques to determine the biophysical and surface properties of M-CFU-cs. Cell suspensions were either allowed to adhere to plastic tissue culture dishes, or filtered through glass beads or nylon wool columns. Between 50%–70% of cells adhered to plastic dishes, 30%–40% to glass bead columns, and 35%–40% to nylon wool.

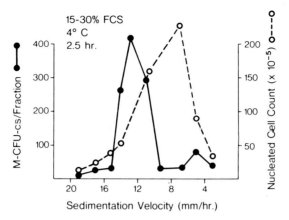

Fig. 7. Results of a typical sedimentation-velocity separation of cells from a bone marrow of a myeloma patient in relapse. Cells (1 × 10⁸) were sedimented through a 15%–30% fetal calf serum (FCS) gradient at 4°C for 150 min. The dotted line represents the nucleated cell profile; the unbroken line the M-CFU-c profile. Reproduced from J Clin Invest [4] with permission of the publishers.

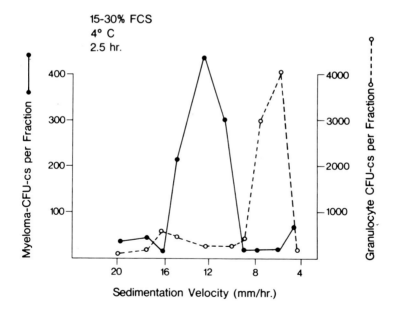

Fig. 8. Results of a sedimentation velocity separation of cells from a bone marrow of a second myeloma patient in relapse. Cells were sedimented as described, fractions were split, and cells were grown in either a granulocyte or myeloma culture assay. The unbroken line represents the number of M-CFU-cs per fraction; the dotted line represents the number of granulocyte colony-forming cells (granulocyte CFU-cs) per fraction. Reproduced from J Clin Invest [4] with permission of the publishers.

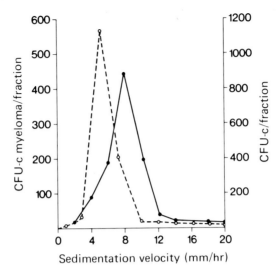

Fig. 9. Velocity sedimentation of peripheral blood leukocytes from the same patient. Cells were sedimented as described in Figure 7. The fractions were then split, and the cells were grown in either a granulocyte or a myeloma culture assay. The unbroken line represents the CFU/c myeloma per fraction; the broken line represents the number of CFU/c per fraction. Mean colony count from four replicate cultures. Number of CFU/c myeloma per fraction (●–●); number of CFU/c per fraction (○–○). Reproduced from J Cell Physiol [20] with permission of the publishers.

M-CFU-cs were enriched in populations that failed to adhere to plastic tissue culture dishes or glass beads (Table VI). In contrast, M-CFU-cs adhered to nylon wool.

Immunological Studies

Although M-CFU-cs were enriched in all fractions enriched for morphologically recognizable plasma cells, we were unable to identify the myeloma colony-forming cell. Although M-CFU-cs were enriched 20-fold in the Staput experiments, plating efficiencies were too low to make definitive statements as to the nature of the myeloma colony-forming progenitor. Thus, although these separation techniques provide useful information on biophysical properties of M-CFU-cs, they are not adequate for purification. Therefore, we explored the use of an antiserum to a human-myeloma cell line. Potentially, the use of a specific antiserum would permit the selective staining of myeloma progenitors with the fluorescein-labeled antibody and the sorting of them with a fluorescent-activated cell sorter, or the preparation of immunoadsorbants for myeloma-cell purification. Therefore, the effect of the rabbit antiserum on colony formation was studied. Control experiments of colony formation by the 8226 myeloma cell line indicated that 50% inhibition of colony formation could be observed at an antiserum dilution of 1:32.

TABLE VI. Myeloma-Colony Formation by Adherent and Nonadherent Cell Populations*

Separation technique	Experiment no.	Number of M-CFU-cs per 5×10^5 cells plated		
		Control	Adherent	Nonadherent
Glass bead column	1	29 ± 4	6 ± 3	86 ± 11
	2	16 ± 8	10 ± 4	42 ± 9
	3	106 ± 24	20 ± 11	254 ± 20
Nylon wool	1	22 ± 10	66 ± 9	8 ± 3
	2	10 ± 5	210 ± 12	10 ± 5
Plastic dish	1	25 ± 3	2 ± 2	64 ± 4
	2	40 ± 11	15 ± 6	245 ± 44

*Reproduced from J Cell Physiol [5] with permission of the publishers.

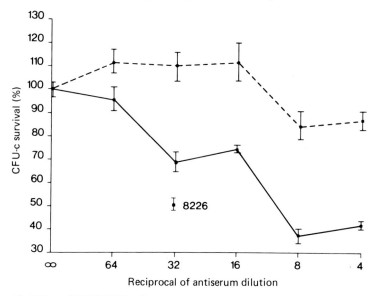

Fig. 10. Effect of RPMI 8226 cell antiserum on myeloma and granulocyte colony growth. The solid line represents the number of myeloma colonies; the dotted line the number of granulocyte colonies; "8226" represents antibody dilution at which 8226 colony formation was inhibited 50%. Each point represents the mean of four plates ± SE. Reproduced from J Cell Physiol [5] with permission of the publishers.

As determined by trypan blue exclusion, the antiserum was not cytotoxic to the majority of bone marrow cells from myeloma patients even in the presence of complement. Incubation of myeloma bone marrow cells with antiserum at dilutions of 1/2 to 1/128 with or without rabbit complement for one hour at 37°C did not result in more than 10% cytotoxicity.

Figure 10 shows the result of addition of anti-RPMI 8226 serum to a culture of bone marrow cells from a myeloma patient in relapse. The bone marrow sample contained 30% plasma cells. At a dilution of 1:8, the antiserum reacted with 15% of cells as determined by indirect immunofluorescence on fixed smears. The results of the colony assay showed a 63% inhibition of colony formation at an antiserum dilution of 1:8. In contrast, the antiserum reduced granulocyte formation from the same marrow 10% at this dilution. Normal rabbit serum at dilutions of 1:4 and 1:8 resulted in a slight depression (5%) of both myeloma colony-forming cells and granulocyte colony-forming cells. Clearly, the antiserum to the myeloma cell line is cytotoxic to the myeloma colony-forming units within random myeloma cell populations from various myeloma patients. It therefore does recognize myeloma-associated antigens that are expressed on myeloma stem cells. However, in view of the results with immunofluorescence, it is likely that this antigen is shared by both clonogenic and nonclonogenic human myeloma cells.

SUMMARY

This chapter outlines our development of an in vitro soft agar assay for detection of human myeloma colony-forming cells. Growth was induced with either 0.02 ml of human type O erythrocytes or 0.25 ml of medium conditioned by the adherent spleen cells of mineral-oil-primed BALB/c mice. A maximum plating efficiency of 0.1% was obtained. The number of myeloma colonies was proportional to the number of cells plated between concentrations of 10^5-10^6/ml. Morphological, histochemical, and functional criteria showed the colonies to consist of immature plasmablasts and mature plasma cells. 60%–80% of cells picked from colonies contained intracytoplasmic monoclonal immunoglobulin. Tritiated thymidine suicide studies provided evidence that for most myeloma patients, a very high proportion of myeloma colony-forming cells were actively in transit through the cell cycle. Using biophysical and immunologic studies, we were able to further characterize myeloma stem cells and obtain partial enrichment of the colony-forming cells. Increased numbers of myeloma colonies were seen when the marrow was depleted of CSF, elaborating adherent cells before plating. Antibody to granulocyte colony-stimulating factor, which did inhibit granulocyte colony formation, did not reduce the number or size of the myeloma colonies. This bioassay has subsequently served as the basis for studies of in vitro biological behavior of multiple myeloma, and for measurement of drug sensitivity.

The general methodology which we first developed for myeloma appears to have general applicability not only to monoclonal plasma cell disorders, but also to many other tumor types as well. Detailed biological studies and anlaysis of culture conditions (similar to those we have carried out in myeloma) will no doubt prove important in understanding the biology and drug sensitivity of various forms of human cancer.

REFERENCES

1. Park CH, Bergsagel D, McCulloch E: Mouse myeloma tumor stem cells: A primary cell culture assay. J Natl Cancer Inst 46:411–416, 1971.
2. Metcalf D: Colony formation in agar by mouse plasmacytoma cells: Potentiation by hemopoietic cells and serum. J Cell Physiol 81:397–410, 1973.
3. Park C: Studies of the growth characteristics of myeloma in vitro. PhD dissertation University of Toronto 31–61, 1971.
4. Hamburger AW, Salmon SE: Primary bioassay of human myeloma stem cells. J Clin Invest 60:846–854, 1977.
5. Hamburger AW, Salmon SE: The nature of cells generating human myeloma colonies in vitro. J Cell Physiol 98:371–376, 1979.
6. Durie B, Salmon SE: A clinical staging system for multiple myeloma. Cancer 36: 842–854, 1975.
7. Pike B, Robinson W: Human bone marrow colony growth in vitro. J Cell Physiol 76:77–81, 1970.
8. Namba Y, Hanoka M: Immunocytology of cultured IgM-forming cells of mice. J Immunol 109:1193–1200, 1972.
9. Salmon SE, Buick RN: Preparation of permanent slides of intact soft agar colony cultures of hematopoietic and tumor stem cells. Cancer Res 39:1133–1136, 1979.
10. Iscove N, Till J, McCulloch E: The proliferative state of mouse granulopoietic progenitor cells. Proc Soc Exp Biol Med 134:33–37, 1970.
11. Messner H, Till J, McCulloch E: Interacting cell populations affecting granulopoietic colony formation by normal and leukemic human marrow cells. Blood 42:701–710, 1973.
12. Miller R, Phillips R: Separation of cells by velocity sedimentation. J Cell Physiol 73:191–201, 1969.
13. Julius M, Simpson E, Herzenberg LA: A rapid method for the isolation of functional thymus-derived murine lymphocytes. Eur J Immunol 3:645–649, 1973.
14. Handwerger B, Schwartz R: Separation of murine lymphoid cells using nylon wool columns and recovery of the B cell enriched population. Transplantation (Baltimore) 18:544–547, 1974.
15. Shortman K, Williams W, Jackson H, Russell P, Bynt P, Diener E: The separation of different cell classes from lymphoid organs. IV. The separation of lymphocytes from phagocytes on glass bead columns, and its effect on subpopulations of lymphocytes and antibody forming cells. J Cell Biol 48:566–597, 1971.
16. Krueger R, Staneck L, Boehlecke J: Tumor-associated antigens in human myeloma. J Natl Cancer Inst 46:711–715, 1976.
17. Zucker-Franklin D, Grusky G: The identification of eosinophil colonies in soft agar cultures by differential staining for peroxidases. J Histochem Cytochem 24:1270–1272, 1976.
18. Potter M: Immunoglobulin-producing tumors and myeloma proteins of mice. Physiol Rev 52:631–719, 1972.
19. Izaquirre C, Minden M, McCulloch EA: A colony assay for normal and malignant B-lymphocyte progenitors. Blood 54(Suppl 1):172a (abstr 448), 1979.
20. Shadduck R, Metcalf D: Preparation and neutralization characteristics of an anti-CSA antibody. J Cell Physiol 86:247–252, 1975.
21. Hamburger AW, Kim MB, Salmon SE: In Baum SJ, Ledney G (eds): "Experimental Hematology Today." New York: Springer-Verlag, 1979, pp 155–162.

4
Soft-Agar Cloning of Cells From Patients With Lymphoma

Anne W. Hamburger, Stephen E. Jones, and Sydney E. Salmon

Lymphomas are relatively common hematologic malignancies, and substantial information has been observed concerning their immunologic features. Tumor is often quite accessible for biopsy from superficial lymph nodes or bone marrow. Additionally, tissue is often available from an initial staging laparotomy.

Although we initially observed tumor-colony growth in a few cases of lymphoma using conditions developed for myeloma, the optimal conditions favoring growth of lymphoma colonies appeared to differ from those promoting growth of multiple myeloma and ovarian cancer colonies. We therefore studied a series of laboratory conditions that would more reliably promote growth of lymphoma colonies in vitro. Our initial results in lymphoma were recently reported [1].

MATERIALS AND METHODS

Patient Population

After obtaining informed consent, we studied 65 patients with non-Hodgkin lymphoma. All patients had advanced disease (stages III or IV by the Ann Arbor classification), and all lymphomas were diagnosed according to the Rappaport histological classification [2]. There were nine patients studied prior to treatment, 22 at the time of relapse, and the others while in remission. Of the 65 patients, 27 had nodular lymphoma (lymphocytic, 23; mixed cell type, 2; histiocytic, 2); 38 had diffuse lymphoma (lymphocytic, poorly differentiated, 22; lymphocytic, well-differentiated, 5; histiocytic, 7; lymphoblastic (T cell), 2; undifferentiated, 1; hairy cell leukemia, 1).

Collection of Cells

Bone marrow cells were prepared as described for myeloma in Chapter 3. Lymph nodes or spleen samples were minced with a scalpel and teased apart with 22-gauge needles to form a single-cell suspension. Viability, as determined by exclusion of trypan blue, was more than 90% if the cell suspensions were prepared within 2 hr of surgery. Pleural effusions were collected in heparinized (10 U/ml) vacuum bottles. The cells were centrifuged out at 150g for 10 min, and washed twice in Hank's balanced salt solution (HBSS) and 10% heat-inactivated fetal calf serum (FCS).

Culture Assay

Cells were cultured in the 2-layer agar system as described earlier. Briefly, 1-ml feeder layers containing 0.25 ml of medium conditioned by the adherent spleen cells of mineral-oil-primed BALB/c mice in 0.5% Bacto-agar were prepared in 35-mm plastic Petri dishes. In other experiments, a variety of media (as described in Results) were substituted in this feeder layer. One of the media, "1788 conditioned medium," was prepared as follows. A lymphocyte cell line, ATCC CCL156 (RPMI 1788), was obtained from the American Type Culture Collection, Rockville, MD. The line was originally derived by Dr. George Moore from the peripheral blood cells of a normal volunteer. Cells were cultured in RPMI 1640 medium with 20% FCS and antibiotics in stationary suspension cultures in 250-ml tissue culture flasks. The medium was collected when cells had reached a concentration of 1×10^6 cells/ml. For harvesting, cultures were centrifuged at 200g for 10 min and the supernatant medium collected and filtered through 0.22-μ Naglene filters. The cell-free conditioned medium was then aliquoted and stored at $-20°C$ for up to one month. In later experiments, filtered medium was ultracentrifuged at 67,000g to pellet extracellular EB virus or viral DNA. Lymphoma cells grew equally well in both types of media.

Lymphoma cells to be tested were plated over various feeder layers. They were suspended in 0.3% agar in enriched CMRL 1066 medium. 2-Mercaptoethanol (2-ME) was added to give a final concentration of 50 μM immediately before culture. Each culture received 5×10^5 bone marrow or effusion cells or 10^6 lymph node or spleen cells in 1 ml agar medium mixture. Cultures were performed in quadruplicate. All cultures were incubated at $37°C$ in 7.5% CO_2 in a humidified incubator. Cells growing in colonies were removed from dishes using a fine capillary pipette and routinely stained with Wright–Giemsa and methyl green pyronin for morphology, as well as for peroxidase, nonspecific esterase, and periodic acid, Schiff (PAS) reactivity. In addition, entire culture plates were routinely monitored for contaminating granulocyte-macrophage colonies by the peroxidase method of Zucker-Franklin and Grusky ([3] and Chapter 12).

For detection of surface-membrane immunoglobulin, slides of individual colonies were air dried, incubated for 30 min with 0.025 ml of a 1:6 dilution of fluorescein-conjugated monovalent rabbit antihuman kappa or lambda light-chain antiserum. Slides were then washed in three changes of phosphate-buffered saline (PBS) at $20°C$ and examined in a fluorescent microscope. Alternately, 1 ml of a 1:6 dilution of antiserum was gently pipetted over the culture plates. The antiserum was allowed to diffuse through the plate at $4°C$ for 4 hr. The antiserum was removed and the dishes bathed in PBS for three days at $4°C$, changing PBS twice daily. This washing step reduced background fluorescence significantly.

RESULTS AND DISCUSSION

Clinical and Pathologic Correlations With Putative Stem Cell Colony Growth

In patients whose bone marrows were histologically involved by lymphoma, colony growth was observed in four of eight patients with nodular lymphocytic lymphoma and in seven of ten patients with diffuse lymphocytic lymphoma. Growth was also noted from one other patient with Burkitt lymphoma (diffuse

TABLE I. Relationship of the Growth of Putative Lymphoma Colonies to Bone Marrow Involvement by Lymphoma*

	Lymphoma No. positive/no. cultured (%)	
	Nodular	Diffuse
Lymphoma present[a]		
NLPD	4/8 (50%)	DLPD 5/6 (83%)
		DLWD 2/4 (50%)
		Others 1/3 (33%)
Lymphoma absent[a]		
NLPD	0/11	DLPD 0/14
NM	0/2	DLWD 0/1
NH	0/2	DH 0/6

*Reproduced from Blood [1] with permission of the publishers.
[a]Bone marrows were evaluated by examination of aspirated marrow smears, clot sections, and core biopsies obtained at the same time as cultures.
Abbreviations: NLPD, nodular lymphocytic poorly differentiated; NM, nodular mixed cell type; NH, nodular histiocytic; DLPD, diffuse lymphocytic poorly differentiated; DLWD, diffuse lymphocytic well differentiated; DH, diffuse histiocytic.

TABLE II. Growth of Putative Lymphoma Stem Cell Colonies (No. Colonies/No. Cultured) From Normal and Affected Tissue Other Than Bone Marrow*

	Lymphoma type		
	Nodular	Diffuse	Normal tissue
Lymph nodes	0/2	3/4	0/8
Spleen	0/1	–	0/10
Pleural fluid	1/1	–	–
Soft tissue mass	–	0/1	–
Peripheral blood	–	0/2	0/10
Thymus	–	–	0/3

*Modified from Blood [1] with permission of the publishers.

undifferentiated). Cultures from one patient with hairy cell leukemia and another with a T-cell lymphoma failed to grow. Conversely, no colony growth was observed in 36 cultures from patients whose bone marrows were not involved by lymphoma (Table I). Likewise, ten bone marrows from healthy subjects failed to produce lymphoid colonies.

Table II summarizes our experience with cultures of various tissue shown to be involved by lymphoma. Colony growth was achieved from three of six lymphomatous lymph nodes and from a cytology-positive pleural fluid. We were unable to obtain growth of colonies from a spleen extensively involved by nodular lym-

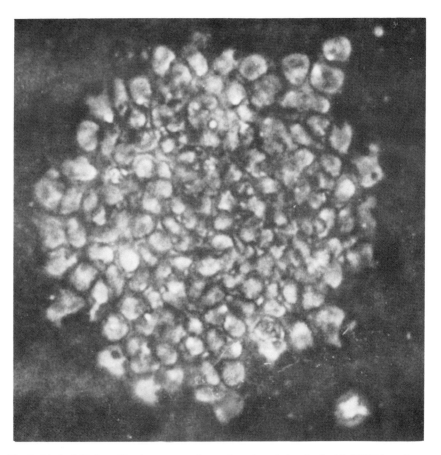

Fig. 1. Typical 10-day-old colony grown from a lymph node involved with DLPD lymphoma. Culture shown is from patient 1 (Table III) using the feeder layer containing 1788-conditioned medium plus 30% horse serum. Reproduced from Blood [1] with permission of the publishers.

phocytic lymphoma or from 30 normal spleens. No growth was observed after plating peripheral blood from two patients with leukemic phases of lymphoma. No growth was obtained from ten adult or two fetal thymuses tested. No growth was obtained from peripheral blood samples of 20 normals tested at cell concentrations of 5×10^5 cells per plate (Table II).

Characteristics and Identification of Colonies

The colonies stimulated by conditioned medium from adherent spleen cells of mineral-oil-primed BALB/c mice (BALB/c CM) consisted of tightly packed clumps of cells of uniform size regardless of cell source.

Cell doublings were usually observed within 24 hr of plating, and colonies of more than 40 cells appeared four days after plating. There was a progressive in-

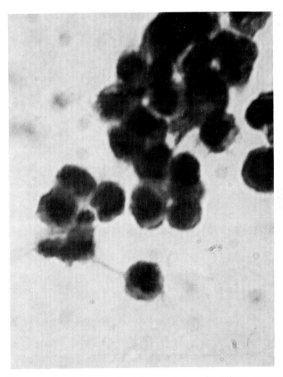

Fig. 2. Wright–Giemsa stain of cells plucked from a 7-day-old colony grown from a mediastinal lymph node histologically involved by diffuse lymphocytic lymphoma (intermediate differentiation). Colony growth was supported by the 1788-conditioned medium (patient 2 of Table III). Reproduced from Blood [1] with permission of the publishers.

crease in the number of cells that commenced proliferation during the first week of incubation. These colonies reached a peak size seven days after plating and degenerated after three weeks in culture. Cultures were not refed, and no attempts were made to serially subculture the clones.

The number of cells per colony sometimes reached several thousand, with the average number being about 800. A typical colony from a patient with a diffuse histiocytic lymphoma appears in Figure 1. Individual cells varied considerably in size from patient to patient but averaged about 10 μ in diameter. Cells picked from colonies had lymphoid morphology on Wright–Giemsa stain and were negative for peroxidase and nonspecific esterase activity (Figs. 2–4). Most, but not all, cells exhibited block positivity with the PAS stain. Cells from colonies from cultures from two patients with nodular lymphoma were tested for cell-surface immunoglobulin and expressed monoclonal light chain.

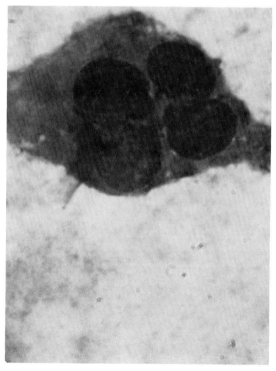

Fig. 3. Multinucleated giant cell (Wright–Giemsa stain) plucked from a colony from a patient with diffuse lymphocytic, poorly differentiated lymphoma.

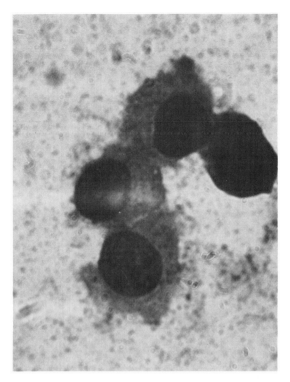

Fig. 4. Plasmacytic lymphoid cells (Wright–Giemsa stain) plucked from a colony from the patient with DLPD also depicted in Figure 3.

The colonies stimulated by 1788-conditioned medium resembled those induced by BALB/c CM but differed somewhat in their growth characteristics. For example, colonies from lymph node samples appeared about six days after plating and reached a peak size ten days after plating. They degenerated after about 14 days in culture. Individual cells were usually small (about 10–15 μ). The number of cells per colony averaged about 500 but could become as large as 1,000.

Overall, the number of colonies reached a maximum of $1,000/10^6$ cells plated, with an average of 150 colonies per plate. Thus, the plating efficiency was maximally 0.1% but often considerably less.

Factors Affecting Cell Growth

We had previously determined that cells from patients with non-Hodgkin lymphoma could be cloned in agar in the presence of 0.25 ml of conditioned medium

TABLE III. Effect of Substitution of Alternative Feeder Layers on Lymphoma-Colony Formation From Cell Suspensions of Lymph Nodes*

Patient	Diagnosis	Feeder layer	No. colonies/10^6 cells plated (mean ± SE)
1	DLPD	0.83 ml 1788-CM + 30% HS	100 ± 28
		0.83 ml 1788-CM + 15% HS	11 ± 6
		10^6 1788 cells	0
		Lysates 10^6 1788 cells	0
		BALB/c CM	9 ± 5
2	DLPD	0.83 ml 1788-CM + 30% HS	245 ± 59
		T cell CM	0
		BALB/c CM	0
3	DHL	0.83 ml 1788-CM + 30% HS	236 ± 9
		0.25 ml 1788-CM + 30% HS	100 ± 28
		BALB/c CM	9 ± 5

*Reproduced from Blood [1] with permission of the publishers.
Abbreviations: CM, conditioned media; HS, horse serum; DLPD, diffuse lymphocytic poorly differentiated lymphoma; DH, diffuse histiocytic lymphoma; 1788-CM, media conditioned by RPMI-1788 cells; BALB/c CM, media conditioned by adherent spleen cells of mineral-oil-primed BALB/c mice.

from adherent spleen cells of BALB/c mice. However, the overall success rate with this technique was less than 50%.

In order to develop a more reliable culture system, substitution of various feeder layers to support growth of lymphoma cells taken from involved tissues led to the following conclusions. Growth was induced by the substitution of 0.83 ml of RPMI-1788-conditioned media to the underlayer and by raising the horse-serum concentration in the CMRL layer to 30%. All of a small series of samples from lymph nodes formed colonies under these conditions (Table III). Two of these patients had diffuse lymphocytic, poorly differentiated lymphoma and one had diffuse histiocytic lymphoma. The addition of smaller volumes (ie, 0.25 ml) of 1788-CM with either 15% or 30% horse serum did not support colony growth (Table III).

Ten normal lymph nodes and 30 normal spleens failed to grow under these conditions. Peripheral blood cells from 18 normal volunteers failed to grow when plated at concentrations of 5×10^5 cells/ml. Galbraith et al [4], however, obtained growth of normal T lymphocytes in soft agar from peripheral blood cells stimulated by 1788-conditioned medium. However, they used a cell concentration of 10^6/ml. Cells from peripheral blood of two out of ten normal volunteers grew when we repeated their conditions.

Several other conditions failed to support colony growth (Table III). Addition of 10^6 cultured amniotic fluid cells to the underlayer of each culture failed to sup-

TABLE IV. Relationship of Clinical Status, Bone Marrow Findings, and Stem Cell Colony Growth in 25 Serial Studies in 12 Patients*

	Bone marrow No. positive bioassay/no. cultured (%)	
Clinical status[a]	Involved	Uninvolved
Before treatment	4/4 (100%)	0/3
Partial response	1/2 (50%)	0/3
Complete response	–	1/9
Relapse	0/1	0/3
Totals	5/7 (71%)	1/18 (6%)

*Reproduced from Blood [1] with permission of the publishers.
[a]Partial response, greater than 50% regression of disease on chemotherapy; complete response, no clinical, laboratory, radiographic, or histologic evidence of lymphoma.

port growth. Media conditioned by those cells failed to stimulate colony growth at dilutions of 1:4 and 1:8. Neither 10^6 freshly excised unfractioned thymus cells nor supernatants from three-day-old cultures of these cells stimulated lymphoma-colony growth. Addition of 0.02 ml of B pertussis antigen failed to stimulate lymphoma growth. Media conditioned for seven days by normal human spleen cells did not support lymphoma cell growth. Finally, 10^6 1788 cells added to underlayers failed to stimulate lymphoma growth. Lysates of the cells obtained by freeze thawing also failed to stimulate lymphoma cell growth.

The evidence suggesting that the growth observed was due to proliferation of lymphoma stem cells in vitro was as follows. First, cells picked from colonies were peroxidase and nonspecific-esterase negative and had lymphoid morphology with Wright–Giemsa stain. Second, most (but not all) colonies were PAS positive. Third, colonies exhibited monoclonal surface immunoglobulin characteristic of the tumor of origin. Fourth, the clinical conditions under which colony growth from tissues was observed is extremely strong evidence in support of the colonies being derived from lymphoma stem cells (Table IV). For example, from bone marrow samples, colony growth was noted in 11 (61%) of 18 bone marrows involved by either nodular or diffuse lymphocytic lymphoma. In contrast, only one of 49 pathologically negative bone marrows grew lymphoid colonies. This growth of colonies was not a random event of cultivation of normal B cell colonies, but was associated with known involvement of the bone marrow or other tissue by lymphoma.

However, more definitive studies of the identity of the cells in the lymphoid colonies is essential. Despite the excellent correlations between clinical state and colony growth in the non-Hodgkin lymphomas, we still consider the cells that gave rise to these colonies as putative lymphoma stem cells. Perhaps the best evidence for a malignant origin of these colony-forming cells would be demonstration of the chromosome 14 translocations recently reported for a variety of non-Hodgkin lymphomas by Fukuhara and Rowley [5]. In addition to cytogenetics, immunologic studies — including cell surface immunofluorescent and rosette analyses — would be essential to demonstrate colony cells were derived from the tumor of origin [6]. In addition, cells should be examined for the presence of EBNA and VCA antigens to demonstrate that cells growing in colonies were not the descendants of normal EB virus transformed B lymphocytes.

SUMMARY

Growth of cells from patients with lymphoma was promoted by feeder layers containing medium conditioned by adherent spleen cells of mineral-oil-primed BALB/c mice or by cells from a human B lymphocyte line (RPMI 1788). Sixty-five patients with all histologic types of non-Hodgkin lymphoma were studied. Lymphoid colony growth was obtained in 61% of bone marrows and 50% of lymph nodes histologically involved by lymphocytic lymphoma. Conversely, colony growth was observed in only a single instance from 49 bone marrows without overt lymphoma and was not observed in cultures of normal lymph nodes, spleens, bone marrow, peripheral blood, or thymuses. Colonies appeared within four days of plating and reached peak size in 7–10 days. Plating efficiency ranged from 0.001% to 0.1%. Morphological, histochemical, and immunological studies of cells from the colonies identified them as lymphoid, and sufficient evidence is available to designate the colony-forming units as putative lymphoma stem cells.

REFERENCES

1. Jones SE, Hamburger AW, Kim MB, Salmon SE: Development of a bioassay for putative human lymphoma stem cells. Blood 53:294–303, 1979.
2. Byrne GE: Rappaport classification of non-Hodgkin's lymphoma. Histologic features and clinical significance. Cancer Treat Rep 61:935–944, 1977.
3. Zucker-Franklin D, Grusky G: The identification of eosinophil colonies in soft agar culture by differential staining for peroxidase. J Histochem Cytochem 24:1270–1272, 1976.
4. Galbraith R, Goust J, Fudenberg H, Hure CJ: Induction of colonies by conditioned media from human lymphoid cell lines. J Exp Med 46:182–186.
5. Fukuhara S, Rowley JD: Chromosome 14 translocations in non-Burkitt's lymphoma. Int J Cancer 22:14–21, 1978.
6. Jaffe EX, Braylan RC, Namba K, et al: Functional markers: A new perspective on malignant lymphomas. Cancer Treat Rep 61:953–962, 1977.

5
Blast Colony Formation in Myelogenous Leukemia: Drug Sensitivity and Renewal Capacity of the Leukemic Clone

Ronald N. Buick

Over the last fifteen years methods have been developed to allow analysis of the hemopoietic cell-renewal system by assessment of culture clonogenicity. These techniques all depend on the use of viscid (methylcellulose) or semi-solid (agar) media, which allow the progeny of a single cell to remain localized as a colony; the presence in the culture media of specific stimulators have proven obligatory. The first such assay to be described was for the committed granulopoietic progenitor (CFU-C) in the mouse [1, 2]. These procedures were readily adapted to human granulopoiesis [3, 4], and now colony assays exist also for human erythropoiesis [5, 6], megakaryocytopoiesis [7, 8], lymphopoiesis [9, 10], and most recently, for a more primitive multipotent stem cell [11, 12].

Assays for committed progenitors have been applied to leukemic hemopoiesis in acute myeloblastic leukemia (AML) but thus far have failed to account numerically [13] for the major cell population in AML, the leukemic blasts. This has led a number of groups to study culture methods directed specifically to the blast cell population and its maintenance. These techniques have in common their use of phytohemagglutinin (PHA). Leukemic cells have been stimulated to grow either by preincubation with PHA [14], by culturing in suspension over agar with PHA in both phases [15], or by continuous feeding with conditioned medium from PHA-treated cells [16]. The colony assay on which this particular study is conducted is also based on stimulation by PHA-LCM [17, 18]. The characteristics of this assay have been reported; briefly, the cells of origin of the colonies are believed to be part of the blast population, since 1) a highly significant correlation was found between concentrations of morphologically identifiable blast cells and colony-forming cells in the peripheral blood of patients at diagnosis [19], 2) chromosomal abnormalities identified in the marrow or peripheral blood of certain patients were also found in cells of colonies [18, 20] and the cells were blasts by light and electron microscopy [21] and failed to express any markers of lymphoid cells [18].

The results of these initial studies were taken to mean that a minority subpopulation of leukemic blasts have sufficient proliferative capacity to be clonogenic and that, in fact, these cells represented a part of that population maintaining the tumor in situ. As such, these cells should be the critical population in terms of tumor response to therapy, and they also should display the biological characteristic of self-renewal (a necessary function for maintenance of the tumor, and the defining property of a stem cell). Accordingly, we have performed two studies, one to establish conditions for measurement of drug sensitivity of the clonogenic cells [22]; the other, to measure the self-renewal capacity of the leukemic clone(s) [23].

For the determination of drug dose survival curves, peripheral blood blast cells were exposed to various concentrations of the drug, washed extensively, and plated for colony formation. The dose response curves so constructed were found to be simple negative exponentials extending at least 2 logs, so they could be characterized by a D_{10} value. These values were found to vary from patient to patient over a 30-fold range. However, only weak correlation was found between D_{10} values for Adriamycin and the outcome of remission induction with an Adriamycin-containing regimen [24]. The simple negative exponential response of human leukemic clones to such cycle nonspecific drugs as in distinct contrast to the more complex curves observed with solid tumors, as illustrated in Chapters 16–18.

The procedures developed to study self-renewal are based on replanting colonies after a primary growth phase and assessing secondary colony formation (see Methodology). In contrast to in vitro drug sensitivity, measurement of self-renewal capacity in a series of 21 AML patients has shown considerable promise of defining a prognostic indicator [23, 25]. A significant association has been found between low capacity for self-renewal and succesful remission induction ($0.05 > P > 0.025$; Spearman's rank sum test).

The purpose of this chapter is to describe some preliminary experiments in which these two characteristics of the leukemic clone (drug sensitivity and self-renewal capacity) have been combined experimentally in an attempt to derive a value for the drug sensitivity of the self-renewal process.

METHODOLOGY

Colony growth, in vitro drug sensitivity, and self-renewal capacity were assessed as previously described [17, 22, 23]. Briefly, cells from the peripheral blood of AML patients are cultured in methylcellulose and growth medium supplemented by the addition of PHA-LCM. Small colonies (40–200 cells) of blast-like cells can be enumerated after five to ten days in culture. Routinely, peripheral blood mononuclear cells are depleted of T-lymphocytes prior to this colony assay [26].

Drug sensitivity of primary colony formation is assessed by constructing fractional survival curves for colony-forming cells versus increasing drug dose. Drug is administered as a pulse (ten minutes, 37°C, subsequent washing) or as continuous contact by incorporating the drug at the appropriate concentrations in the growth culture [22].

Self-renewal during colony formation can be considered to have occurred when one can demonstrate a cell(s) within a colony capable of yielding new colonies when re-plated. For the assessment of self-renewal of blast progenitors, cells are pooled from many colonies, rendered into a single cell suspension, and replated under the identical conditions as for primary colony formation. If the number of colonies contributing to the pooled cell suspension is sufficiently large, the value of PE2 is a measure of the progenitor's property of self-renewal (as opposed to the initial plating efficiency (PE1), which is a minimum estimate of progenitor frequency). Drug sensitivity of self-renewal has been assessed by quantitating the PE2 for pooled colonies surviving drug treatment. Primary colony formation is assessed in a dose-response curve to increasing quantities of drug in continuous contact. Surviving colonies are pooled at individual doses, washed, and rendered into a cell suspension. These cells are then replated at a concentration of 10^4 cells/ 0.1 ml medium in microwell plates, and secondary colony formation is assessed after an additional five to ten days in culture.

RESULTS

A feature of in vitro derived drug dose-response curves for leukemic blast colony formation is the fact that, almost invariably, they can be drawn as single negative exponetials extending into the third log kill. An example of such curves for one patient tested against pulse exposure to Adriamycin or m-AMSA is shown in Figure 1. The consistency of this finding is in contrast to the experience of dose-response curves derived from solid tumor clonogenic cells (Chapters 16 and 18). In such cases, marked heterogeneity has been seen in terms of the shape of the dose-response curves, and such extensive cell kill is rare. It is possible to conclude that such differences are reflective of the homogeneity of the clonogenic populations under assessment; diversity due to clonal evolution is much more apparent in solid tumor populations than in the leukemias.

Adriamycin, cytosine arabinoside, and methane sulfonamide, N-[4-9-acridinyl-amino)-3-methoxyphenyl] (m-AMSA) have been tested for their effect on the self-renewal capacity of blast progenitors. The dose-response curve for the effect of the drugs on primary colony formation was measured, and surviving colonies at a variety of doses were pooled (for an individual dose) and replated for measurement of PE2. Representative data for an Adriamycin experiment are shown in Table I. Primary colonies at all levels of survival had an apparently equal probability of self-renewal during development. However, for cytosine arabinoside (Table II) a frequent finding was a decrease in self-renewal in colonies surviving 10^{-6} to 10^{-7} M drug. For AMSA, the reverse situation was apparent (Table III). Clono-

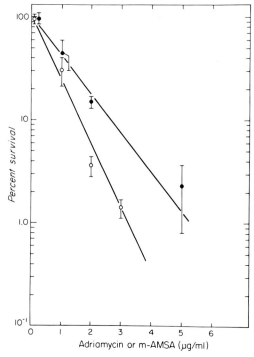

Fig. 1. Adriamycin and m-AMSA survival curves for blast precursors: leukemic blasts (T cell-depleted) were preincubated with drug concentrations, as shown, for ten minutes at 37°C, washed twice, recounted, and plated for colony formation. ○—○, Adriamycin; ●—●, m-AMSA.

TABLE I. Self-Renewal Capacity of Clonogenic Cells Surviving Adriamycin In Vitro (Pulse Exposure)*

Adriamycin concentration (μg/ml)	PE1 (colonies/10^5 cells)	PE2 (colonies/10^4 cells)
0	193 ± 8	37.5 ± 2.5
0.25	160 ± 12	49.6 ± 8.2
0.5	99.5 ± 4.5	38.0 ± 2.6
1.0	64 ± 11.5	46.5 ± 11.5
2.5	16.5 ± 4.5	32.0 ± 8.0

*Results are mean ± SE of quadruplicate plates.

TABLE II. Self-Renewal Capacity of Clonogenic Cells Surviving Cytosine Arabinoside In Vitro (Continuous Exposure)*

ARA-C concentration (M)	PE1 (colonies/10^5 cells)	PE2 colonies 10^4 cells
0	95.5 ± 40	26.5 ± 4.0
5×10^{-8}	170 ± 45	32.25 ± 6.0
10^{-7}	78 ± 5.0	26.5 ± 5.5
5×10^{-7}	34 ± 2.0	4.25 ± 2.0
10^{-6}	44 ± 0.5	9.25 ± 3.0
5×10^{-5}	2.5 ± 1.0	Insufficient cells
10^{-5}	0	–

*Colony counts are expressed as mean ± SE of quadruplicate plates.

TABLE III. Self-Renewal Capacity of Clonogenic Cells Surviving in m-AMSA In Vitro (Continuous Exposure)*

Dose AMSA (μg/ml)	Primary colony formation/10^5 cells (A)	Secondary colony formation/10^4 cells (B)	Ratio B/A
0	1186* ± 252	24 ± 2.4 (A)	1.0
10^{-5}	1085 ± 65	47 ± 2.5	1.95
10^{-4}	798 ± 77	53.7 ± 6.9	2.23
10^{-3}	398 ± 169	30 ± 3.8	1.25
10^{-2}	326 ± 66	22.7 ± 9.4	0.945
10^{-1}	213 ± 26	23.7 ± 5.3	0.98
1	11 ± 3.1	Insufficient cells	

*Colony counts are expressed as mean ± SE of 4 replicates.

genic cells surviving AMSA treatment had a higher (two- to threefold) probability of renewal during clonal expansion. This effect was also dose-dependent, being most apparent at 10^{-4} to 10^{-5} μg/ml.

The generality of these results can be seen in Table IV, in which the experience with cells from a number of patients is summarized. Cells of all six patients tested have shown no effect of Adriamycin on self-renewal; three of five patients have shown lower self-renewal probability in survivors of in vitro cytosine arabinoside treatment; and cells of all four patients treated with m-AMSA have shown increased self-renewal in the surviving fraction.

TABLE IV. Summarized Experience of Drug Effects on Self-Renewal (SR) and Primary Clonogenicity (PC)

Drug	Comparative effect of drug on SR or PC
Adriamycin	SR = PC (6)[a]
Cytosine arabinoside	SR < PC (3) or SR = PC (2)
m-AMSA	SR > PC (4)

[a]Numbers in parentheses indicate the number of patients studied.

DISCUSSION

The study described depends on the ability to measure the property of self-renewal independently of in vitro drug sensitivity. This can be achieved because practical measurement of self-renewal is derived from secondary colony formation and because drug sensitivity is a measure of reduction of primary colony formation (although the actual process of renewal does take place during primary colony development). There is no relationship between primary plating efficiency (PE1) and self-renewal capacity (PE2) [23]. Thus, the two progenitor properties can be measured independently. However, they should not necessarily be considered as biologically independent variables; preliminary analysis [27] has shown that Adriamycin sensitivity and self-renewal capacity are inversely related. Thus blast cell populations, which might be less agressive as judged by low self-renewal capacity in culture, are less responsive to chemotherapy. The biological basis for this relationship is under study.

The importance of studying the self-renewal process is underlined by the significant relationship with remission induction [23]. Such a relationship might be expected on a theoretical basis if one considered the parameter of self-renewal as related to the "aggressiveness" of the leukemic clone. The lack of prognostic significance of in vitro drug sensitivity measurements is in sharp contrast to the ability of such measurements to predict response in myeloma and ovarian cancer [28] (Chapter 18). Clearly, remission induction in AML is not governed solely by inherent drug sensitititivy. In this case, it seems that more prognostic significance will derive from studies aimed at biological characteristics of clonogenic cells, such as self-renewal. Additionally, in AML, other prognostic factors also may have significance in relation to the induction of complete remission (eg, sepsis, major bleeding, advanced age). The incorporation of such prognostic factors into a general equation that includes in vitro measures is discussed by Dr. Moon in Chapter 17. However, self-renewal appears to be an independent variable that does not correlate with these clinical variables [23]. In the long run, in vitro drug sensitivity, self-renewal, and clinical factors may all be important in a multifactorial analysis of the variance in complete remission induction.

The preliminary experiments presented in ths chapter point to a number of important areas for future development. The apparent selectivity of cytosine arabinoside in inhibiting self-renewal suggests a key biological role for this agent in successful treatment of AML. This would seem to be reflected in the regular inclusion of this agent in remission induction protocols. Adriamycin, however, would appear to have its function in a generally cytoreductive mode. It remains to be seen whether the patient-to-patient heterogeneity with respect to effect of cytosine arabinoside on self-renewal (Table IV) will be related to the outcome of attempted remission induction. From the point of view of the future selection of agents potentially useful in treatment of AML, on a theoretical basis it would be advantageous to choose agents that had an inhibitory effect on self-renewal. m-AMSA, as can be seen from Tables III and IV, consistently allows increased self-renewal in survivors of treatment, whereas on a molar basis, m-AMSA seems as efficient as Adriamycin in terms of cytoreduction. These characteristics would be consistent with a drug that was efficient in inducing remission, but that would be associated with a short duration of remission. The clinical experience with m-AMSA is sufficient to suggest that the drug will induce remissions in AML [29], but no information is yet available on remission duration.

This chapter provides an example of the complexity of concepts to be considered when studying leukemic hemopoiesis in AML. Colony assays such as the one described provide the specificity to allow dissection of these concepts into quantifiable laboratory parameters.

ACKNOWLEDGMENTS

This work was supported by a grant from the National Leukemia Association.

REFERENCES

1. Pluznik DH, Sachs L: The cloning of normal "mast" cells in tissue culture. J Cell Comp Physiol 66:319–324, 1965.
2. Bradley TR, Metcalf D: The growth of mouse bone marrow cells in vitro. Aust J Exp Biol Med Sci 44:287–299, 1966.
3. Senn JS, McCulloch EA, Till JE: Comparison of the colony forming ability of normal and leukemic human marrow in cell culture. Lancet 2:597–598, 1967.
4. Brown CH, Carbone PP: In vitro growth of normal and leukemic human bone marrow. J Natl Cancer Inst 46:989–1000, 1971.
5. Tepperman AD, Curtis JE, McCulloch EA: Erythropoietic colonies in cultures of human marrow. Blood 44:659–669, 1974.

6. Iscove NN, Sieber F, Winterhalter K: Erythroid colony formation in cultures of mouse and human bone marrow: Analysis of the requirement for erythropoietin by gel filtration and affinity chromatography on agarose concanavalin A. J Cell Physiol 83:309–320, 1974.
7. Nakeff A, Daniels-McQueen S: In vitro colony assay for a new class of megakaryocyte precursor: Colony forming unit megakaryocyte (CFU-M). Proc Soc Exp Biol Med 151: 587–590, 1976.
8. Vainchenker W, Breton-Gorius J: Megakaryocyte colony formation by blood and marrow precursors. Blood 52(Suppl 1):234, 1978.
9. Rozenszajn LA, Shoham D, Kalechman I: Clonal proliferation of PHA-stimulated human lymphocytes in soft agar culture. Immunology 29:1041–1054, 1975.
10. Radnay J, Goldman I, Rozenszajn LA: Growth of human B-lymphocyte colonies in vitro. Nature 278:351–353, 1979.
11. Fauser AA, Messner HA: Granulo-erythropoietic colonies in human bone marrow, peripheral blood and cord blood. Blood 52:1243–1248, 1978.
12. Fauser AA, Messner HA: Identification of megakaryocytes, macrophages and eosinophils in colonies of human bone marrow containing neutrophilic granulocytes and erythroblasts. Blood 53:1023–1027, 1979.
13. Lan S, McCulloch EA, Till JE: Cytodifferentiation in the acute myeloblastic leukemias of man. J Natl Cancer Inst 60:265–269, 1978.
14. Dicke KA, Spitzer G, Ahearn MJ: Colony formation in vitro by leukemic cells in acute myelogenous leukaemia with phytohaemagglutinin as stimulating factor. Nature 259: 129–130, 1976.
15. Lowenberg G, Hagemeijer H: Attempts at improving colony methods for human leukemic cells. J Supramol Struct 2:182, 1978.
16. Park C, Savin MA, Hoogstraten B, et al: Improved growth of the in vitro colonies in human acute leukemia with the feeding culture method. Cancer Res 37:4594–4601, 1977.
17. Buick RN, Till JE, McCulloch EA: Colony assay for proliferative blast cells circulating in myeloblastic leukaemia. Lancet 1:862–863, 1977.
18. McCulloch EA, Buick RN, Till JE: Cellular differentiation in the myeloblastic leukemias of man. In Saunders GF (ed): "Cell Differentiation and Neoplasia." New York: Raven Press, 1978, pp 211–221.
19. Till JE, Lan S, Buick RN, et al: Approaches to the evaluation of human hemopoietic stem cell function. In Clarkson B, Marks PA, Till JE (eds): Differentiation of Normal and Neoplastic Hematopoietic Cells. Cold Spring Harbor, New York: Cold Spring Harbor Laboratory, 1978, pp 81–92.
20. Izaguirre CA, McCulloch EA: Cytogenetic analysis of leukemic clones. Blood 52(Suppl): abstr 287, 1978.
21. McCulloch EA, Howatson AF, Buick RN, et al: Acute myeloblastic leukemia considered as clonal hemopathy. Blood Cells 5:261–282, 1979.
22. Buick RN, Messner HA, Till JE, et al: Cytotoxicity of Adrimaycin and daunorubicin for normal and leukemia progenitor cells of man. J Natl Cancer Inst 62:249–255, 1979.
23. Buick RN, Minden MD, McCulloch EA: Self-renewal in culture of proliferative blast progenitor cells in acute myeloblastic leukemia. Blood 54:95–104, 1979.
24. Buick RN, McCulloch EA: A colony assay for human leukemic blast cells used to measure anthracycline sensitivity in culture. Proceedings of the American Association for Cancer Research, Washington, D.C. Cancer Res 19:90, 1978.
25. McCulloch EA: Abnormal myelopoietic clones in men. J Natl Cancer Inst 63:883–891, 1979.

26. Minden MD, Buick RN, McCulloch EA: Separation of blast cell and T-lymphocyte progenitors in the blood of patients with acute myeloblastic leukemia. Blood 54:186–195, 1979.
27. McCulloch EA, Buick RN, Minden MD, et al: Differentiation programmes underlying cellular heterogeneity in the myeloblastic leukemias of man. In Golde DW, Cline MJ, Metcalf D, et al: "Hematopoietic Cell Differentiation." New York: Academic Press, 1978, pp 317–333.
28. Salmon SE, Hamburger AW, Soehnlen BJ, et al: Quantitation of differential sensitivity of human tumor stem cells to anticancer drugs. N Engl J Med 298:1321, 1978
29. Von Hoff DD, Howser D, Gormley P, et al: Phase I study of methanesulfonamide, N-[4-(9-aeridinylamino)-3-methoxyphenyl]-(m-AMSA) using a single dose schedule. Cancer Treat Rep 62:1421–1426, 1978.

6
Development of a Bioassay for Ovarian Carcinoma Colony-Forming Cells

Anne W. Hamburger, Sydney E. Salmon, and David S. Alberts

Epithelial tumors of the ovary are now the most common fatal gynecological cancer in the United States [1]. Despite this, the growth of ovarian tumor cells in vitro had not been extensively studied until recently [2–5]. We selected ovarian carcinoma for early application of our clonogenic assay for tumor colony-forming cells (TCFUs) [6] because tissue was readily available (from ascites) and because repeated biopsies were feasible. This facilitated systematic study of culture conditions for a variety of solid tumors. We have recently reported successful cloning of more than 85% of tumors from patients with ovarian cancer [5]. The study of these tumors in primary culture could provide much-needed information on the biology of ovarian cancers. Additionally, application of this assay to measurement of individual patient sensitivity to anticancer drugs and new-drug development appeared to be obvious practical applications of the methodology. Subsequent to our report, Von Hoff's group in San Antonio (Chapter 10) and Ozols, Young, and associates from the National Cancer Institute (Chapter 19), and Epstein in San Francisco (Chapter 21) have also investigated ovarian tumor-colony formation using this assay system.

MATERIALS AND METHODS

Patient Studies

Patients with well-documented ovarian carcinomas were selected for study. The international system for clinical staging of ovarian cancer was utilized [7]. Patients with ovarian cancer all had epithelial-type cancers. All patients had been previously treated with either radiation, chemotherapy, or a combination of the two.

Collection of Cells

Malignant effusions were collected in heparinized vacuum bottles (100 U/ml). Heparin was essential for obtaining usable cultures. After centrifugation at 150g for 10 min, the cells were collected and washed twice in Hank's balanced salt solution (HBSS) with 10% heat-inactivated fetal calf serum (FCS). The addition to 10 μg/ml of DNAase was often, but not invariably, helpful in eliminating clumps of tumor cells. The viable nucleated cell counts, determined in a hemocytometer with trypan blue, were routinely more than 90% if samples were used within 24 hr of collection. All effusions were stored at 4°C prior to plating.

Tumor nodules obtained immediately after surgery were mechanically dissociated under aseptic conditions in a laminar-flow hood. Tumors were minced with a scalpel and then teased apart with needles. Cells were filtered through sterile gauze to remove cell clumps, passed through 25-gauge needles, and then washed by centrifugation as described previously. Cell viability, as determined by trypan blue, was more than 90% if samples were obtained within two hours of surgery. Cultures were usually not successful if the viability of the cell suspension was less than 30%. Differential counts were performed on slides prepared with a cytocentrifuge and stained by the Papanicolaou and/or Wright-Giemsa methods [8].

Culture Assay for Ovarian Colony-Forming Cells

Cells were cultured as described earlier in this volume. One-ml underlayers containing 0.25 ml of Millipore-filtered medium conditioned by the adherent spleen cells of mineral-oil-primed BALB/c mice in 0.5% agar were prepared in 35-mm, plastic Petri dishes. In some instances, the same medium-serum combination was used without prior conditioning. Cells to be tested were suspended in 0.3% agar in enriched CMRL 1066 medium with 15% horse serum. 2-Mercaptoethanol (2-ME) was not added to the ovarian cultures (see Results). Each culture received 2×10^5 cells in 1 ml agar-medium mixture. Cells were incubated at 37°C in a 7.5% CO_2, humidified atmosphere.

Determination of Percentage of Cells in DNA Synthesis by H^3Tdr Suicide

The suicide method was used to measure the proportion of ovarian colony-forming cells in the DNA synthetic phase of the cell cycle [9]. Samples of 2×10^6 cells suspended in HBSS and 10% FCS were added to 1.0 ml of HBSS containing 40 μCi of H^3Tdr (23 Ci/mmole). Cell suspensions were incubated for 30 min at 37°C with shaking and washed twice with 50 ml of cold HBSS containing 100 μg/ml of cold thymidine and 10% FCS. Each suicide and control suspension was cultured in eight replicate plates at a concentration of 5×10^5 cells/plate.

Removal of Phagocytic Cells

One $\times 10^7$ cells obtained from effusions were incubated in 15 ml of McCoy's 5A medium with 10% heat-inactivated FCS and 40 ml of dry-heat sterilized carbonyl iron in 250-ml Falcon flasks. After incubation at 37°C for 45 min in a shaking water bath, the flasks were placed flat down on a magnet and supernatant cells were carefully decanted. This step was repeated as often as necessary (generally four to five times) to remove all the iron powder and iron-laden phagocytic cells. Cells were examined in a hemocytometer to determine that all iron-laden cells were excluded. The cells in the supernatant were considered nonphagocytic. Cells were washed twice in HBSS and 10% heat-inactivated FCS. This procedure removed between 30%–60% of cells.

RESULTS AND DISCUSSION

Development and Identification of Colonies

Cultures were examined in an inverted phase microscope immediately after the agar had hardened to determine if cell clumps were present. Any experiments in which a majority of plates contained clumps of more than eight cells were discarded. In general, when good single-cell suspensions were obtained, cell doublings were observed within 24 hr of plating and clusters of eight to 20 cells appeared within three to eight days. Individual cells appeared to increase in size within 24 hr of plating. Colonies (collections of 30 or more cells) appeared 10 to 14 days after plating. Cell lysis generally occurred 21 days after plating. During the first five days of incubation, there was a progressive increase in the number of cells that commenced proliferation. Colonies consisted of 30 to several hundred large (30μ) round cells (Fig. 1). Many cells were vacuolated with a characteristic signet-ring appearance. Cells in large colonies were tightly packed without the free-cell layer at the periphery that is characteristic of macrophage colonies. The morphology of the tumor colonies did not vary with histological type. Overall, tumor colony formation has been observed in 85% of the more than 100 ovarian cancer patients tested thus far.

The number of colonies ranged up to 2,000/plate, yielding a maximum plating efficiency of 1%. A linear relationship was obtained between the number of nucleated cells plated and the number of colonies found after 14 days (Fig. 2). Drug-sensitivity studies were valid only when such a linear relationship was obtained.

When cells from individual colonies were plucked from agar and stained by the Wright-Giemsa and Papanicolaou methods, they appeared to have the same morphological characteristics as did tumor cells in the original suspension.

In general, cells from colonies grown from patients with serous adenocarcinoma were oval with pale-staining cytoplasm. The nucleus was pleomorphic with one or more nucleoli. Multinucleated cells were occasionally present. Cells usually contained a moderate amount of PAS-positive material and occasional oil-red O-positive granules (Fig. 3).

Mucinous adenocarcinoma cells were often larger, polyhedral, and frequently contained secretory vacuoles. The nuclei were often small, irregularly shaped, and hyperchromatic when pushed to the periphery by the secretory vacuoles (Fig. 4).

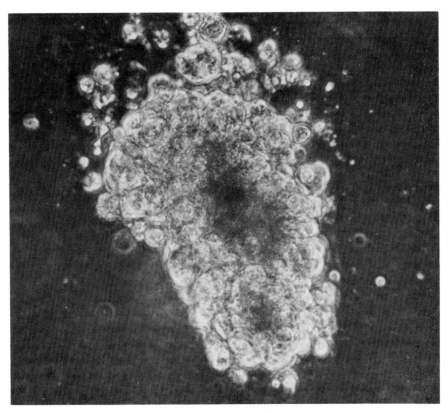

Fig. 1. Phase-contrast view of a typical ovarian cancer colony from an 11-day-old culture grown from a patient with serous adenocarcinoma in relapse. A pseudoacinar pattern was apparent (×435). Reproduced from Cancer Res [5] with permission of the publishers.

Endometroid cancer cells often showed PAS-positive material and were large, with large nuclei and prominent nucleoli.

Poorly differentiated cells, although pleomorphic with atypical nuclei, still showed faintly staining, PAS-positive material. However, many cells contained vacuoles that did not stain for either oil-red O or PAS. Multinucleated cells were often seen.

However, morphological features of cells from different patients with the same histological type varied. In addition, the appearance of cells from the same patient varied from colony to colony to a lesser degree. Moreover, the morphology and secretory activities of the tumor cells changed under culture conditions. Quite often, cells from patients with serous adenocarcinoma were more positive for PAS or oil-red O stains after culture. Individual cells appeared larger and more vacuolated than did cells in the original suspension.

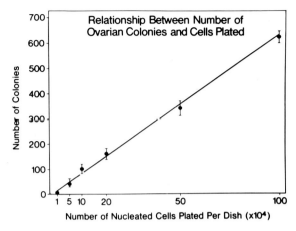

Fig. 2. Linear relationship between colony formation and the number of nucleated effusion cells plated. Points, mean of four dishes ± SE. These are the results of a typical experiment. Three other trials on different patients have yielded similar results. Reproduced from Cancer Res [5] with permission of the publishers.

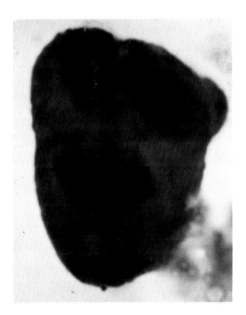

Fig. 3. Three cells plucked from an ovarian tumor colony from a patient with a mucinous adenocarcinoma of the ovary. The cytoplasm of the cells was strongly PAS positive ($\times 2,680$). Reproduced from Cancer Res [5] with permission of the publishers.

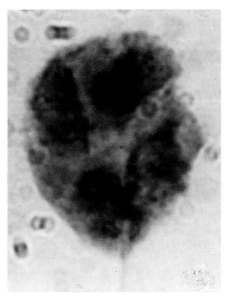

Fig. 4. Four cells plucked from an ovarian tumor colony from a patient with a serous adenocarcinoma of the ovary. The cytoplasm of these cells stained lightly with PAS (×2,680). Reproduced from Cancer Res [5] with permission of the publishers.

The use of a battery of morphological and histochemical stains thus indicated that cells picked from colonies had the same features as did the original tumor cells. However, routine staining techniques are often inadequate for the recognition of malignant cells in pleural and peritoneal effusions [10]. Even the use of histochemical stains for ovarian tumor cells is not completely specific [11]. The development and use of an ovarian tumor-specific antiserum, such as that raised by Ioachim et al [12] and Bhattacharya and Barlow [13] would be of use in confirming the malignant nature of the colony cells. In addition, cytogenetic analysis may reveal the presence of marker chromosomes in both the original tumor cell suspension and the cultured cells.

Effect of Histological Type on Cell Growth

The potential influence of histopathology on in vitro colony formation has been assessed on malignant effusions and solid ovarian tumor biopsies from 31 patients. Ovarian colony growth has been achieved from 75% of patients with serous adenocarcinoma and 100% of patients with mucinous, endometroid, or undifferentiated types (Table I). All patients but one were in relapse from treatment with multiple drugs or radiotherapy.

TABLE I. Effect of Tumor Type on Ovarian Colony Growth*

Tumor type	No. of successful cases/ total no. of cases		% Successful	Colony range
	Effusion	Solid tumor		
Serous adenocarcinoma	10/15	5/5	75	127–500
Mucinous adenocarcinoma	3/3	1/1	100	12–2,549
Endometrial adenocarcinoma	2/2	–	100	144–5,000
Undifferentiated carcinoma	5/5	–	100	20–220

*Reproduced from Cancer Res [5] with permission of the publishers.
Cells from patients with ovarian cancers of different histological types were plated in soft agar, and results were compared.

Cells from nonmalignant effusions (consisting of macrophages, lymphocytes, mature granulocytes, and mesothelial cells) obtained from patients in cardiac failure failed to form colonies at concentrations of 5×10^3 to 5×10^5 cells/plate.

Factors Affecting Cell Growth

Unlike myeloma, cells from the majority of patients with ovarian cancer did not require the presence of the BALB/c spleen-cell-conditioned medium for growth. For cells that did require a conditioning factor, substitution of type-O washed human RBC for the spleen-cell-conditioned medium permitted cell growth in all but one case (Table II). Substitution of medium conditioned by adherent spleen cells of CD-1, DBA/2, or BALB/c mice that were not oil primed decreased the number of ovarian colonies. Conditioned media from WI-38 cells, MA-184 cells (human bone marrow fibroblasts), or primary explants of human skin fibroblasts appeared cytotoxic. However, it is possible that such supernatants contained Mycoplasma or endotoxin that was responsible for the cytolytic effect.

Ovarian tumor cells (in contrast to myeloma cells grown in this system) did not require the presence of 2-ME for cell growth (Table III).

Cell suspensions were routinely plated over underlayers containing either BALB/c spleen-cell-conditioned medium, O RBC, or McCoy's medium alone. Such plates can be stored in the incubator up to one week. Requirements of tumor cells from individuals can be routinely determined this way.

Proliferative State of Ovarian Colony-Forming Cells

Incubation of cells with high-specific-activity H^3Tdr reduced colony formation to as little as 35% of control. No suicide effect was seen in two patients with undifferentiated cancer. We were unable to correlate suicide effects with tumor mass (Table IV).

TABLE II. Effect of Substitution of Alternative Feeder Layers on Ovarian Colony Growth*

Substituted material	Trial	No. of colonies/5 × 10⁵ cells		
		Control (oil-primed conditioned media)	Experimental	Media alone
RBC	1	306 ± 35	320 ± 40	80 ± 6
RBC	2	2,549 ± 39	833 ± 15	–
RBC	3	420 ± 30	300 ± 50	99 ± 20
RBC	4	380 ± 19	350 ± 30	–
Effusion	5	99 ± 9	29 ± 8	–
Heat	5	99 ± 9	38 ± 2	–
Pronase	5	99 ± 9	50 ± 4	–
CD-1 conditioned media	6	580 ± 27	320 ± 40	220 ± 30
DBA/2 conditioned media	6	580 ± 27	240 ± 36	220 ± 30
BALB/c conditioned media	6	580 ± 27	109 ± 9	220 ± 30
MA-184 conditioned media	7	420 ± 50	0	–
WI-38 conditioned media	7	420 ± 50	0	–

*Reproduced with modifications from Cancer Res [5] with permission of the publishers. Various growth factors were added to underlayers at a dilution of 1:4. Ovarian cells were then plated over feeder layers.

TABLE III. Effect of 2-ME on Ovarian Colony Growth*

Trial	No. of colonies/200,000 cells	
	+ 2-ME	– 2-ME
1	280 ± 14[a]	306 ± 48
2	183 ± 16	195 ± 21
3	91 ± 1	99 ± 9
4	100 ± 11	146 ± 35

*Reproduced from Cancer Res [5] with permission of the publishers.
[a]Mean ± SE.
Ovarian tumor cells were cloned in the presence or absence of 50 μM 2-ME.

Effect of Depletion of Phagocytic Cells

The number of ovarian colonies seen after depletion of phagocytic cells was decreased in every case. The addition of 50 μM 2-ME only partially restored ovarian cell growth (Table V). A number of authors have noted the stimulatory effect of macrophages on malignant cell growth [14–17]. Nathan and Terry [14] have demonstrated that DNA synthesis of many (although not all) mouse lymphomas was stimulated by normal mouse peritoneal macrophages. Similarly, Namba and Hanoka [15] reported that phagocytic cells or a glycoprotein released by them was re-

TABLE IV. Effect of H³Tdr on Human Ovarian Colony Formation*

Trial	Stage at diagnosis	Estimated tumor bulk	Source of tissue	Control (no. of colonies/500,000 cells)	% Survival of colony-forming cells H³Tdr
1	4	>1 kg	Tumor nodule	54 ± 6[a]	—
2	3	>1 kg	Tumor nodule	78 ± 8	—
3	3	>2 kg	Ascites	500 ± 19	65
4	3	>1 kg	Tumor nodule	183 ± 16	36
5	3	>1 kg	Ascites	233 ± 40	46
6	3	>1 kg	Ascites	47 ± 12	83
7	3	>1 kg	Ascites	20 ± 5.7	120
8	3	>1 kg	Ascites	120 ± 23	39

*Reproduced from Cancer Res [5] with permission of the publishers.
[a]Mean ± SE.
Ovarian cancer cells were incubated with high specific activity H³Tdr for 1 hour prior to plating.

TABLE V. Effect of Depletion of Macrophages on Ovarian Tumor Cell Growth*

		No. of colonies/500,000 cells	
Trial	Control	— Macrophages	— Macrophages + 2-ME
1	157 ± 30[a]	31 ± 6	44 ± 8
2	146 ± 35	17 ± 8	25 ± 14
3	99 ± 9	5 ± 3	40 ± 8
4	147 ± 20	58 ± 10	87 ± 7

*Reproduced from Cancer Res [5] with permission of the publishers.
[a]Mean ± SE.
Cell populations in malignant effusions from patients with ovarian cancer were depleted of phagocytic macrophages by ingestion of carbonyl iron prior to plating. The potential growth restorative effect of 2-ME (50 µM) was also tested.

quired for growth of MOPC-104E mouse myeloma cells in spinner culture. Calderon and Unanue [16] have also found a cell-proliferation factor for L1210 leukemic cells in the dialyzed supernatants from mouse peritoneal macrophages. Most recently, Hewlett et al [17] have found that L1210 cells, depleted of macrophages prior to subcutaneous injection in DBA/2 mouse, proliferate more slowly than control cells.

Whether the stimulatory effect seen here was due to direct contact between macrophages and tumor cells or to a macrophage-derived factor could not be deduced from these experiments. We speculate that the reason ovarian cancer often fails to spread outside peritoneal or pleural cavities may be related to the dependence of tumor cells on macrophages in such exudates for growth. General hypotheses on the potential "feeder layer effect" of macrophages as promoters of tumor growth have been formulated [18; and Chapter 23] in relation to our studies. Substantial additional efforts toward clarifying the relationship between endogenous macrophages and clonogenic tumor cells have therefore been pursued in our laboratories and are summarized in Chapter 11 by Buick et al.

As reported in Chapters 18, 19, 21, and 22, the ovarian tumor-colony system is quite suitable for in vitro sensitivity studies involving both cytotoxic anticancer drugs and biological response modifiers such as interferon. Human ovarian cancer may prove to be a "model tumor" for detailed studies of clonogenicity and drug sensitivity.

SUMMARY

We have reviewed the application of our in vitro assay for human tumor stem cells to the cloning of human ovarian adenocarcinoma cells in soft agar. Tumor colonies grew from both effusions and biopsies from 85% of more than 100 ovarian cancer patients tested. Up to 2,000 colonies appeared after 10 to 14 days in culture, yielding a maximum plating efficiency of 1%. Cells from nonmalignant effusions did not form colonies under these conditions. The number of tumor colonies was proportional to the number of cells plated between concentrations of 10^4 to 10^6 cells/dish.

Morphological and histochemical criteria showed that the colonies consisted of cells with the same characteristics as those of the original tumor. H^3Tdr suicide indices provided evidence that in most cases a high proportion of ovarian tumor-colony-forming cells were actively in transit through the cell cycle. Removal of phagocytic cells with carbonyl iron markedly reduced the plating efficiency, and 2-mercaptoethanol could only partially substitute for macrophages. Spleen cell-conditioned medium from oil-primed BALB/c mice was not required. Endogenous macrophages within the tumor may provide the conditioning factor or factors required for in vitro growth.

Thus, this assay is proving extremely useful for studying the biology and drug sensitivity of human ovarian cancer.

REFERENCES

1. Young RC: Chemotherapy of ovarian cancer: past and present. Semin Oncol 2:267–276, 1975.
2. DiSaia P, Morrow M, Kanabus J, Piechal W, Townsend DE: Two new culture lines from ovarian cancer. Gynecol Oncol 3:215–219, 1975.
3. Zaharia M, Samorlescu M: In vitro growth peculiarities of some human ovarian tumors. Morphol Embryol 21:283, 287, 1975.

4. Hecker D, Saul G, Wolf G: Untersuchungen zum einsatz von zellund organkultur bei zytostatika sensibilitatistestungen air menschlichen ovarialkarzinomen. Arch Geschwulstforch 46:34–43, 1976.
5. Hamburger AW, Salmon SE, Kim MB, Trent JM, Soehnlen BJ, Alberts DS, Schmidt HJ: Direct cloning of human ovarian carcinoma cells in agar. Cancer Res 38(10):3438–3444, 1978.
6. Hamburger AW, Salmon SE: Primary bioassay of human tumor stem cells. Science 197:461–463, 1977.
7. Technical Bulletin 47: Chicago: American College of Obstetrics and Gynecology, 1977.
8. Luna L (ed): "Manual of Histologic Staining Methods." New York: McGraw-Hill, 1968, pp 70–158.
9. Iscove N, Till J, McCulloch E: The proliferative state of mouse granulopoietic progenitor cells. Proc Soc Exp Biol Med 134:33–37, 1970.
10. Soendergaard K: On the interpretation of atypical cells in pleural and peritoneal effusions. Acta Cytol 21:413–416, 1977.
11. Long M, Hosannah Y: Comparative histochemical studies of l-leucyl-β-naphthylamidase in ovarian epithelial tumors. Cancer 19:909–925, 1965.
12. Ioachim H, Sabbath M, Anderson B, Barber HK: Tissue cultures of ovarian carcinomas. Lab Invest 31:381–390, 1974.
13. Bhattacharya M, Barlow J: Immunologic studies of human serious cystadenomas of the ovary. Demonstration of tumor-associated antigens. Cancer 31:588–595, 1973.
14. Nathan CF, Terry WD: Differential stimulation of murine lymphoma growth in vitro by normal and BCG-activated macrophages. J Exp Med 142:887–902, 1975.
15. Namba Y, Hanoka M: Immunocytology of cultured IgM-forming cells of mice. J Immunol 109:1193–1200, 1972.
16. Calderon J, Unanue E: Two biological activities regulating cell proliferation found in cultures of peritoneal exudate cells. Nature 253:359–361, 1975.
17. Hewlett G, Opitz HG, Schlumberger HD, Lemke H: Growth regulation of a murine lymphoma cell line by a 2-mercaptoethanol or macrophage-activated serum factor. Eur J Immunol 7:781–785, 1977.
18. Salmon SE, Hamburger AW: Immunoproliferation and cancer: A common macrophage derived promoter substance. Lancet 1(8077):1289–1290, 1978.

7
Soft Agar-Methylcellulose Assay for Human Bladder Carcinoma

Thomas H. Stanisic, Ronald N. Buick, and Sydney E. Salmon

As detailed elsewhere in this volume, the assessment of clonogenicity of human tumor cells has wide-reaching implications for the study of tumor biology and for the development of effective therapeutic modalities [1]. We describe here our initial experience of applying the concepts of short-term clonogenic assays to human transitional cell carcinoma (TCC) of the bladder. Although TCC is a moderately common human cancer, relatively few studies have been performed to analyze the biology of the tumor. Limited kinetic data are currently available [2, 3] and, recently, tissue culture in collagen-coated sponge has shown promise of defining important cellular characteristics [4].

The efficacy of treatment of this form of cancer has progressed very slowly and until recently has been limited to surgical and radiotherapeutic approaches. While locally recurrent superficial TCC can often be managed for relatively long periods of time, the invasive form of the disease is highly malignant and frequently fatal. Recently, chemotherapeutic approaches have been initiated both as primary therapy and in an adjuvant setting. However, knowledge with respect to this form of treatment is in its infancy. TCC, therefore, is a human tumor in which knowledge of the clonogenic population might have a profound effect on the development of more effective therapy. In addition, it will become apparent that the urothelium is a unique tissue for biological investigations because of the access of the investigator to cellular material using minimally invasive procedures.

METHODOLOGICAL APPROACHES

Patient Samples

Tissue and cell specimens were obtained by one of us (T.H.S.) from patients undergoing routine clinical care (from July 1978 to present) on the urology service at the Arizona Health Sciences Center and Veteran's Administration Hospital, Tucson, AZ. Bladder washings were obtained by irrigating the bladder with sterile 0.9% NaCl solution at the time of cystoscopy, prior to tumor resection. The recovered cell suspension was immediately placed on ice. Samples of solid tumor were obtained by transurethral biopsy under anesthesia. Generally, using 0.9% NaCl solution as the irrigating fluid to distend the bladder, early in our work tumors were biopsied transurethrally using cold-cup biopsy forceps to minimize tissue trauma. Later, we began using the resectoscope and a pure cutting current and glycine as an irrigating solution for tumor resection for study without any apparent effect on tumor viability or growth in vitro. After biopsies were obtained, the residual tumor was resected in the usual fashion using standard urological techniques and water as an irrigating solution. Pieces of tumor were immediately placed in sterile, iced Hank's balanced salt solution (HBSS).

Preparation of Cell Suspensions

Pieces of solid tumor were minced to approximately 1 mm in size and then teased apart using the tips of 21-gauge needles. The resulting suspension was passed through needles of decreasing size to 23-gauge. Cells were washed once with McCoy's Medium 5A containing 10% heat-inactivated (57°C, 1 hr) fetal calf serum (FCS) (Flow Laboratories, McLean, VA). Cells obtained by bladder irrigation were collected by centrifugation (1,250 rpm, 10 min) and then washed once with the same medium. Cell counts were performed by hemocytometer, and viable counts were assessed by trypan blue exclusion.

Assay for Colony Formation

The basic agar procedure used was that of Hamburger and Salmon [5] except that conditioned medium was not required [6]. Briefly, an underlayer of 0.5% agar in enriched McCoy's Medium 5A containing 10% FCS was prepared (1 ml in a 35-mm Falcon plastic Petri dish). Cells to be tested for colony formation were suspended in a plating layer of 0.3% agar in enriched CMRL Medium 1066 with 15% horse serum (Flow Laboratories, Rockville, Maryland). In certain experiments, the plating layer consisted of 1 ml of McCoy's Medium 5A containing the same enrichments and 0.8% methyl cellulose (Methocel; Dow Chemical Co., Midland, Michigan). This plating layer was used either on top of the conventional agar underlayer or alone. Cells were routinely plated at a concentration of 5×10^5 cells/ml in the 1-ml plating layer. Cultures were incubated at 37°C in a 7.5% CO_2, humidified atmosphere of air. Cultures were examined serially with an inverted microscope ($\times$ 100) and scored 7–28 days after plating. Colonies were defined to be aggregates of $>$ 40 cells and clusters as aggregates of 8–40 cells. For rapid evaluation by light microscopy, Papanicolaou stains of dried colony-containing plating layers were prepared on microscope slides ([7] and Chapter 12).

Electron Microscopy

The agar plates were prepared as detailed in Chapter 12, with glutaraldehyde fixation, following by thickening of the agar by warming, and dissection of the colonies followed by placement in 0.1 M cacodylate buffer (pH 7.4) containing 7% sucrose [8]. The blocks were postfixed in 2% osmium tetroxide for 1 hr at 4°C. After a rinsing in 0.1 M cacodylate buffer, the blocks were dehydrated in a series of increasing concentrations of ethanol and embedded in Epon [9]. Thin sections (600–900 Å) were prepared and stained with 8% aqueous uranyl acetate and lead citrate [10] and examined on a Philips 300 electron microscope.

Cytogenetic Analysis

Bladder colonies selected for cytogenetic analysis were harvested as described by Trent et al (Chapter 14).

LABORATORY STUDIES

Characteristics of Clonogenicity

Morphological characteristics of tumor colonies forming in agar from human bladder transitional cell carcinoma cells are shown in Figure 1. By light microscopy (Fig. 1, top left), the colonies have a very characteristic symmetrical spherical appearance with a clearly defined outer perimeter. Histological profiles of colonies were studied by means of toluidine blue-stained, 1-μ thick sections [6]. Cells appeared uniformly ovoid to cuboid, displaying large nuclei and prominent nucleoli. Well-defined cytoplasmic boundaries with prominent surface processes were observed. Findings of transmission electron microscopy of colonies were consistent with cells of epithelial origin. The colony surface (Fig. 1, bottom left) was uncovered on all sides by numerous surface microvilli. Additionally, elaborate folding and interdigitating of cell membranes was occasionally observed. High-power electron micrographs revealed cell junctions characterized by the presence of desmosomes (Fig. 1, top right).

Colony size varied from approximately 40 cells to greater than 1,000 cells, and the time to maximal growth varied from 7 to 28 days. Plating at cell concentrations between 1×10^4 and 2×10^6 cells/ml provided evidence of linearity between the number of nucleated cells plated and the number of colonies [6]. Single-cell origin of colonies could be directly observed by serial inverted microscopic studies of the cultures.
subsequent cellular manipulation. This procedure is much more easily achieved after growth in methyl cellulose rather than agar. We have, therefore, compared the clonal properties of the biopsy material in the conventional two-layer agar system, agar underlayer with methyl cellulose plating layer, and methyl cellulose

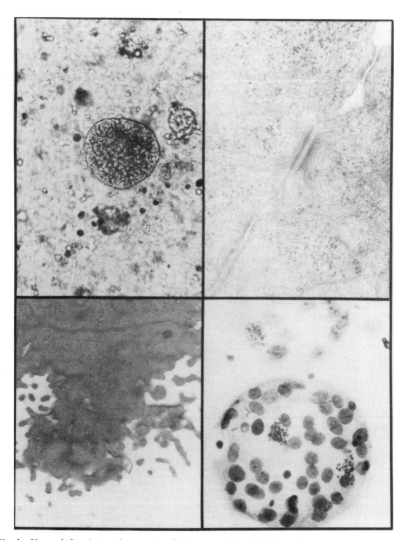

Fig. 1. Upper left: photomicrograph of 7-day transitional cell carcinoma colony as seen through inverted phase microscope. Upper right: electron micrograph showing a tight junction between two cells within a bladder colony (× 150,000). Lower left: electron micrograph demonstrating microvilli on the surface of a bladder cancer colony (× 20,000). Lower right: photomicrograph of a bladder colony treated for metaphase visualization (Giemsa × 400) (see appendix 3, this volume).

alone [6]. In the plates containing only methyl cellulose (0.8%) as the semisolid support, clonal growth was slightly inhibited in one case and severely limited in the other. The morphology of the colonies, however, was markedly altered, with the cells now forming flat, spreading colonies on the bottom of the dish. In the plates containing an agar underlayer with a methyl cellulose plating agar, colony growth was quantitatively and qualitatively similar to that obtained in the two-layer agar system, but the colonies could be readily harvested (by dilution of medium).

Cytogenetic assessment of tumor cells is greatly enhanced by a study of these in developing bladder colonies. Colony architecture can be maintained throughout the harvesting procedures, and this allows visualization of the number and location of proliferating cells within individual colonies, as shown in Figure 1, bottom right. Results of the cytogenetic analysis of 451 mitotic figures from two cases of TCC revealed one tumor with a modal number of 68 and a second containing a bimodal distribution at 41 and 45. G-banding analysis of the latter case revealed only numerical deviation, with consistent loss of a C-group chromosome in those cells with 45 chromosomes and loss of C- and G-group chromosomes in cells displaying 41 chromosomes. No "normal" mitotic figures containing 46 chromosomes were observed in either case.

Quantitation of clonogenic cells from solid tumor biopsies was described previously [6]. Of note is the fact that neither number or size of colonies was related to tumor grade or stage. However, as discussed in Chapter 2 of this volume, such in-vitro-derived parameters are inherently inaccurate. Of more relevance to the question of grade and stage of tumor may be the biological characteristics of the clonogenic cells.

CLINICAL STUDIES

We have applied the laboratory methods described above to the study of two clinical problems: 1) characterization of the biologic potential of the urothelium; 2) in vitro drug-sensitivity studies.

Urothelial Biologic Potential

The urothelium of patients undergoing cystoscopy for a variety of clinical indications was studied using the bladder barbotage technique described above and elsewhere [11]. The patient population studied was divided on the basis of history, cystoscopic examination, biopsy and cytology findings into three categories: 1) control patients were those with no historical, cystoscopic, or cytologic evidence of current or previous bladder tumor; 2) a second group of patients had histologic confirmation of current bladder cancer based on tumor or random bladder biopsies; 3) finally, a third group of patients are best termed "suspicious." These individuals had no definite histologic or cytologic evidence of current malignancy but were suspicious because of either a past history of bladder tumor or current cytology or biopsy evidence of severe "atypia or dysplasia."

Eighty percent of 26 barbotage specimens from patients with confirmed bladder cancer formed clusters or colonies in agar. The mean number of clusters/colonies formed in this group was 171/500,000 cells plated. The range was 2–1,700 colonies per plate. In contrast, only 27% of 45 "control" patients demonstrated clonal growth with colony number ranging from 15 to 500 colonies per plate, with a mean of 94 colonies per plate. Seventeen "suspicious" patients were studied, and 70% formed clusters or colonies with a mean of 60 colonies/500,000 cells plated and a range of 5–350 colonies.

Based on the above data, we feel that urothelial samples obtained from patients with bladder cancer and from "control" patients with nonmalignant urologic disease exhibit differential clonogenic capacity in agar. The data from the suspicious group suggest that clonogenicity in this system may serve as a sensitive, early indicator of future neoplastic degeneration. Three of our suspicious group studied a year ago have subsequently developed evidence of frank urothelial malignancy [11]. However, the fact that 25% of specimens from control patients also cloned in agar indicates that colony formation per se is *not* a specific indicator of such urothelial malignant potential. Based on clinical experience, 25% of such patients do not go on to develop urothelial malignancies. Perhaps in such individuals we are measuring a hyperplastic reparative response to nonspecific urothelial injuries. Currently, we are carrying on studies of CEA surface antigens and electron micrographic studies of clonogenic cells in an effort to define their biologic characteristics more precisely.

In Vitro Drug-Sensitivity Studies

As discussed in Chapter 18 of this volume, considerable data are accumulating from studies of a variety of tumor types that in vitro drug sensitivity, as measured by clonogenic assay techniques, correlates well with in vivo drug sensitivity in the patient. Tumors best suited for studies of this type should be readily accessible for biopsy with minimum patient morbidity, easily disruptible into a single-cell suspension, and grow readily in soft agar. Bladder cancer satisfies each of these criteria. Tumor cell specimens are easily obtainable transurethrally by biopsy or bladder barbotage. Most transitional cell tumors are highly cellular and contain minimal fibrous stroma and, as such, are easily separated mechanically into a single cell suspension. Urothelial samples obtained by bladder barbotage are almost a single-cell suspension when obtained and require even less mechanical manipulation prior to plating. Finally, bladder cancer cells obtained by bladder biopsy or bladder barbotage grow readily in soft agar. Of 17 specimens obtained by transurethral bladder tumor biopsy, approximately 70% have formed 30 or more colonies/500,000 cells plated — growth sufficient to perform meaningful drug-sensitivity studies in vitro. Fifty percent of bladder barbotage specimens obtained from bladder cancer patients have formed more than 30 colonies/ 500,000 cells plated. Of special interest is a group of five patients with carcinoma in situ, a form of bladder cancer theoretically ideally suited to intravesical chemotherapy. All five patients studied to date have grown at least 30 colonies/plate.

Recently, we have begun to examine in vitro drug sensitivity of clonogenic bladder cancer cells in a small group of patients. Figure 2 illustrates a typical pattern of a patient's response to six cytotoxic agents. As might be expected, from data derived from clinical drug trials, five of the six agents tested fail to achieve significant cell kill in concentrations tested. At a concentration of 1.0 μg/ml of adriamycin, only 20% of clonogenic cells survived, however. Figure 3 demonstrates the spectrum of response seen in specimens from eight patients exposed in vitro to mitomycin-C, an agent used both systemically and intravesically in the treatment of bladder cancer. Drug-sensitivity studies of bladder cancer are just beginning, and we have not as yet attempted any in vitro/in vivo drug-sensitivity correlations. However, in vitro sensitivity patterns seen to date resemble

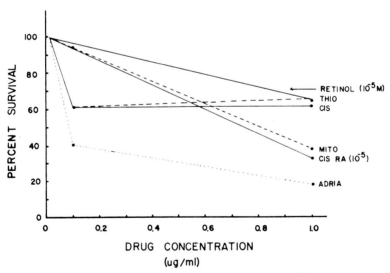

Fig. 2. In vitro response to clonogenic cells, from a patient with bladder cancer to cytotoxic agents. Cells were exposed for 1 hr to each cytotoxic agent as described by Salmon et al [12]. Each point plotted represents the mean counts from three separate plates. Minimal variation between plates for each point was noted, and standard deviation was not plotted in this figure or in Figure 3.

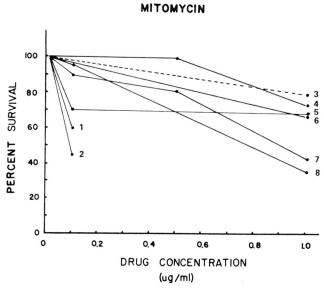

Fig. 3. In vitro response to mitomycin-C of clonogenic cells from eight different bladder cancer patients. Drug exposure was for 1 hr as described by Salmon et al ([12] and Chapters 16–18).

those seen in other tumors that have been studied in more detail. Chapter 18 summarizes the very positive in vitro/in vivo correlations, which have been observed with these other tumor types. Hopefully, in vitro/in vivo correlation with TCC will be similar as well. If so, this may open a new era for predictive testing and new drug screening specifically for bladder cancer.

REFERENCES

1. Steel GG: In "Kinetics of Human Tumors." Oxford: Clarendon Press, 1977, pp 216–267.
2. Fulker MJ, Cooper EH, Tanaka T: Proliferation and ultrastructure of papillary transitional cell carcinoma of the human bladder. Cancer 27:71–82, 1971.
3. Hainau B, Dombernowsky P: Histology and cell proliferation in human bladder tumors. Cancer 33:114–126, 1974.
4. Leighton J, Abaza N, Tchao R, Geisinger K, Valentich J: Development of tissue culture procedures for predicting the individual risk of recurrence in bladder cancer. Cancer Res 37:2854–2859, 1977.
5. Hamburger AW, Salmon SE: Primary bioassay of human tumor stem cells. Science 197:416–463, 1977.
6. Buick RN, Stanisic TH et al: Development of an agar-methylcellulose clonogenic assay for cells in transitional cell carcinoma of the human bladder. Cancer Res 39:5051–5056, 1979.

7. Salmon SE, Buick RN: Preparation of permanent slides of intact soft agar colony cultures of hematopoietic and tumor stem cells. Cancer Res 39:1133–1136, 1979.
8. Zucker-Franklin D, Grusky A: Ultrastructural analysis of hematopoietic colonies derived from human peripheral blood. J Cell Biol 63:855–863, 1974.
9. Luft JH: Improvements in epoxy resin embedding methods. J Biophys Biochem Cytol 9:409–414, 1961.
10. Reynolds ES: The use of lead citrate at high pH as an electron opaque stain in electron microscopy. J Cell Biol 17:208–212, 1963.
11. Stanisic TH, Buick RN: An in vitro clonal assay for bladder cancer: Clinical correlation with the status of the urothelium in 33 patients. J Urol 124:30–33, 1980.
12. Salmon SE, Hamburger AW et al: Quantitation of differential sensitivities of human tumor stem cells to anticancer drugs. N Engl J Med 298:1321, 1978.

8
Human Melanoma Colony Formation in Soft Agar

Frank L. Meyskens, Jr.

Advanced human malignant melanoma is an aggressive malignancy that responds poorly to available therapeutic manipulations. Our inability effectively to control this disease reflects in large part a fundamental gap in our knowledge about the biology of the disease and the lack of effective chemotherapeutic or hormonal agents. We have examined melanoma colony formation on fresh biopsies from 38 patients, and the results of these studies form the basis of this chapter.

RESULTS AND DISCUSSION

Preparation and Determinants of Assay

The standard mechanical cell preparation technique summarized in Chapters 4 and 6 was employed. In general it was not difficult to obtain single cell suspensions from melanotic metastases. Dispersion of amelanotic melanomas into single cells was somewhat more difficult, and cell recovery was usually lower than that obtained from melanotic melanomas. The use of enzymes in amelanotic melanoma needs to be explored.

The viability of the initial cell inoculum as determined by trypan blue exclusion was not related to tumor colony formation (Table I), a result different from that noted in other cell types. This suggests either that trypan blue is not an accurate measure of viability of melanoma clonogenic cells or that clonogenic melanoma cells may be more hardy than the general melanoma cell population.

The standard double-layer agar technique was used, and conditioned medium was not required for melanoma colony formation in vitro. To date 58% (22 of 38) of all histologically positive biopsies have resulted in melanoma colony formation. Details of the characteristic of melanoma colony formation in these patients are shown in Tables I and II. The source of the melanoma tissue does not appear to play a role in contributing to successful growth (Table I), although more samples from visceral sites are

TABLE I. Growth of Human Melanoma Cells in Soft Agar

Tissue site	Growth[a]	No growth[b]
Subcutaneous/skin	9	8 (3A)
Node	7	6 (1A)
Liver	1	1 (1A)
Lung	1	0
Brain	3	0
Vulva	1	1
	22	16

[a]Viability, 10–85%; [b]viability, 30–90%.
A = amelanotic on standard microscopic histopathology.

needed. Five of six visceral (brain, lung, liver), seven of 13 lymph node, and nine of 17 skin metastases formed colonies. Also, the source of the tissue was not related to the suicide index, cloning efficiency, or the type of colony formed (Table II).

An important practical consideration is whether tumor cells can be cryopreserved and tumor colony-forming units (TCFUs) successfully recovered. A limited number of studies suggest that melanoma colony-forming units can be recovered from melanoma cells frozen at $-80°C$ at 5×10^6 cells/ml in F10 medium containing 20% serum and 8% DMSO for up to three months. Seven of eight samples subjected to cryopreservation gave rise to melanoma cluster/colonies after thawing. However, the frequency of colony formation was slightly reduced in all instances (25–50%), and the cryopreserved clonogenic cells gave rise to smaller colonies (four cases) or formed only clusters (three cases). It will be important to develop cryopreservation techniques that will allow recovery of intact clonogenic cells with similar or identical biological and chemosensitivity responses to fresh clonogenic cells. The development of effective cryopreservation methodology will facilitate serial and systematic study of specific patients' melanomas.

Characteristics of Melanoma Colony Formation in Soft Agar

Two morphologic variants of colony formation were identified (Fig. 1): 1) groups of ten to 40 large, light (LL) cells (Fig. 1 A and B) and 2) groups of 15 to 50 small, dark (SD) cells. (Fig. 1A). The major morphologic variant in any patient did not seem to be related to the source of the tissue or to the cloning efficiency, which ranged from 0.003 to 0.02% for different patients (Table II). Suicide indices obtained with a one-hour exposure to 76 µg/ml hydroxyurea ranged from 0.10 to 0.87, and there was no apparent relationship to cloning efficiency, suggesting that the ability to clone in soft agar is not dependent on cell cycle status of the melanoma TCFU at the time of plating. However, as discussed in Chapter 13, additional studies with ^{3}H-thymidine suicide will be required to substantiate this conclusion.

TABLE II. Characteristics of Melanoma Colony Formation in Soft Agar

Patient	Tissue site[a]	Viability[b] (%)	Colony description[c]	SI (% survival)[d]	Cloning efficiency ($\times 10^{-5}$)
1	LN	—	—	—	0.8
2	SS	90	S*	—	2.0
3	SS	85	L, S	63	2.2
4	LN	80	S	—	4.2
5	LU	85	S	57	5.0
6	LN	—	—	10	5.6
7	BR	—	—	—	7.4
8	LN	25	L	74	6.2
9	SS	—	—	—	8.0
10	LN	60	S	—	8.1
11	SS	40	L	—	9.0
12	LU	—	—	—	9.8
13	SS	10	L, S	29	9.8
14	BR	—	—	—	10.0
15	LN	40	L	—	10.1
16	LN	—	—	32	10.5
17	SS	50	S	54	12.0
18	BR	60	—	—	17.0
19	SS	—	—	—	20.2
20	LN	—	L, S	87	20.2
21	LN	—	S	86	22.2

[a]LN, lymphe node; SS, skin/subcutaneous; LU, lung, BR, brain.
[b]Determined by trypan blue exclusion.
[c]L, 12–30 large, relatively amelanotic cells; S, 15–40 small, melanotic cells; S*, small, amelanotic cells.
[d]Suicide index (SI) determined by incubation of cells in 76 μg/ml hydroxyurea for 1.0 hour at 37°C, followed by standard plating.

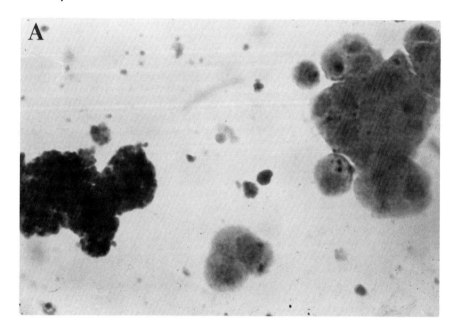

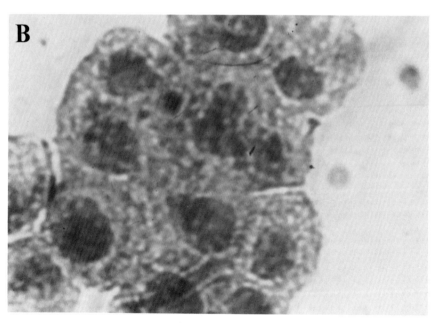

Fig. 1. Melanoma colony formation. Photomicrograph of cluster/colonies dried down onto slides and stained with Papanicolaou using the dried slide technique. A) Light, large cell colony variant and dark, small cell colony variant (× 100). B) Light, large cell colony in A (× 400).

Histological properties of melanoma colonies were assessed after preparation using the dried slide technique (Chapter 12). Colonies were stained with Panaicolaou. The presence of melanin was determined with either the Lilly melanin stain (positive in 60% of samples tested) or with Mishima's modification [1] of the Masson's ammoniated silver nitrate method for tyrosinase, premelanin, and melanin sites (positive in 92% of samples tested) (Fig. 6 in Chapter 12).

Effect of Host Cells on Melanoma Colony Formation

As detailed elsewhere in this volume (Chapter 11), host cells play an important role in determining tumor colony formation. In melanoma additional complex interactions may depend upon the colony variant present and the presence of host cells, including macrophages and lymphocytes. For example, in one patient (Fig. 2A) the number of colonies formed was linear between 10^4 and 10^6 cells plated in the initial cell inoculum. In a second patient (Fig. 2B) the number of colonies formed was nonlinear between one and 10^6 cells, and when phagocytic cells in the initial cell inoculum were removed by magnetic depletion after ingestion of carbonyl iron, colony formation was markedly decreased. It was fortuitous that, in this first patient, the major colony variant was the light, large cell type. We have subsequently analyzed the effect of magnetic depletion after carbonyl iron ingestion in an additional five patients and have determined that *only* the expression of the hypomelanotic, large cell colony variant appeared to be affected. Small cell, dark colonies were unaffected by this procedure, and in fact, the clonogenicity increased slightly at lower numbers of plated cells in the two cases in which this colony variant was predominant. A particularly interesting case is demonstrated in Figure 2C, in which control plates contained countable numbers of both colony variants at all cell numbers plated. Phagocytic depletion caused a marked decrease in the large, light colony variant, whereas no effect was seen on the small, dark colony variant. These results suggest that colony formation may be dependent on a factor released by phagocytic host cells (?macrophages) which selectively affects different melanoma progenitor cells.

A role of lymphocytes in melanoma growth and regression has long been suspected. We have recently turned our attention to this area and have investigated the effect of E-rosette depletion on melanoma TCFUs in four patients. In two patients in whom the major colony variant was large, light, a marked decrease in TCFUs occurred. In the two other patients the small, dark colony variant predominated in control plates, and after E-rosette depletion a twofold increase in TCFUs was seen. These preliminary results with phagocytic and E-rosette depletion of host cells in melanoma suspensions suggest that a complex immunological regulation of the proliferation of melanoma TCFUs exists and can be assessed with the in vitro clonogenic assay.

Effects of Different Conditions on Colony Formation

Conditioned media. Several conditioned media were also tested (Fig. 3). Media conditioned with Balb/C spleen cells, placental cells, and fibroblasts obtained from children and adults manifest some degree of inhibition of melanoma colony forma-

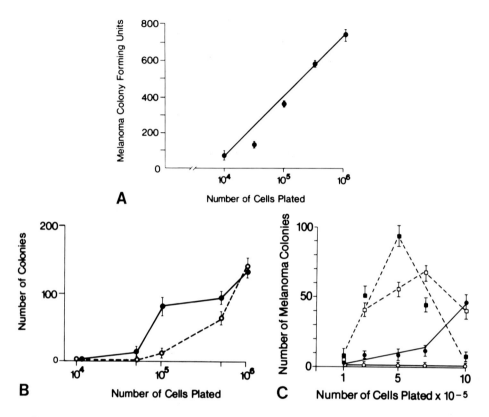

Fig. 2. Relationship of number of cells plated and number of colonies. The indicated number of cells was plated in the 0.3% agar overlayer. Source of cells was A) a large ulcerative supraclavicular nodule from patient 3: histological examination showed less than 5% nonmelanocytes; B) a large non-matted axillary lymph node from patient 2; C) a large subcutaneous metastasis from patient 1. Figure 2A shows the linear relationship between the total number of cells plated and the number of colonies formed. Figure 2B depicts the effect of carbonyl iron depletion of phagocytic cells on large light colony formation. ●——● untreated; □------□ phagocytes depleted. Figure 2C depicts similar data for a patient whose cultures contained both large light and small dark colonies. ■------■ small dark untreated, □------□ small dark, phagocytes depleted; ●——● large light, untreated, ○——○ large light, phagocytes depleted. Phagocyte depletion was carried out by shaking the cells with 40 mg carbonyl iron/10^7 cells for 2 hours at $37°C$ followed by magnetic removal of the iron-laden cells. The non-phagocytic cells were collected, recounted and plated at the standard numbers.

tion in vitro. Further studies with medium conditioned by host cells such as macrophages, lymphocytes, and fibroblasts that may play a role in normal and transformed melanocyte regulation should be informative.

Effects of growth factors on melanoma colony formation. Clinical observations and studies of animal melanoma suggest that the growth of this tumor is affected by hormonal compounds [2]. As part of our initial attempts to investigate culture con-

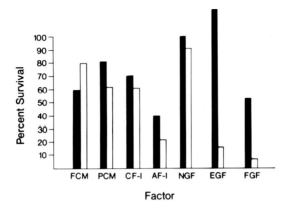

Fig. 3. Effect of continuous exposure of selected conditioned media and purified factors on melanoma colony formation. Conditioned media (used at 20% concentration in underlayer): FCM – mineral oil primed BalbC/mice (see Chapter 3 and Appendix 1); PCM – obtained from cultivation of 12-week placenta for two weeks; CF-1 – medium obtained from cultivation of neonatal forskin fibroblasts; AF-1 – medium obtained from cultivation of adult forearm skin fibroblasts. Purified factors (collaborative research): NGF – 5 units/ml (concentration in underlayer); EGF – 50 ng/ml; FGF – 100 µg/ml. ■ patient 8; □ patient 1.

ditions for human melanoma clonogenic cells, we have examined the effect of various growth factors on human melanoma colony formation in medium containing inactivated fetal calf serum (Figs. 3–5, Table III). We have tested the effects of epidermal growth factor (EGF) and fibroblast growth factor (FGF) on TCFU on biopsies from two patients (Fig. 3). Both FGF and EGF inhibited colony formation at the concentration tested. Nerve growth factor (NGF) (5 units/ml) appeared to have little effect on colony formation in the first two cases tested (Fig. 3). However, we have done dose-reponse curves in three subsequent patients in whom over 95% of the colonies in the control plates were the hypomelanotic large cell variant. The findings were similar in all three cases, and one case is presented in Figure 4. At low concentrations of NGF (< 1 ng/ml) colony formation was inhibited, whereas at higher concentrations the clonogenic frequency increased, and only the dark, small cell colony variant was expressed. In two additional cases in which the small, dark colony variant was predominant, NGF had little effect on colony formation. These results suggest that FSH, melatonin, and NGF modulate proliferation and differentiation of melanoma TCFU. Additional experiments will be required to clarify this phenomenon.

Effects of neuroendocrine hormones on melanoma colony formation. The melanocyte is embryologically derived from the neural crest, and we have therefore investigated the effect of single concentrations of melatonin, follicle stimulating hormone (FSH), and melanocyte stimulating hormones (MSH) on melanoma TCFUs from a

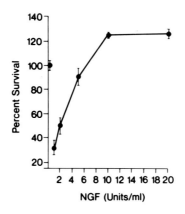

Fig. 4. Effect of nerve growth factor on human melanoma colony formation by cells from patient 8. Low doses of NGF inhibited proliferation relative to the control, whereas concentrations of greater than 10 units/ml restored proliferation or produced slight stimulation.

series of patients. Heterogeneity of response to these compounds was seen, and the results are summarized in Table III. The TCFUs from different patients exhibited differential responsiveness to the six neurohormones. MSH (2×10^{-7} M) reduced (3, 61, 65, 81%) TCFU in three melanomas and had no effect on one. FSH (2.5×10^{-9} M) reduced (10, 33, and 41%) TCFUs in three patients and increased (536%) TCFUs in one patient. Melatonin (10^{-6} M) reduced 10, 35, 61, 68, 69, and 76%) TCFUs in six patients and increased (24 and 54%) TCFUs in two patients. In those patients in which the number of TCFUs was affected by greater than 50% (either reduced or increased) and in which the LL colony variant was predominant, a shift to the SD colony variant occurred.

These results with single concentrations of neurohormones suggested that dose-response investigations would be of interest. Dose-response studies in three patients with predominantly the LL colony variant were performed. Both FSH (0.25–25 nM) and melatonin (10^{-5}–10^{-15} M) reduced TCFUs without affecting the colony variant in two patients, but in one patient (Fig. 5 A and B) a shift to the small, dark colony variant occurred and at optimal concentrations of hormone total TCFUs increased [FSH (53%), melatonin (950%)]. These results indicate that proliferation and differentiation of human melanoma TCFUs can be modulated by neurohormones and suggest that the disease process may be susceptible to neurohormonal manipulations.

Effects of retinoids on melanoma colony formation. Retinoids have been reported to alter the ability of carcinogen-exposed cells to express neoplastic potential and also to modulate the growth of transformed cells in animals in vitro. We have examined the effects of various retinoids (retinol, β-trans-retinoic acid, 13-cis-retinoic acid, and aromatic retinoic acid ethyl ester) on melanoma colony formation and have found

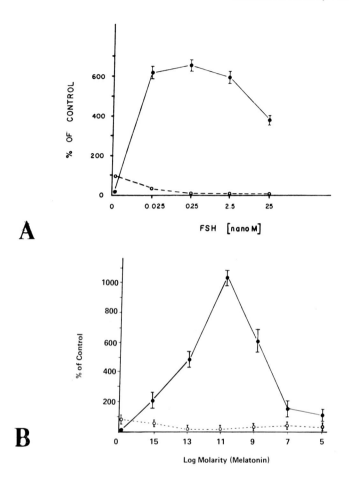

Fig. 5. Effect of neurohormones on human melanoma colony variants. The compound under study was present at the indicated concentration. A) Follicle stimulating hormone; B) melatonin. ●——● dark, small cell colony variant, ○------○ large light cell variant. FSH inhibited proliferation of the light, large cell colony variant, while proliferation of the dark, small cell variant markedly increased. Melatonin also inhibited the light, large cell colony variant. The proliferation of the dark, small colony variant was affected in a biphasic concentration-dependent manner.

that these compounds had variable inhibitory effects (Table IV). A more detailed report of the effect of retinoids on melanoma colony formation has recently been published [3]. These data with retinoids have provided us with a reasonable basis for a clinical trial of retinoids in advanced melanoma patients. The heterogeneous responses to retinoids suggest that combining these agents may also be fruitful.

TABLE III. Effects of Neurohormones on Melanoma Colony Formation*

Patient	Factor: MSH	Percent of Control FSH	Melatonin
A	19		24
B	100		41
C	35		39
D	36		31
E	21	100	–
F	–	64	82
G	31	–	52
H	–	130	588
I	–	62	75

*The conditions of the experiment were as in Materials and Methods except that the factor under study was included in the underlayer at the following concentrations: MSH (2×10^{-7} M), melatonin (1×10^{-6} M), and FSH (0.25 nM).

TABLE IV. Inhibition of Melanoma Colony Formation by Short-Term Incubation With Retinoids

Retinoid[a] Patient	Retinol	Percent of control 13-cis RA	β-trans RA	Aromatic RA (R-10-9359)
3	–	30	35	50
5	30	100	70	60
10	–	88	90	80
17	18	100	85	25
21	–	90	–	95

[a]Respective retinoid incubated at a concentration of 10^{-9} M with 5×10^5 cells. The treated cells are then washed twice and plated in the overlayer.

A theory of neurohormonal and immunological regulation of melanoma TCFUs. Our studies of biological factors on melanoma colony formation have led me to propose that major regulators of melanoma clonogenic cells include immune modulators and neurohormones. This theory is diagrammatically represented in Figure 6. It is reasonable to consider that neurohormones may influence clonogenicity of melanoma TCFUs since both melanocytes and most brain tissues are derived from the neural crest. A seldom appreciated fact is that the epithelial portion of the thymus

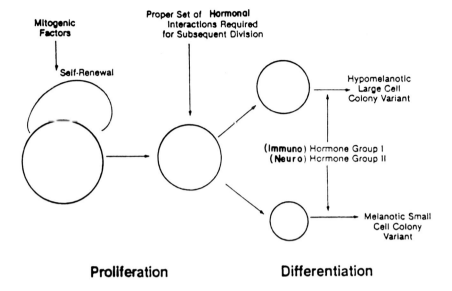

Fig. 6. Modulation of melanoma clonogenic cells by neurohormonal and immunological factors. Progenitor melanoma cell differentiation and proliferation is proposed to be regulated by neurohormonal and immunological agents. These stimulators may influence the expression of different pigmentary variants.

is also derived from neural crest tissue, and so involvement of thymic epithelial-produced factors or receptors that respond to such factors also needs to be studied. Further evaluation of control mechanisms influencing melanoma cell proliferation may justify clinical investigations of these agents in patients with melanoma.

Experience With In Vitro Therapeutic Modalities

Chemotherapy. Chemotherapy of metastatic melanoma is of limited effectiveness. The in vitro system may have several roles in this regard: 1) to suggest effective regimens of currently available drugs, 2) to detect sensitivities in particular patients to Phase II compounds, and 3) to screen new agents for possible clinical trials.

Using the drug sensitivity testing system outline in Chapter 18, we have obtained 37 in vivo/in vitro drug correlations for 26 patients. Although overall correlations are delineated in that chapter, it is clear that clonogenic melanoma cells are resistant to most anticancer drugs (also see [4]). Nevertheless, some interesting responses have been observed. In vitro sensitivity to cis-platinum was detected in a woman with vaginal melanoma. She was treated with this agent and responded with marked regression of skin lesions for three months. We also found that dacarbazine (DTIC) was active in vitro on some patients TCFUs, suggesting that liver microsomal activation of this agent may not be needed for in vivo effect. We took care to avoid light activation of

DTIC in these studies, and therefore conclude that the drug can be activated by cells present in melanoma suspensions. In many patients an intermediate sensitivity (or resistance) to melphalan and hydroxyurea was noted. We have initiated a trial to test the effectiveness of this combination in patients who have failed DTIC/actinomycin D. One of three patients treated with these three drugs had a partial response of bulk nodal disease for greater than six months. Future protocols will be based on a detailed analysis of cross-resistance tables as described in Chapter 18.

We have recently completed an "in vitro Phase II trial" in malignant melanoma [5]. The approach and some of our results are outlined in Chapter 22. These studies indicate that melanoma clonogenic cells are resistant to most Phase II agents, although vindesine and AMSA appear promising. A particularly useful aspect of this assay system is as a screen for new experimental agents. One such compound is methyl ester DOPA, which as a false DOPA analogue was proposed to affect melanoma growth. Our preliminary dose-finding studies against clonogenic melanoma cells (Fig. 7) indicates potential activity of this DOPA congener. Continuous contact of methl ester DOPA in the agar appears to be more effective than the standard one-hour exposure. Therefore, preclinical toxicology data will need to examine closely prolonged exposure to this agent. Should this agent be advanced to the clinic for Phase I-II study in melanoma, continuous infusions will be of greatest interest.

Heat. A considerable number of clinical studies have suggested that melanoma is susceptable to elevated temperatures. We have begun to investigate the effect of hyperthermia on melanoma TCFUs in an attempt to quantify this phenomenon. Our initial studies measured the effect of 42°C for one hour on the tumor cell suspension before plating. Heterogeneity of response of different patients' melanoma TCFUs was detected (Table V). These results are encouraging inasmuch as they suggest that the effect of hyperthermia can be quantitated and studied in our system.

Radiation therapy. Most clinical studies of melanoma confirm that this tumor is usually radioresistant. Several investigations suggest that this apparent resistance may be related to a broad shoulder on the survival curve. We have therefore tested the effect of high doses of radiation on melanoma TCFUs in 11 patients (Fig. 8). A greater than 70% decrease in survival of melanoma TCFUs was seen in four of 11 cases at 500 rads. The in vitro radiation results have encouraged us to test the effect of high fraction dose radiation in patients with locally aggressive melanoma.

Biological Heterogeneity of Melanoma Clonogenic Cells

The results that we have detailed here show that melanoma TCFUs are heterogeneous inasmuch as they exhibit differential responses from patient to patient to chemotherapy, host cell factors, endocrine agents, and retinoids. An important area of investigations will be to evaluate simultaneously responses of clono-

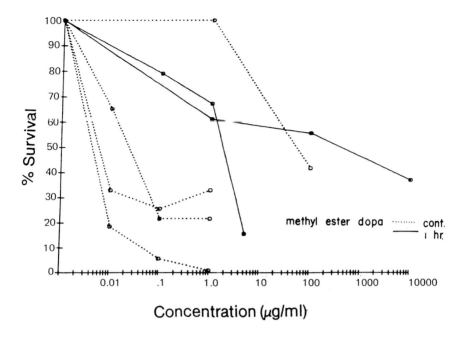

Fig. 7. Effect of methyl ester DOPA HCL on melanoma clonogenic cells from a series of patients. Survival of melanoma TCFUs appeared to be affected more by continuous incubation with the drug by one-hour exposure. Controls were not exposed to this agent.

TABLE V. Inhibition of Melanoma Colony Formation by Hyperthermia*

Patient	Percent survival
J	100
K	100
L	76
M	62
N	53
O	49
P	42
Q	25
R	20
S	9

*Melanoma tumor cell suspensions are incubated at 42°C for one hour and then plated. Simultaneously, controls for calculating percent survival for each patient were incubated at 37°C for one hour.

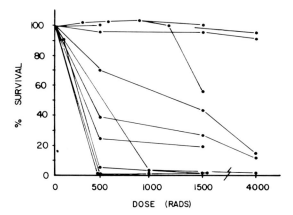

Fig. 8. Reduction of melanoma TCFUs by radiation. The tumor cell suspensions were irradiated at a rate of 1 rad/sec to the indicated total dose and then plated. Melanoma TCFUs from several patients appeared to be strikingly radioresistant, whereas others were relatively sensitive.

genic cells obtained from several biopsy sites in the same patient. We have recently begun such investigations. Melanoma may serve as a model for neoplasms expressing clonal heterogeneity. Definition of the cellular basis for such heterogenity may be of crucial importance in deining fundamental properties of neoplastic growth.

SUMMARY

Human malignant melanoma can form tumor colonies in soft agar, and at least two major morphological variants can be identified. Host cells play an important role in human melanoma colony formation, and their effects on the colony variants are selective. Neuroendocrine compounds also modulate expression of the melanoma TCFU pigmentary variants. This system appears suitable for evaluation of the response of melanoma TCFUs to chemotherapy, retinoids, heat, and radiation.

ACKNOWLEDGMENTS

I thank Reea Rodriguez for skilled secretarial assistance. Methyl ester DOPA HCl was provided by Dr. Leonard H. Kedda of the Drug Synthesis and Chemistry Branch, Division of Cancer Treatment, National Cancer Institute, Bethseda, Maryland. The 13-cis-retinoic acid (RO-43780) and aromatic retinoic acid ethyl ester analog (RO-10-9359) were kindly provided by Dr. E. Miller, Hoffman-LaRoche, Inc., Nutley, New Jersey.

REFERENCES

1. Mishima Y: Modification of combined DOPA-premelanin reaction: New technique for comphrehensive demonstration of melanin, and tyrosinase sites. J Invest Deratol 34:355–360, 1960.
2. Mather JP, Sato GH: The growth of mouse melanoma cells in hormone-supplemented, serum free medium. Exp Cell Res 120:191–200, 1979.
3. Meyskens FL, Salmon SE: Inhibition of human melanoma colony formation by retinoids. Cancer Res 39:4055–4057, 1979.
4. Meyskens FL, Salmon SE: Application of a tumor clonogenic survival assay to the study of drug sensitivity in human malignant melanoma. Clin Res 28:56A, 1980.
5. Salmon SE, Meyskens FL, Alberts DS, Soehnlen BJ, Young L: New drugs in ovarian cancer and malignant melanoma: In vitro phase II screening with the human tumor stem cell assay. Cancer Treat Rep (in press).

9
Cloning of Human Neuroblastoma Cells in Soft Agar

James T. Casper, Jeffrey M. Trent, and Daniel D. Von Hoff

The survival rate in neuroblastoma has not changed markedly over the last 20 years in spite of advances in treatment of other childhood tumors. This lack of survival improvement possibly is related to the fact that about 70% of neuroblastoma patients have widespread disease at the time of diagnosis and that effects of chemotherapy are variable and transient [1–3]. Clearly different approaches to this disease are needed to make an impact on survival.

The purpose of the present study was to employ a recently developed human tumor stem cell assay system in an attempt to grow neuroblastoma from a variety of clinical sources. This stem cell assay system has been reported to be selective for tumor growth [4–10; and numerous chapters in this volume]. In addition, it has been shown to be useful as a predictor for response of an individual patient's tumor to chemotherapy [9, 10; and Chapters 10 and 18].

We recently reported our use of this stem cell assay system for the study of neuroblastoma [11]. This chapter summarizes the growth of the cells in vitro, their morphology, karyology, and secretion of a tumor marker.

MATERIALS AND METHODS

Patient Populations

Only patients with pathologically documented neuroblastoma were chosen for this study. From October 1978 to October 1979, 18 patients with neuroblastoma were treated at the three institutions participating in this study. At the time of diagnosis 11 of these patients had stage IIIC disease, three were IIB, and two were stage IIA according to the St. Jude's staging classification [2]. The median age was two and one-half years; one of the 18 patients was less than one year of age. All of the 18 patients had abdominal primary tumors. Nine were female and nine were male. Marrow involvement was documented pathologically in 11 of the patients.

After obtaining informed consent according to federal regulations, a total of 71 bone marrow aspirations (~2.0 cc) were performed on the 18 patients. One patient had an open liver biopsy, one, a laparotomy for a solid tumor, and three had lymphadenectomy of superficial lymph nodes. These procedures were all performed as part of routine diagnostic work-up and follow-up studies. One-half of each specimen was sent for routine pathology studies, and the other half was sent from one of the investigator hospitals (JC) to the stem cell laboratory of the principal investigator (DVH). These specimens were sent blindly, without clinical information or results of pathologic examination of the aspiration or biopsy specimen known to the principal investigator. The average delay from sample collection to placement in culture was 48 hours. At the end of the study, patient identity was made known, and pathologic, clinical, and stem cell correlations were performed.

Collection of Cells

After informed consent, bone marrow cells were obtained by iliac puncture. Cells were aspirated into a syringe containing preservative-free heparin (100 units/ml). After centrifugation at 150g for ten minutes, the cells in the buffy coat were harvested with a Pasteur pipette and washed twice in Hanks' balanced salt solution (Grand Island Biological Co., Grand Island, NY) with 10% heat-inactivated fetal calf serum (Grand Island Biological Co.). Lymph nodes obtained immediately after surgery were mechanically dissociated under aseptic conditions. Nodes were minced with a scalpel, teased apart with needles, passed through 20-, 22-, and 25-gauge needles, and then washed by centrifugation as previously described [6]. The viability of both bone marrow and lymph node specimens was determined in a hemocytometer with trypan blue. Viability was routinely more than 90% for marrow specimens and 46–50% for the solid tumor tissue.

Culture Assay for Tumor Colony-Forming Cells

Cells were cultured as described by Hamburger and Salmon [4–7; and this volume]. Cells to be tested were suspended in 0.3% agar in enriched CMRL 1066 medium (Grand Island Biological Co.) supplemented with 15% horse serum, penicillin (100 μ/ml), streptomycin (2 mg/ml), glutamine (2 mM), $CaCl_2$ (4 mM), insulin (3 units/ml). Just prior to plating, asparagine (0.6 mg/ml), DEAE-dextran (0.5 mg/ml) (pharmacia Fine Chemical, Division of Pharmacia, Inc., Piscataway, NJ), and freshly prepared 2-mercaptoethanol (final concentration 50 μm) were added to the cells. One milliliter of the resultant mixture was pipetted onto 1 ml feeder layers in 35 mm plastic Petri dishes. The final concentration of cells in each culture was 5×10^5 cells in 1 ml of agar-medium mixture. One milliliter feeder layers were used in this study. The feeder layer consisted of McCoy's 5a medium plus 15% heat-inactivated fetal calf serum and a variety of nutrients as described by Pike and Robinson [12]. Immediately before use, 10 ml of 3%

tryptic soy broth (Grand Island Biological Co.), 0.6 ml asparagine, and 0.3 ml DEAE dextran were added to 4.0 ml of the enriched medium. Agar (0.5% final concentration) was added to the enriched medium, and underlayers were poured in 35 mm Petri dishes.

After preparation of both bottom and top layers, cultures were incubated at 37°C in a 7.5% CO_2 humidified atmosphere.

Scoring and Identification of Colonies of Cultures

Cultures were examined with a Zeiss inverted-phase microscope at ×20, ×100, and ×200. Colony counts were made five, ten, 15, and 20 days after plating. Aggregates of 50 or more cells were considered colonies. To identify more specifically colony morphology slides were prepared from the entire top layer by embedding the entire two layers of agar in paraffin after fixation according to the method described by Salmon and Buick [8; and Chapter 12]. Sections were cut with a microtome and stained with hematoxylin and eosin. These sections of colonies were compared to sections taken from the original tumor.

Marker Studies

Culture plates with neuroblastoma colonies were selected on days 7, 14, 21, and 28. Two milliliters of Hanks' balanced salt + 10% fetal calf serum were added to the top of the plates, and the plates were incubated at 37°C for 24 hours. The supernatant was recovered and sent for determination of catecholamines and VMA by standard colorimetric techniques [13].

Cytogenetic Analysis

Cytogenetic analysis of agar cultures was performed as described by Trent [14; Chapter 14 and Appendix 3]. Briefly, cultures were overlaid with 0.1 μm colchicine, the agar upper layer was removed and centrifuged, the upper layer was exposed to hypotonic treatment, tumor colonies were recentrifuged, and cells were fixed. Following preparation of air-dried slides, standard Giemsa and G-banding studies [15] were performed.

RESULTS

Development and Identification of Colonies

Cell doublings were usually observed within 12 hours of plating, and clusters of 8–20 cells usually appeared within three to five days. Colonies (collections of 50 or more cells) appeared eight to ten days after plating. Cell lysis generally occurred 35–40 days after plating. Cultures were not refed, and no attempt was made to subculture the colonies. The appearance of a typical culture plate on day 8 is shown in Figure 1. It is noteworthy that colonies developed at different

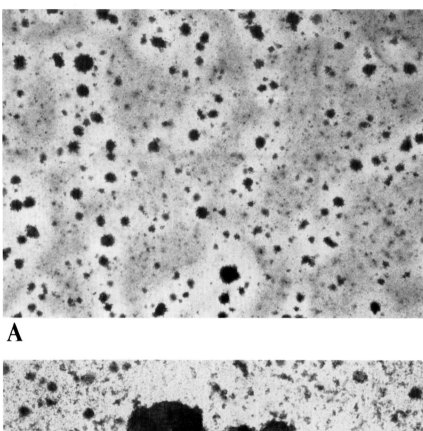

A

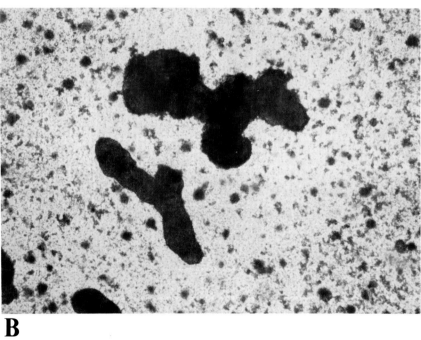

B

rates. Colonies were tightly packed and consisted of 50 to several hundred small uniform round cells (15 μm).

The number of colonies that grew from histologically positive bone marrows ranged from six to 7,214 per 500,000 nucleated cells plated, yielding a plating efficiency of 0.0012–1.44% (median plating efficiency = 0.005%; mean, 0.087%). A linear relationship was obtained between the number of nucleated cells plated and the number of colonies found after 15 days. The tumor plating efficiencies in two lymph node specimens were 0.0238% and 0.047% versus 0.004% and 0.1404%, respectively, in the bone marrow of the same patients.

To prove that the colonies growing in culture were indeed composed of neuroblastoma cells and not fibroblasts or granulocytes and macrophages, three approaches were used, including light microscopy, tumor marker determination, and karyotyping.

Light microscopy studies were performed with a variety of stains, including hematoxylin and eosin, Wright-Giemsa, Papanicoulau, peroxidase, and nonspecific esterase [16]. Hematoxylin and eosin stains demonstrated that colonies were composed of collections of cells slightly larger than lymphocytes that had deeply staining nuclei (Fig. 2). Mitoses could be seen in several of the colonies. Peroxidase stains of entire plates revealed an average of four colonies (range 0–5) per plate that were peroxidase positive. These probably represented granulocyte macrophage colonies and usually demonstrated the "starburst" pattern characteristic of those colonies.

Marker studies were performed on cultures from three different patients who had elevated levels of urinary vanillylmandelic acid (VMA), metanephrines, and catecholamines. Figure 3 demonstrates that the level of catecholamines in the tissue culture media increased with time in culture as compared to control plates with ovarian cancer colonies growing in them. Cytogenetic analysis was attempted on four tumors grown in soft agar, with three specimens yielding sufficient mitoses for cytogenetic analysis. Results from these tumors were derived from direct counts of 300 of the over 2,000 observed mitotic cells. In all cases, tumor colony mitoses were found to be near-diploid with minimal chromosome change. One patient displayed a hyperdiploid mode at 50 (with 10% polyploidy), the remaining two evidenced hypodiploidy, the first with a bimodal distribution at 43–44, the second with a modal number of 45. Interestingly, one patient displayed a "giant" marker chromosome approximately 1.8× the size of chromosome 1. This giant marker was present in approximately 80% of the 179 cells analyzed for modal number. None of the tumors sampled contained either double minute bodies (dm's), or homogeneously staining regions

---

Fig. 1. A. Typical culture plate demonstrating neuroblastoma colonies growing on day 8. Note heterogeneity of sizes of clusters and colonies (30× inverted microscope). B. Culture plate demonstrating neuroblastoma colonies growing together by day 27 in culture (30× inverted microscope). Reproduced from Cancer Res [11] with permission of the publishers.

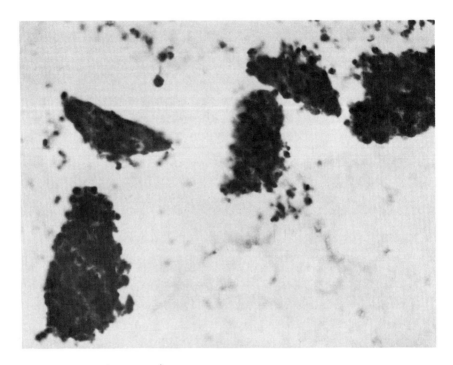

Fig. 2. Hematoxylin and eosin stain of neuroblastoma colony on day 21. Note colony is composed of a collection of cells that are slightly larger than lymphocytes and have deeply staining nuclei (300×).

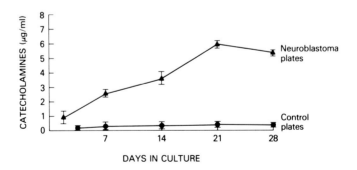

Fig. 3. Secretion of catecholamines into the supernatant by neuroblastoma colonies in culture. The quantity increases consistently up to 21 days in culture. Control plates contain ovarian cancer colonies growing under the same conditions.

(HSR's). Although dm's and HSR's are common elements of neuroblastoma cell lines, direct tumor samples have evidenced dm's or HSR's in less than 20% of cases (17). Also, the finding of near-diploid mitoses in direct samples of neuroblastoma is apparently a common feature of these cancers (17). Despite the lack of either dm's or HSR's in these three tumors, the presence of apparently stable clones with aneuploid chromosome numbers is suggestive of the malignant origin (rather than origin form normal cellular contaminants) of colony forming cells in vitro.

Sensitivity and Specificity of Method for Diagnosing Neuroblastoma

In all, there were 76 specimens (71 marrow, three lymph node, two abdominal tumors) for which both routine histologic examination and soft agar culturing results were obtained. Table I details the significant correlation between histologic and soft agar assay results ($P < 0.001$, chi square test corrected for continuity). There were 38 instances in which histology of bone marrow (34 specimens), lymph node (three specimens), or primary tumor (one specimen) demonstrated neuroblastoma cells. The agar culture showed colony growth (≥ 5 colonies per plate) in 30 of these instances (26 bone marrow, three lymph node, and one solid tumor) (true positives) whereas in eight instances the soft agar system did not show colony growth (false negatives). There were 38 specimens that were histologically negative for neuroblastoma. Thirty of these 38 specimens also showed no growth in the agar system (true negatives). However, eight histologically negative specimens formed colonies in the soft agar system (median of five colonies per plate; range 5–9). These eight histologically negative specimens were from six different patients. It is possible that the histologic examinations of these eight specimens in the six patients were falsely negative. Three of the six patients had histologically documented relapse of tumor of the bone marrow on subsequent marrow examinations within two months of the histologically negative studies. Two of the six patients with growth of colonies in the soft agar system and histologically negative marrows had had prior histologically positive

TABLE I. Correlation Between Histologic Findings and Soft Agar Colony Growth of Neuroblastoma*

		Colony +	Growth −
Histology	+	30	8
	−	8[a]	30

*Reproduced from Cancer Res [11] with permission of the publishers.
[a]These eight specimens were obtained from six patients. Five of the six patients did have positive histologies on prior or subsequent marrow examinations.

bone marrows two months before. It is possible the tumor had been present but was not detected histologically in these five patients. One of the patients with growth of colonies in the soft agar system never had nor ever developed a histologically positive marrow, and that patient's specimen is considered a definite false positive for the soft agar culture system.

The relationship between histologic bone marrow status and the number of colonies growing in culture is demonstrated in Figure 4. It should be noted that of the 38 histologically negative specimens only eight formed more than five colonies per plate in the soft agar system (an average of 1.8 colonies for all 38 specimens, range = 0–8, median = 0.8 colonies/plate), whereas for the 38 histologically positive specimens, 30 formed more than five colonies (an average of 404 colonies,

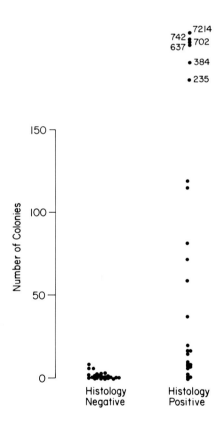

Fig. 4. The relationship between histologic findings in bone marrow (negative or positive for neuroblastoma) and the number of colonies growing in culture. See text for discussion. Reproduced from Cancer Res [11] with permission of the publishers.

range 0–7,314, median = 19 colonies/plate). There is a highly significant association between the number of colonies per plate and histologic status ($P < 0.005$, Wilcoxon rank-sum test).

DISCUSSION

These studies have demonstrated that tumor cells from patients with neuroblastoma can form colonies in soft agar. These colonies grew even from bone marrow that had been in the mail 48 hours after harvesting. The neuroblastoma colonies have light microscopic characteristics of the parent neuroblastoma. Tumor cells in the colonies secreted catecholamines in culture as they did in the patient. The levels of the catecholamines increased with time. This phenomenon correlates well with the recently described secretion of the tumor markers in the human tumor stem cell assay system [18; and Chapter 10]. Cytogenetic analysis revealed a high modal chromosome count in the cultured cells. Growth of one of five neuroblastomas in a soft agar system has been reported by McAllister and Reed [19], and growth of neuroblastoma from one patient was also reported by Hamburger and Salmon [4] and is illustrated in Figure 1 of Chapter 1.

One recurring question is whether the medium used in the stem cell system is specific for growth of tumor cells. Histochemical stains of slides made from the soft agar revealed that only an average of four granulocyte/macrophage colonies were present on each plate (5×10^5 nucleated cells plated). More important, this blind study has demonstrated that there is a very good correlation between histologic evidence of marrow involvement by neuroblastoma and growth of tumor colonies in the tumor stem cell assay system.

When compared to histologic examination of the marrow, the accuracy of the stem cell system for a positive diagnosis of neuroblastoma in the marrow was

$$\left(\frac{\text{true positive}}{\text{true positive + false negative}}\right) = 0.79 \text{ (sensitivity)}$$

and the accuracy of the soft agar system for finding a marrow free of tumor was also 0.79 (specificity). There were eight samples from six patients in which the soft agar assay grew ≥5 colonies per plate but the marrow was histologically negative (called a false-positive soft agar assay). In five of the six patients, the marrow had histologically documented neuroblastoma in the bone marrow within two months before or after the histologically negative marrow. Thus in these five cases it is conceivable that growth in the soft agar system was more sensitive than the histologic examination of the bone marrow. In some patients it might have been predictive of subsequent tumor relapse.

The eight instances in which the soft agar assay did not grow tumor from histologically positive marrows (false-negative stem cell assay) are bothersome. Reasons for this failure to grow could be related to recent chemotherapy (administered on the same day in three instances), plus sampling or technical problems. It is also possible that, even though tumor cells were seen on histologic examination, these tumor cells lacked ability to replicate because of chemotherapy-induced damage or lack of appropriate conditions in the culture system.

Since approximately 70% of children with neuroblastoma have marrow involvement at the time of diagnosis [3], tumor is accessible for sampling in the majority of patients. Because of tumor accessibility, studies using a soft agar culture system can serve as the basis for selection of the appropriate chemotherapy for an individual patient with neuroblastoma (as described in Chapter 10).

The ability to grow neuroblastomas from bone marrow, lymph node, and solid tumor specimens in a soft agar system has thus clearly been documented. Careful prospective clinical testing of this soft agar technique appears warranted to determine its utility in diagnosis, selecting the most effective drug for an individual patient with neuroblastoma, and for following the clinical course of patients with neuroblastoma.

ACKNOWLEDGMENTS

This work was supported in part by BRSG grant 507 BR-05654, awarded by Biomedical Research Support Grant Program, Division of Research Resources, National Institutes of Health; American Cancer Society institutional grant IN-116B; American Cancer Society grant CH162; National Cancer Institute research grant CA26226; and the MACC Fund (Midwest Athletics Against Childhood Cancer).

REFERENCES

1. Delta BG, Pinkel D: Bone marrow aspiration in children with malignant tumor. J Pediatr 64:542–546, 1964.
2. Green AA, Hustu HD, Palmer R, Pinkel D: Total-body sequential irradiation and combination chemotherapy for children with disseminated neuroblastoma. Cancer 38:2250–2257, 1976.
3. Maurer HM: Solid tumors in children. N Engl J Med 299:1345–1348, 1978.
4. Hamburger AW, Salmon SE: Primary bioassay of human tumor stem cells. Science 197:461:463, 1977.
5. Hamburger AW, Salmon SE: Primary bioassay of human myeloma stem cells. J Clin Invest 60:846–854, 1977.
6. Hamburger AW, Salmon SE, Kim MB, Trent JM, Soehnlen BJ, Alberts DS, Schmidt HT: Direct cloning of human ovarian carcinoma cells in agar. Cancer Res 38:3438–3443, 1978.
7. Buick RN, Stanisic TH, et al: Development of an agar-methylcellulose clonogenic assay for cells in transitional cell carcinoma of the human bladder. Cancer Res 39:5051–5056, 1979.

8. Salmon SE, Buick RN: Preparation of permanent slides of intact soft agar colony cultures of hematopoietic and tumor stem cells. Cancer Res 39:1133–1136, 1979.
9. Salmon SE, Hamburger AW, Soehnlen BJ, Durie BGM, Alberts DS, Moon TE: Quantitation of differential sensitivity of human tumor stem cells to anticancer drugs. N Engl J Med 298:1321–1327, 1978.
10. Salmon SE, Soehnlen BJ, Durie BGM, Alberts DS, Meyskens FL, Chen HSG, Moon TE: Clinical correlations of drug sensitivity in tumor stem cell assay. Proc Am Assoc Cancer Res, Am Soc Clin Oncol 20:340, 1979.
11. Von Hoff DD, Casper J, Bradley E, Trent JM, Hodach A, Reichert C, Makuch R, Altman A: Direct cloning of human neuroblastoma cells in agar: A potential assay for diagnosis, response and prognosis. Cancer Res (in press).
12. Pike B, Robinson W: Human bone marrow colony growth in vitro. J Cell Physiol 76:77–81, 1970.
13. Jacobs SL, Sobel C, Henry RJ: Specificity of the trihydroxyindole method for determination of urinary catecholamines. J Clin Endocrinol Metab 21:305, 1961.
14. Trent JM, Salmon SE: Human tumor karyology: Marked analytic enhancement via short-term agar culture. Br J Cancer (in press).
15. Sun NC, Chu EH, Chang CC: Staining method for the banding patterns of human mitotic chromosomes. Mammalian Chromosomes Newsletter 14:26–27, 1973.
16. Luna L: "Manual of Histologic Staining Methods." New York: McGraw-Hill, 1968, pp 70–158.
17. Brodeur GM, Sekhon GS, Goldstein MN: Chromosomal aberrations in human neuroblastomas. Cancer 40:2256–2263, 1977.
18. Von Hoff DD, Johnson GE: Secretion of tumor markers in the human tumor stem cell system. Proc Am Assoc Cancer Res, Am Soc Clin Oncol 20:51, 1979.
19. McAllister RM, Reed G: Colonial growth in agar of cells derived from neoplastic and non-neoplastic tissues of children. Pediatr Res 2:356–360, 1968.

10

Initial Experience With the Human Tumor Stem Cell Assay System: Potential and Problems

Daniel D. Von Hoff, Gary J. Harris, Gloria Johnson, and Daniel Glaubiger

Work on the cell kinetics of normal tissues has led to the concept that for every renewal tissue in an adult there is a subpopulation of stem cells. These stem cells are defined as cells that can reproduce themselves (capacity of self-renewal) and also give rise to a differentiating line of mature and functional cells [1]. In many tissues the identity and properties of stem cells have not been elucidated, but for the bone marrow and intestinal epithelium this has been an area of intense investigation [1–3]. It has been shown that the marrow stem cells make up a very small percentage (1%) of the marrow population [2]. Under normal circumstances they proliferate rather slowly. In intestinal epithelium, most of the cells in the crypts of Lieberkühn proliferate, rapidly migrating as a sheet onto the intestinal villi. Under normal circumstances only cells near the base of the crypts are the "effective" stem cells, because all of the progeny cells that are higher up on the villi are eventually lost by exfoliation [1].

Bearing in mind that many tumors retain some of the structural and morphologic characteristics of the tissue of origin, it is possible to conceptualize that tumors may also have stem cell populations. As discussed in Chapter 2, part of the definition of a tumor stem cell is that the cell gives rise to a large number of tumor-specific progeny cells and still has the ability to renew itself (make other stem cells). These tumor stem cells may be only a small proportion of the total number of tumor cells and they may be kinetically different from the majority of tumor cells [1].

It should be pointed out that at present the view that only a small proportion of cells in primary tumors are potential stem cells is only a hypothesis (Chapter 2). However, it is an attractive hypothesis that is gaining some scientific basis.

Cloning or stem cells may be detected in animal tumors by a variety of transplantation techniques, including the end-point dilution method [4], the spleen colony or lung methods [5, 6] regrowth assays [7], and in vitro cloning [8, 9]. It is uncertain whether these various techniques measure the same population of cells [1]. Because of recent developments outlined below, methods are now available to grow human tumor stem cells directly from biopsies in a clonogenic assay in semisolid media.

About 13 years ago, Bruce and colleagues at the Ontario Cancer Institute (OCI) demonstrated the potential for studying tumor stem cells from transplantable murine neoplasma by using a spleen colony assay [6]. Subsequently, Park et al from OCI developed and tested and tested an in vitro agar colony assay for transplantable BALB-C mouse myeloma in which irradiated tumor-inoculated spleen cells were used as a feeder layer [10]. They further showed that the results obtained for drug assays against the tumor in vitro were predictive for in vivo results [11]. Unfortunately, primary explanation of human tumors for colony formation has met with little success, the major problem being the creation of an environment that gives tumor cells a selective advantage over normal cells [12]. Two groups of investigators did, however, have some success in obtaining colony growth in a soft agar system with pediatric solid tumors (rhabdomyosarcomas and hepatoblastomas) [13, 14].

The major breakthrough in culturing progenitor cells of human tumors came with the work of Hamburger and Salmon [13, 15–19; and this text]. They devised a system using soft agar (with a bottom layer containing conditioned media from spleens of BALB/c mice primed with mineral oil) for assay of human myeloma stem cells [13, 15, 17; and Chapter 3]. Using this method, they have been able to grow colonies from 75% of 70 patients with multiple myeloma or related monoclonal disorders [13, 15; and Chapter 3]. The number of colonies that grew was proportional to the number of cells plated (making a quantitative test of drug sensitivity a possibility). Morphologic, histochemical, and functional criteria (including the presence of intracytoplasmic immunoglobulin) showed that the colonies growing in the agar were myeloma cells.

Using the same system, Hamburger and Salmon attempted to grow a variety of metastatic cancers with some success, including oat cell carcinoma of the lung, non-Hodgkin lymphoma, adenocarcinoma of the ovary, melanoma, and neuroblastoma [3]. They have recently extended their observations in ovarian carcinoma with 85% of 31 ovarian cancer biopsy and effusion specimens forming tumor colonies in vitro [16; and Chapter 6]. Morphologic and histochemical criteria confirmed that the colonies consisted of cells with the same characteristics as the original tumor. Results of cytogenetic studies were also consistent with a malignant origin for the tumor colonies.

The human tumor stem cell system has also been more extensively studied in non-Hodgkin lymphoma (Chapter 4). Lymphoid colony growth was obtained in 11 (61%) of 18 bone marrows microscopically involved by tumor and in three (50%) of six lymph nodes histologically involved by lymphocytic lymphoma. Conversely, colony growth was observed in only a single instance from 49 bone marrows without overt lymphoma and was not observed in cultures of four normal lymph nodes, two normal spleens, ten bone marrows, or six peripheral blood specimens.

The most important clinical development with the in vitro stem cell assay came from Salmon and colleagues with their report of using the system to quantitate the differential sensitivity of human tumors to various anticancer agents [19]. Using the in vitro stem cell assay, they performed 32 retrospective or prospective clinical studies in nine patients with myeloma and nine with ovarian cancer. These patients were treated with standard anticancer drugs that were also tested in vitro. Each tumor was cultured using the stem cell assay technique after incubation of the single cell suspension with various drug concentrations for one hour. The number of colonies that eventually grew out on drug-treated specimens was compared to the number of colonies on control plates. The data were expressed as colonies surviving versus drug concentration. In eight cases of myeloma and in three cases of ovarian carcinoma in vitro sensitivity* corresponded with in vivo sensitivity, whereas in one case of myeloma it did not. In vitro resistance correlated with clinical resistance in all five comparisons in myeloma and all 15 in ovarian cancer. These investigators concluded that the assay warranted larger scale testing to determine its efficacy for selection of new agents and for individualization of cancer chemotherapy regimens.

Their experience was recently updated [20]. There have been 22 in vitro-in vivo correlations available for 20 patients with ovarian cancer and 16 patients with multiple myeloma. Sixteen correlations demonstrated sensitivity in vitro and in vivo, eight showed sensitivity in vitro and resistance in vivo (false-positive test), one was resistant in vitro and sensitive in vivo (false-negative test), and 67 showed both in vivo and in vitro resistance. Overall, then, in myeloma and ovarian cancer the false-positive rate for the system is 8/92 (9%) and the false-negative rate is 1/91 (1%). A further updating of these data is summarized in Chapter 18 and the data are similar. These early correlations are impressive and certainly warrant rapid well-designed follow-up studies.

Over the past nine months our laboratory has been studying the human tumor stem cell assay system. This research has taken four major directions, which are discussed in the paragraphs that follow:

1) Types of tumors cultured
2) Confirmatory evidence that tumor is growing, rather than fibroblasts or granulocyte/macrophage colonies
3) Difficulties encountered with the system
4) Clinical applications of the human tumor stem cell system

*As defined by the area under the drug concentration curve.

INITIAL EXPERIENCE

Types of Tumors Cultured

Over the past nine months a total of 120 specimens have been received. All specimens were plated using the technique described by Hamburger and Salmon [13, 15], with the exception that no conditioned media were used in the bottom layer of agar. We defined cultures as positive when at least 5 tumor colonies grew in the plate. Table I describes our experience to date.

As can be seen in Table I, a rather large variety of tumor types have formed colonies using the soft agar technique. Overall, 295 of 401 tumors plated have grown (70%). Both adult and pediatric malignancies have been grown. Figure 1 depicts a typical ovarian cancer colony growing in soft agar on day 21.

Table II details the success rates of growing tumor colonies from the various sources of clinical specimens. As noted below, the solid tumor and lymph node specimens may be more difficult to grow because of technical problems in maintaining viability while making single cell suspensions.

TABLE I. Growth of Tumor Stem Cell Colonies From Various Human Neoplasms

Type of tumor (source of sample)	Number of patients with + culture/total tested
Ovarian	54/70
Neuroblastoma	42/52
Breast	14/19
Melanoma	20/32
Colorectal	14/20
Lung cancer	
Small cell	22/28
Squamous	24/26
Adenocarcinoma	22/24
Head and neck	14/32
Testicular	15/21
Multiple myeloma	6/10
Osteogenic sarcoma	6/10
Rhabdomyosarcoma	6/12
Endometrial	6/6
Pancreatic	3/3
Cervical	6/6
Ewing sarcoma	4/4
Renal	4/6
Hepatoma	4/4
Prostate	6/10
Thyroid	2/4
Wilms	1/2
Normal marrow	0/38

Our first conclusion, then, is that the human tumor stem cell system does grow a variety of types of tumors from a variety of sources (ascites, pleural fluid, solid tumor, marrow). The success of growing the tumor types is variable and depends *at least* upon both the type of tumor and source of the tumor specimen.

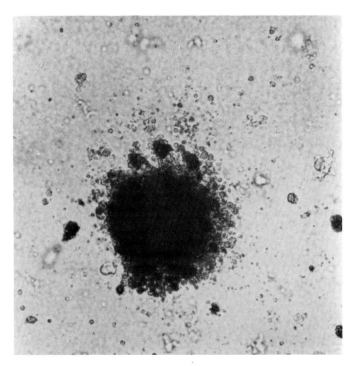

Fig. 1. Human ovarian cancer colony growing in soft agar (inverted microscope, 160 × magnification).

TABLE II. Growth of Tumor Stem Cell Colonies Categorized by Source of Specimen

Source of specimen	Number of patients with + culture/total tested (%)
Ascites	54/60 (90)
Solid	29/61 (48)
Marrow	46/53 (86)
Pleural effusion	17/21 (82)
Pericardial effusion	3/ 3 (100)
Lymph node	7/11 (66)

Confirmation That Colonies in Agar Are Human Tumors

We have used two approaches to the problem: histology, and tumor markers.

Histology. A technique was devised by Salmon and Buick [22; and Chapter 12] to make slides for histologic examination of tumor colonies in the soft agar. We have modified this technique and embedded the whole agar layer in paraffin and cut sections in the usual manner with a microtome. By histologic studies, we have confirmed that histologies of the tumor colonies in vitro are similar to the tumor in man for colon and breast carcinoma, melanoma, ovarian carcinoma, and neuroblastoma (see Chapter 9).

Secretion of tumor markers by cells in the stem cell system.

Melanogens. The tumors of two patients who had clinical melanosis actually turned the agar black after 12 days in culture. The agar contained melanogens (by the Thromalin test). None of the tumors in culture from patients without clinical melanosis secreted melanogens into the agar.

Catecholamines. We have cultured neuroblastoma from 18 patients (Chapter 9). Seven of these patients had elevated levels of urinary vanillylmandelic acid (VMA) and total catecholamines on several determinations. Tumor colonies from these seven patients secreted catecholamines into the agar. As shown in Chapter 9 (Fig. 1), the amount of catecholamines in the culture actually increased with time, whereas it did not do so in control plates of ovarian cancer.

E-rosettes. Cells from a T-cell lymphoma growing in culture formed colonies, and these colonies were plucked on day 21 and the cells dispersed. Each of the cells formed remarkably good E-rosettes with sheep red blood cells.

From the preliminary studies, there is good evidence that the tumor cells grown in culture are producing the same tumor markers that they are producing in the patient. This information provides additional evidence that the human tumor stem cell culture system reflects the in vivo situation. In addition, the system provides an attractive model for detailed studies of human tumor markers (ie, secretory rates, clonal differences, etc).

Problems With the System

Not all tumors grow in the system. As noted in Table I, it is clear that, despite considerable experience, we are not able to grow every patient's tumor nor every type of tumor. This problem may be one of improper growth conditions, lack of appropriate conditioned media, etc. At present efforts are being directed toward variations in media and plating methods to increase the success rate with which tumors can be grown.

Specificity. As noted in Table I, we attempted to grow colonies from six normal bone marrow specimens in the human tumor stem cell assay system. We did not have colony growth in any of the bone marrow cultures. However, other investigators using other soft agar techniques have grown colonies from hyperplastic tissue [13; and Chapter 7]. Therefore, work is being directed at helping to define the specificity of the system.

Low plating efficiency (PE). Table III details the plating efficiencies for five of the types of tumors cultured to date. As noted in Table III, the plating efficiency (number of colonies/number of nucleated cells plated) varies considerably both in terms of median plating efficiency from tumor type to tumor type and in terms of range of PE within a tumor type.

TABLE III. Plating Efficiencies in the Human Tumor Stem Cell System*

Tumor type	Median number of colonies/dish (range)	Median % plating efficiency (range)
Neuroblastoma	81 (12–20,000)	0.01 (0.002–4.0)
Ovarian	54 (6–650)	0.01 (0.001–0.1)
Breast	50 (4–60)	0.01 (0.0008–0.01)
Melanoma	32 (10–210)	0.006 (0.002–0.04)
Colorectal	26 (10–118)	0.005 (0.002–0.02)

*Based on 500,000 nucleated cells plated per dish.

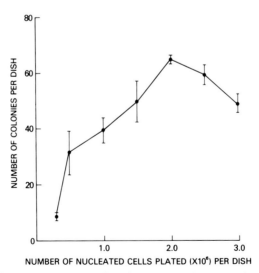

Fig. 2. Effect of increasing number of nucleated cells plated per dish on the number of colonies of breast cancer cells per dish that are present on day 21. Note that after more than 2.0×10^6 nucleated cells/cc plated, the number of colonies that eventually appears decreases.

The low plating efficiency may, in fact, reflect a small cell population in a tumor. However, it may also reflect a large number of mononuclear, nontumor cells in the inoculum. The variation within a type of tumor is of interest. It could reflect differences in stage of tumor, time since prior treatment, area of tumor sampled, etc. It is important to have at least 100 colonies per plate so that counting errors can be minimized and so that log kills of colonies can be discerned. We have attempted to increase the number of colonies per plate by increasing the number of nucleated cells per plate. This improves the PE, but as noted in Figure 2 for breast cancer, as one increases cell number above 2.0×10^6 cell/plate, there is actually a decrease in the number of colonies/plate (probably a nutrient competition/waste build-up problem). This phenomenon is also observed for ovarian and colorectal carcinoma.

Time and resource consumption. This technique is time- and resource-consuming. To set up one tumor in culture for a marker study takes about four hours (20 plates), if one starts with an effusion. It can take six to eight hours if one starts from a solid tumor specimen. If drug studies are carried out for ten drugs (approximately 200 plates), the set-up time is about ten hours.

The other major time commitment is for counting the colonies in an experiment. In a 120-plate experiment, it can take 14 hours to count the number of colonies on all of the plates. Experimental data indicate that one observer must count the entire experiment to ensure consistency in counting. Automatic counters are being explored (Chapter 15).

Difficulties for clinical applications. 1) The tumor colonies take two to three weeks to grow to the point where they can be reliably counted. This delay may be sufficiently long to preclude a selection of drugs in some clinical situations.

2) The optimal exposure time of the tumor cells to the various drugs has not been elucidated. A one-hour drug incubation was an arbitrary choice. This exposure time may not be optimal for cell cycle phase-specific agents such as the antimetabolites.

What are the appropriate drug concentrations? What do sensitivity and insensitivity mean in the system? The drug concentrations utilized in the stem cell assay are usually based on clinical pharmacology data. For anticancer agents this information has gradually accumulated (see Chapter 16 and Appendix 4). The quantitation of drug sensitivity and insensitivity is a difficult area. Figure 3 shows a typical drug-sensitivity curve.

The method of Salmon et al [19; and Chapters 16–18] calculates the area under the curve to determine whether a tumor is sensitive to a drug. If the area is less than a certain cut-off value, then the tumor is considered sensitive to the drug. The determination of this "cut-off" point is based on past patient experience. For some tumors that exhibit a plateau of the curve, this calculation becomes very difficult.

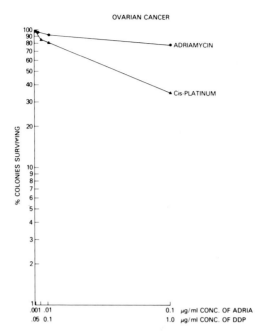

Fig. 3. Typical drug-sensitivity (survival) curves. These curves represent in vitro survival of ovarian carcinoma stem cell colonies after separate incubations with Adriamycin and cis-platinum (DDP). The patient did not respond to either drug.

Early Clinical Applications

Early clinical applications of the stem cell assay have included
1) A series of drug-sensitivity studies on tumors from 435 patients (two-thirds of whom had had prior chemotherapy). Two hundred ninety-one of these tumors grew in vitro. Of these 291 tumors, 105 had greater than 30 colonies in control plates and enough cells to plate at least one drug. Five hundred thirty-five in vitro drug trials have been conducted against these tumors (about five drugs per tumor). An example of the drug-sensitivity information produced in these studies is shown in Figure 3.

Of the 535 in vitro drug trials, only 123 have been duplicated in the patient. Using a retrospective chart review and the definition of in vitro sensitivity/resistance per Salmon and co-workers [19] (area under the curve) and standard clinical response criteria, we have the following in vitro-in vivo clinical correlations to date (see Table IV).

There have been a total of 123 in vitro-in vivo correlations in 101 patients (all retrospective). Seven patients were sensitive to a drug in vitro and responded

TABLE IV. In Vitro-In Vivo Correlations Using the Human Tumor Stem Cell Assay System

Number of patients	Sensitive in vitro and in vivo	Sensitive in vitro/ resistant in vivo[a]	Resistant in vitro/ sensitive in vivo[b]	Resistant in vitro and in vivo	Total number of correlations
101	15	6	2	100	123

[a] False positive.
[b] False negative.

clinically (particularly impressive was an ovarian tumor patient refractory to all standard agents who responded to m-AMSA with a partial response. This response was predicted by the in vitro stem cell assay).

To date all patients, except two, who were resistant in vitro, were also clinically resistant to the drugs tried. Essentially, we have demonstrated that technically the stem cell assay system can be used for in vitro drug studies.

2) We have done serial drug sensitivity studies in patients with ascites secondary to colon cancer. At the National Cancer Institute there is an on-going trial using 5-fluorouracil (5-FU) intraperitoneally via a peritoneal dialysis catheter. Patients receive 5-FU in a series of peritoneal dialysis exchanges with maintenance of 10^{-3} molar concentrations of 5-FU in the abdomen over a 36-hour period [23]. This treatment regimen has been repeated every two to three weeks. Before each treatment we have cultured the patient's ascites using the stem cell assay system. We have also tested 5-FU (both one-hour and 36-hour exposures) against the tumor in vitro. Figure 4 shows that as a patient is treated with 5-FU the tumor becomes more resistant in vitro. This may be due to selection of 5-FU resistant clones. This information is of interest and has potential clinical usefulness.

CONCLUSIONS

As noted in the above discussion, it is clear that a large spectrum of tumor types can be grown from a variety of sources using the human tumor stem cell assay system designed by Hamburger and Salmon. Not all tumors of a given type of tumor will grow, however, and additional work to improve success rates is needed.

From the preliminary studies presented above, there is good evidence that the tumor cells growing in culture are producing the same markers that they are producing in the patient. This information provides additional evidence that the human tumor stem cell assay system closely reflects the in vivo situation.

The questions of specificity for tumor growth, low plating efficiency, time and resource consumption, and difficulties for clinical application have all been

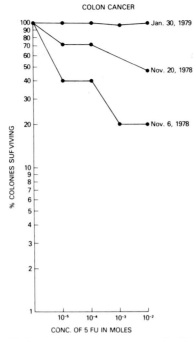

Fig. 4. Serial drug-sensitivity studies on tumor stem cells (from ascites) from a patient with colorectal carcinoma. Patient received 5-fluorouracil intraperitoneally every two to three weeks. Note how a progressively larger percentage of colonies survived incubations with 5-FU as the patient was exposed to intraperitoneal 5-FU.

discussed in detail. At this point, none of the problems seems insurmountable, given additional targeted research in those areas.

Finally, two samples of possible clinical applications of the human tumor stem cell assay system are detailed above. These examples showed the potential utility of the system for predicting for the response or lack of response of an individual patient's tumor to a particular drug. The system was also able to demonstrate the selection of a drug-resistant clone of cells as a patient received treatment with a drug. Both of these examples demonstrate the potential clinical usefulness of the human tumor stem cell assay system.

In summary, the initial experience of our laboratory with the human tumor stem cell assay system has been encouraging. The potential clinical and basic science applications of the technique are enormous.

REFERENCES

1. Steel GG: Cell kinetics and cell survival. In Bagshane KD (ed): "Medical Oncology — Medical Aspects of Malignant Disease." Oxford: Blackwell Scientific Publications, 1975, p 49.

2. Pike BL, Robinson WA: Human bone marrow colony growth in agar-gel. J Cell Physiol 76:77, 1970.
3. Metcalf D: In vitro cloning techniques for hemopoietic cells: Clinical applications. Ann Intern Med 7:483, 1977.
4. Hewitt HB, Wilson CW: A survival curve for mammalian leukemic cells irradiated in vivo. Br J Cancer 13:69, 1959.
5. Hill RP, Bush RS: A lung-colony assay to determine the radio-sensitivity of the cells of a solid tumor. Int J Radiat Biol 15:435, 1969.
6. Bruce WR, Meeker BE, Baleriote FA: Comparison of the sensitivity of normal and transplanted lymphoma colony-forming cells to chemotherapeutic agents administered in vivo. J Natl Cancer Inst 37:233, 1966.
7. Wilcox WS, Griswold DP, Laster WR, Schabel FM, Skipper HF: Experimental evaluation of potential anticancer agents. XVII. Kinetics of growth and regression after treatment of certain solid tumors. Cancer Chemother Rep 47:27, 1965.
8. Brown CH, Carbone PP: In vitro growth of normal and leukemic human bone marrow. J Natl Cancer Inst 46:989, 1971.
9. Hermans AF, Barendsen GW: Changes of cell proliferation characteristics in a rat rhabdomyosarcoma before and after x-irradiation. Eur J Cancer 5:173, 1969.
10. Park CH, Bergsagel DE, McCulloch EA: Mouse myeloma tumor stem cells: A primary cell culture assay. J Natl Cancer Inst 46:411, 1971.
11. Ogawa M, Bergsagel DE, McCulloch: Chemotherapy of mouse myeloma: Quantitative cell culture predictive of response in vivo. Blood 41:7, 1973.
12. Hamburger AW, Salmon SE: Primary bioassay of human tumor stem cells. Science 197: 461, 1977.
13. McAllister RM, Reed G: Colony growth in agar of cells derived from neoplastic and non-neoplastic tissues of children. Pediatr Res 2:356, 1968.
14. Altman AJ, Crussi FG, Rierden WJ, Baehner RL: Growth of rhabdomyosarcoma colonies from pleural fluid. Cancer Res 35:1809, 1975.
15. Hamburger AW, Salmon SE: Primary bioassay of human myeloma stem cells. J Clin Invest 60:846, 1977.
16. Hamburger AW, Salmon SE, Kim MB, Trent JM, Soehnlen BJ, Alberts DS, Schmidt HJ: Direct cloning of human ovarian carcinoma cells in agar. Cancer Res 38:3538, 1978.
17. Hamburger AW, Kim MB, Salmon SE: The nature of cells generating human myeloma colonies in vitro. J Cell Physiol 98:371, 1979.
18. Jones SE, Hamburger AW, Kim MB, Salmon SE: Development of a bioassay for putative human lymphoma stem cells. Blood 53:294–303.
19. Salmon SE, Hamburger AW, Soehnlen BJ, Durie BGM, Alberts DS, Moon TE: Quantitation of differential sensitivity of human tumor stem cells to anticancer drugs. N Engl J Med 298:1321, 1978.
20. Salmon SE, Soehnlen BJ, Durie BGM, Alberts DS, Meyskens FL, Chen HSG, Moon TE: Clinical correlations of drug sensitivity in tumor stem cell assay. In Mathe (ed): "Recent Results in Cancer Research." New York: Springer Verlag (in press).
21. Von Hoff DD, Johnson GE: Secretion of tumor markers in the human tumor stem cell system. Proc Am Assoc Cancer Res, Am Soc Clin Oncol 20:51, 1979.
22. Salmon SE, Buick RM: Preparation of permanent slides of intact soft agar colony cultures of hematopoietic and tumor stem cells. Cancer Res 39:1133, 1979.
23. Speyer JL, Collins TM, Dedrick RL, Brennan MF, Londer H, DeVita VT Jr, Myers CE: Phase I and pharmacological studies of intraperitoneal (i.p.) 5-fluorouracil (5-FU). Proc Am Assoc Cancer Res, Am Soc Clin Oncol 20:352, 1979.

III. Specialized Studies Relevant to In Vitro Tumor Cloning of Human Tumors

11

Variables in the Demonstration of Human Tumor Clonogenicity: Cell Interactions and Semi-Solid Support

Ronald N. Buick and Sydney E. Salmon

The principles involved in the determination of in vitro clonogenicity of human tumor cells are not understood. As discussed in Chapter 2, it seems highly probable that the extent to which a stem cell can perform its in situ function will depend on cell interactions intrinsic to the structural integrity of the tumor tissue. By destroying that structure during preparation of single cell suspension, it is quite possible that important cell interactions are irreversibly lost, causing culture assessment of clonogenicity to underestimate seriously the frequency and function of clonogenic cells. However, since no in situ assay is feasible for human tumor stem cells, we must carefully evaluate the role of cell interactions in determination of culture clonogenicity both to investigate the basis for in vitro clonogenicity and to ensure that conclusions reached regarding such in vitro clonogenicity are not biased by measurement of a non-representative clonogenic population.

No evidence is available on cell interactions governing the proliferation of freshly explanted primary human tumors. However, a number of reports have been made of the significance of lymphoreticular infiltration in human tumors [1–4] and animal tumors [5]. Studies have generally focused on the cytotoxicity involved in such interactions. A few reports have been made of lymphoreticular stimulation of tumor growth [6–10], and it is clear that the extent to which in vitro cytotoxicity or stimulation is observed is very much dependent on the stoichiometry of any interactions [11]. In prior work in this laboratory, Hamburger et al [12] noted decreased tumor cell clonogenicity of ovarian effusions with a one-step phagocytic procedure. This finding and other data led Salmon and Hamburger [13] to hypothesize that macrophages can stimulate the proliferation of clonogenic tumor cells. Stromal interactions have been very much less studied, although a requirement for stromal tissue has been noted for the establishment of tumor cell lines from solid tumors.

This chapter is concerned with the development of techniques to allow the investigation of the role of cell interactions in human tumor clongenicity. No attempt has been made to consider tumor stromal elements, but efforts have been concentrated on putative tumor–lymphoreticular cell interactions. As a source of cells for such developmental studies we have chosen malignant effusions from patients with carcinoma. Clonogenicity of tumor cells from such sources is well documented quantitatively [12, 14], and from a practical point of view, effusions have the decided advantage that tumor cells exist as an approximation of a single cell suspension in contact with mesothelial cells, monocyte/macrophages, lymphoid cells, and granulocytes. Although it can be argued that such metastatic tumor tissue may not be representative of the primary tumor, we feel that the use of such cells can be supported for the development of techniques applicable to analysis of microenvironmental factors affecting solid tumors.

METHODOLOGY

In vitro culture procedures have been described elsewhere [12, 14; and this volume]. Briefly, cultures are grown in the upper layer of two-layer semi-solid medium; either agar underlayer plus agar overlayer or agar underlayer plus methylcellulose overlayer. Enrichments are as described by Hamburger and Salmon [15]; however, conditioned medium was not employed [14].

Fractionation of cells and reconstitution studies were performed as shown in Figure 1. For adherence, cells were incubated overnight at $37°C$ in a humidified atmosphere of 7.5% CO_2 in air in 150 mm plastic dishes at a cell concentration of 10^6 cells/ml in McCoy's 5A containing 10% fetal calf serum (FCS) (10 ml/dish). Nonadherent cells were removed, and the adherent layer was washed twice with 5 ml volumes of McCoy's 5A plus 10% FCS (10 ml/dish). The washings were pooled with the nonadherent fractions. The washed adherent layer was then removed by mechanical means with a rubber policeman. In cases where the adherent cells were to be used as a feeder monolayer, the adherence procedure was performed in 35 mm plastic dishes that were subsequently used for cell culture without removal of the adherent layer. In such experiments, adherent cell layers were prepared with serial increases in the number of cells permitted to adhere ($10^4 - 10^7$). As determined directly by microscopy, the number of cells adhering in any experiment was linear from any one patient and was found to be a constant percentage of the number of cells plated for the adherence procedure. In view of this linearity, experiments were designed using the number of cells plated, not the number of adherent cells.

To remove phagocytic cells from the nonadherent tumor cell suspension, the nonadherant cells were then incubated at a concentration of 10^6 cells/ml in McCoy's 5A and 10% FCS containing 40 mg dry-heat-sterilized carbonyl iron/10^7 cells in 250 ml

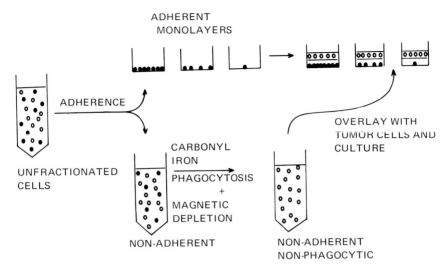

Fig. 1. Schematic representation of experimental procedure for cell fractionation and reconstitution.

Falcon flasks. After an incubation at 37°C for 45 minutes in a shaking water bath the flasks were placed flat on a magnet, and the supernatant cells were carefully decanted. The cells were subjected to this magnetic selection procedure until all iron and iron-laden phagocytic cells were removed.

Density gradient fractionation of cells was performed by bovine serum albumin gradient centrifugation; 5%–23% or 7%–35% BSA gradients were constructed in 12 ml polyallomar tubes by layering 10×1 ml aliquots of solutions of decreasing BSA percentage. Twenty to 100×10^6 cells in approximately 0.5 ml of McCoy's 5A was layered on top, and the tube was spun at 600g for one hour. The gradient giving most optimal fractionation was fractionated by puncturing the bottom of the tube and collecting the drops in a series of sterile tubes.

Cell morphology of fractions was performed on slides prepared with a Shandon centrifuge and stained with Papanicolaou, Wright-Giemsa, and for nonspecific esterase. Cell fractions were also assessed for the ability to ingest latex particles and to synthesize prostaglandin E_2 and $F_2\alpha$ [14].

RESULTS AND DISCUSSION

Information on the basis of clonogenicity can be gained by correlation analysis of the colony-forming process and a possibly important variable. We have reported recently the use of correlation analysis of clonogenicity with the cellular differential of malignant effusions [14]. In a series of 37 effusions from 31 patients with carcinoma, we calculated Spearman rank correlation coefficients beween tumor colony

TABLE I. Spearman Rank Correlation Coefficients of Colonies/ 5×10^5 Cells or Cloning Efficiency (Colonies/Tumor Cell) With Percentage Tumor Cells, Macrophages, or Lymphocytes in 37 Malignant Effusions

	Tumor cells	Macrophages	Lymphocytes
Colonies/5×10^5 cells	0.6178[a]	−0.0784	−0.3544
Cloning efficiency	0.1965	−0.0758	−0.0090

[a]Significant at the 1% confidence limit.

growth or cloning efficiency and the percentage of tumor cells, macrophages, or lymphoid cells in the cell sample (Table I). The only significant correlation occurred between the number of colonies/5×10^5 cells and the percentage of tumor cells (Rs = 0.6178, P < 0.01, n = 37). It is of interest to note that cloning efficiency (ie, that fraction of the tumor cells that was clonogenic) is not significantly correlated with percentage either of tumor cells, macrophages, or lymphocytes. This test simply states that when the cell populations involved are assessed on the basis of gross morphological characteristics, no relationship between clonogenicity and nontumor populations can be demonstrated. However, evidence to be cited below indicates that it is quite possible that nonmorphologically distinct functional subpopulations of these cell types are involved in the determination of clonogencity.

In order to analyze cell subpopulations, we have used two techniques: fractionation by adherence and phagocytosis and fractionation by BSA gradient centrifugation.

Fractionation of malignant effusions by adherence and phagocytic depletion and subsequent reconstitution of fractions was performed as shown in Figure 1. The number of cells adhering was linearly related to the number of cells plated [14], but it was different quantitatively from patient to patient. The analyses of fractions at stages of this procedure (unfractionated, adherent, nonadherent, and nonadherent−nonphagocytic) indicated three things: clonogenic efficiency (ie, colony number per tumor cell) was markedly reduced by fractionation, a slight enrichment was seen in morphologically identifiable macrophages in the adherent cells, and a large enrichment was seen in the adherent layer of cells capable of synthesizing prostaglandins [14]. When adherent layers were recombined with nonadherent−nonphagocytic cells in a two-layer system, clonogenicity could be restored in a dose-response fashion. The optimum number of adherent cells varied from patient to patient, but the shape of the curve was similar for all patients (Fig. 2). A drop in clonogenicity was consistently seen when supraoptimal numbers of adherent cells were employed. Such results clearly indicate cell cooperativity and a dose-response relationship of host cells in the determination of clonogenicity. The fact that changes in cell populations brought about by fractionation were not of similar magnitude to changes in clonogenicity indicates that a morphologically nonrecognizable subpopulation is respon-

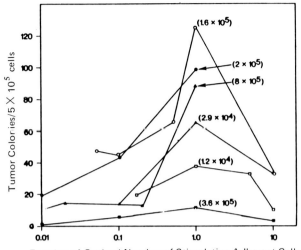

Fig. 2. Ability of autologous adherent cells to reconstitute tumor colony formation by enriched tumor cells in six patients with carcinoma. Inasmuch as the absolute tumor colony modulating activity of adherent cells varied from patient to patient, the reconstitution data have been normalized with respect to the concentration of adherent cells beneath the feeder layer that produced peak stimulation of tumor colony formation. Results are mean of duplicate plates.

sible. Consistent with this hypothesis is the marked enrichment of prostaglandin synthetic capacity of the stimulating cells. Investigations are under way to determine the molecular basis for the observed stimulation of clonogenicity and to characterize better the responsible cell population(s). However, the system we have used does have obvious shortcomings for such studies, since the reconstitution is so obviously cell number-dependent. Thus far we have not been able to predict the number of adherent cells giving optimum stimulation. Therefore experiments investigating the action of modulating agents on the reconstitution process have to be set up "blind" and are therefore cumbersome and impractical in size. On the positive side, it is possible to predict with some certainty the clonogenic potential of the effusion based on tumor cell percentage (Table I). A search is under way for a more consistent and reproducible source of stimulating cells. It should be noted that, although the cell interaction is clearly mediated through a diffusible intermediate, we have thus far been unable to demonstrate activity of conditioned media derived from adherent cell layers. This would suggest that the diffusible substance is either short-lived or unstable, or that the adherent cells are metabolizing an inhibitory factor in the environment.

Further information on the basis of clonogenicity is derived from analysis of density gradient fractionation of effusions. We have used BSA gradients to analyze cell subpopulations. Data from a representative experiment are shown in Figure 3.

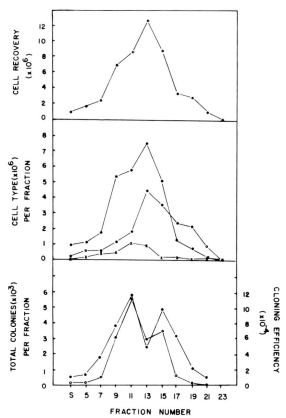

Fig. 3. BSA gradient (5–23% w/v) fractionation of a malignant effusion of patient with breast cancer. A. Cells recovered per fraction. B. Morphological assessment (% of cell types) ●——●; tumor cells, ○——○; lymphocytes, ×——×; macrophages. C. Colony formation of cells from fractions expressed as ○——○; total colonies per fraction or ●——●; cloning efficiency.

Fractions are assessed on the basis of cell number, cellular morphology, and clonogenicity. The clonogenicity data are expressed in two ways; either as number of colonies/fraction or, more important, as clonogenic efficiency (CE) (colonies/tumor cell). It is clear that there are large differences in CE across the gradient. These differences could be due to inherent heterogeneity in the clonogenic capacity of the tumor cells at these various densities or to the cellular composition of the particular fraction in which the tumor cells finds itself. Prior depletion of either macrophage/monocyte or lymphoid cells from the starting cellular suspension should allow differentiation of these two possibilities.

An important role in determination of clonogenicity might be played by the choice of semi-solid conditions. Apart from the obvious requirement of allowing progeny of a single cell to remain localized during growth, the semi-solid medium probably mimicks the in situ surroundings of the clonogenic tumor cell. We have had the opportunity to compare the clonogenic portential of over 30 cell samples in agar/agar or methylcellulose/agar two-layer systems. Using the procedures described for Table I, we have analysed the statistical significance of the relationship between clonogenicity in these two systems and the percentages of tumor cells, macrophages, or lymphocytes. Again, the only significant correlations occurred between the number of colonies and the percentage of tumor cells. However, the correlation coefficients did not differ significantly when growth was measured in the methylcellulose/agar system as compared to the agar/agar system. The finding of equivalent efficacy of methylcellulose/agar is of importance, inasmuch as methylcellulose provides a convenient matrix from which to recover developing colonies for various biological studies.

The experiments described here detail our approach to experimental analysis of the basis of human tumor cell clonogenicity in fresh biopsy samples. Although the evidence is incomplete, the reconstitution experiments suggest that cells of the monocyte/macrophage series are intimately involved in the interactions. Preliminary evidence points to prostaglandin synthesis as being important. The integration of this fractionation/reconstitution procedure (Fig. 1) and analysis of fractions by BSA gradient centrifugation should allow further insight into the interaction of the functional populations present in malignant effusions.

The results presented make it clear that interpretation of numbers of colonies generated in clonogenic assays should be done with care and that experiments designed to analyze surviving clonogenic populations (eg, after in vitro drug-sensitivity testing) may involve changes in the determinants for clonogenicity as well as direct killing of clonogenic cells. In that context, some chemotherapeutic agents thought to act directly on clonogenic tumor cells may have at least a component of their effect on colony formation mediated by actions on other host cells present in the semi-solid cultures.

REFERENCES

1. Gauci CL, Alexander P: The macrophage content of some human tumors. Cancer Lett 1:29–32, 1975.
2. Hersh EM, Mavligit GM, Gutterman JU, Barsales PB: Mononuclear cell content of human solid tumors. Med Pediatr Oncol 2:1–9, 1976.
3. Hayry P, Totterman TN: Cytological and functional analysis of inflammatory infiltrates in human malignant tumors. I. Composition of the inflammatory infiltrates. Eur J Immunol 8:866–871, 1978.
4. Underwood JCE: Lymphoreticular infiltration in human tumors: Prognostic biological implications: A review. Br J Cancer 30:538–548, 1974.

5. Wood GW, Gollahon RA: T-lymphocytes and macrophages in primary murine fibrosarcomas at different stages in their progression. Cancer Res 38:1857–1865, 1978.
6. Norbury KC: In vitro stimulation and inhibition of tumor cell growth mediated by different lymphoid cell populations. Cancer Res 37:1408–1415, 1977.
7. Small M, Trainin N: Separation of populations of sensitized lymphoid cells into fractions inhibiting and factors enhancing syngeneic tumor growth in vivo. J Immunol 117:292–297, 1976.
8. Jee Jeebhoy H: Stimulation of tumor growth by the immune response. Int J Cancer 15:867–878, 1974.
9. Fidler J: In vitro studies of cellular mediated immunostimulation of tumor growth. J Natl Cancer Inst 50:1307–1312, 1973.
10. Evans R: Tumor macrophages in host immunity to malignancies. In Finck MA (ed): "The Macrophage in Neoplasia." New York: Academic Press, 1976, pp 27–42.
11. Prehn RT: Immunostimulation of the lymphodependent phase of neoplasia growth. J Natl Inst 59:1043–1069, 1977.
12. Hamburger AW, Salmon SE, Kim MB, Trent JM, Soehnlen BJ, Alberts DS, Schmidt HJ: Direct cloning of human ovarian carcinoma cells in agar. Cancer Res 38:3438–3444, 1978.
13. Salmon SE, Hamburger AW: Immunoproliferation and cancer: A common macrophage-derived promoter substance. Lancet 2:1289–1290, 1978.
14. Buick RN, Fry SE, Salmon SE: Effect of host cell interactions on clonogenic carcinoma cells in human malignant effusions. Br J Cancer 41:695–704, 1980.
15. Hamburger AW, Salmon SE: Primary bioassay of human tumor stem cells. Science 197:461–463, 1977.

12
Morphologic Studies of Tumor Colonies

Sydney E. Salmon

The development of clonogenic assays in soft agar or other semi-solid media has greatly increased our understanding of the biology of self-renewal compartments in the bone marrow [1]. The recent development of clonogenic assays for human tumor stem cells (as detailed in this text) provides the same potential for understanding neoplastic growth. Microscopic examination of colonies growing in vitro is required for all clonogenic assays.

Serial examination with inverted light or phase-contrast microscopy usually suffices to evaluate growth per se. However, staining and specialized morphologic techniques are required to optimize evaluation of the types of cells that constitute colonies. The purpose of this chapter is to review useful morphologic techniques that have been either specially devised or modified from standard procedures to facilitate evaluation of tumor colonies and distinguish them from other cell types. Two general groups of procedures have proved useful: 1) stains to facilitate identification and enumeration of colonies in the wet agar at the time of colony counting, and 2) techniques applicable to detailed morphologic assessment on cultures that have been subjected to fixation and subsequent processing. In the second category, a rapid and simple method that we developed [2] to prepare permanent slides of intact agar colony cultures is detailed, as it has proved to be suitable for routine staining as well as for immunofluorescent, autoradiographic, and histochemical studies.

"WET" AGAR STAINS

Wet stains are used at the time of colony counting and are intended to facilitate identification of colonies under low power, to distinguish tumor colonies from granulocyte/macrophage colonies, or to identify cellular tumor markers histochemically.

Contrast Enhancement

A properly adjusted inverted microscope generally provides sufficient contrast to make additional stains for enhancement somewhat optional. However, such stains are useful when an inexpensive dissecting microscope is used for colony counting, as such instruments often provide suboptimal illumination.

Coomassie blue G-250 technique. A simple staining procedure that we developed for contrast enhancement uses the protein stain Coomassie blue G-250 [3]. This dye exhibits its leuco (colorless) form in dilute acid but reverses to an intense blue when the dye binds to proteins. This feature eliminates the need for destaining or elution of background dye from the gel. The staining reagent is prepared by dissolving 20 mg of Coomassie brilliant blue G-250 (Sigma Chemical Co., St. Louis, MO) in 3.0 ml of 100% ethanol with stirring for ten minutes. This is followed by adding 6.0 ml of 85% phosphoric acid, with stirring for an additional ten minutes. The resulting solution is diluted to 50 ml with distilled water, stirred for 15 minutes, and then passed through Whatman Number 1 filter paper. The solution has a transparent tan color and can be stored at room temperature for at least several months.

For the staining reaction, 1.0 ml of the reagent is carefully pipetted on to each Petri dish, and the dish is incubated at 37°C for one to four hours. Plates that have previously been fixed in glutaraldehyde (vide infra) also stain satisfactorily. When viewed under a microscope, colonies stain a deep blue, and the background agar (or methylcellulose) remains clear or stains a light blue (depending on its protein concentration).

Tetrazolium salt technique. Tetrazolium salts can be used to stain viable cells, and this will reduce the colorless compound to a water-insoluble formazan product. Schaeffer and Friend [4] reported that the compound 2-(p-iodophenyl)-3(p-nitrophenyl)-5-phenyl tetrazolium chloride (INT) could be reduced to water-insoluble red formazan, which precipitates inside the cells within colonies in an agar system. Bol et al [5] found this protocol to be useful for cultures of hematopoietic cells. The INT (Aldrich Chemical Co., Milwaukee, WI) is kept as a solution of 1 mg INT/ml saline and sterilized by autoclaving. One day prior to termination of the cultures, 0.5 ml of the INT solution is added to the dishes, and they are returned to the incubator. The next day, granulocyte colonies will be bright red.

We have tried the related compound, nitroblue tetrazolium (NBT) (Sigma Chemical Co., St. Louis, MO), for human tumor colonies, with somewhat variable results. Whether INT will be more consistent for various types of tumor colonies remains to be determined. If colonies are not counted immediately, they can be fixed with formaldehyde or glutaraldehyde (vide infra) and stored in a refrigerator for some weeks.

We initially thought that such stains would be necessary for use in conjunction with a video image-analyzer (Chapter 15) to facilitate enumeration of tumor colonies automatically. However, we were pleased to find that optimizing illumination and electronic circuitry provides sufficient contrast for that application, and such staining procedures were, therefore, not required.

Peroxidase Stain

Recognition of peroxidase-containing cells in agar culture is generally used to identify granulocyte/macrophage colonies. Such colonies usually have a diffuse morphology in contrast to the more tightly packed colonies of tumor cells. The peroxidase stain is clearly of value, however, when the tumor specimen for culture is obtained from a marrow aspirate (eg, with myeloma, lymphoma, or metastatic cancer). A very convenient peroxidase stain is that reported by Zucker-Franklin and Grusky [6]. The staining reagent is prepared as follows:

The peroxidase reagent consists of a solution containing 0.3 gm of benzidine dyhydrochloride, 0.132 M $ZnSO_4 \cdot 7\ H_2O$ (1 ml); 1.0 gm of $NaC_2H_3O_2$, 3% H_2O_2 (0.7 ml); and 1.0 N NaOH (1.5 ml) dissolved in a sufficient volume of 30% ethyl alchohol to yield a final volume of 100 ml. Alternatively, the substrate can consist of hydrogen peroxide and diaminobenzidine tetrachloride, which is used for demonstration of peroxidase in electron microscopy. The reaction with either of these reagents is carried out by pipetting 1.5 ml of the reagent onto each Petri dish, followed by a ten-minute incubation without agitation at room temperature. The fluid should then be pipetted off carefully and several changes of tissue culture medium should be made. For preservation and tissue processing (vide infra), the agar is then overlayed with 0.6 ml of 3% glutaraldehyde, fixed for at least ten minutes for light microscopy (or overnight for electron microscopy) according to the processing procedures described later in this chapter.

In order to stain the specimens, a few drops of the reagent are placed on each culture dish and the cells containing peroxidase (usually neutrophiles and their precursors) rapidly convert the substrate and almost immediately begin to exhibit a brown or black color. We have seen occasional peroxidase-positive cells within myeloma colonies that, on fixed and stained preparations, proved to be myeloma cells. Thus, the distinction is not absolute, but remains relatively good as myeloma cells generally are peroxidase negative.

Lactic Dehydrogenase and Other Dehydrogenases

Tumor cells proliferate well under relatively anaerobic conditions and often express increased quantities of lactic dehydrogenase (LDH). A dehydrogenase reaction, using tetrazolium-formazan (nitroblue tetrazolium) (NBT), which is useful for demonstrating a variety of enzymes [7], is compatible with semi-solid culture conditions. We have used this reaction to demonstrate LDH in developing tumor colonies in soft agar and have observed varying expressions of this enzyme from tumor type to tumor type. The reagent is prepared in 30 ml of 0.05 M buffer, pH 7.2, containing 21 mg of nicotine adenine dinucleotide (NAD), 30 mg of DS-NBT, 168 mg of lithium lactate, and 9 mg of phenazine methosulfate. For staining purposes, 0.5–1.0 ml of the reagent is layered on each plate, which is then incubated for one to four hours at 37°C (or overnight) prior to examination. As with other "wet stains" described above, no washing steps are required.

Melanin

Recognition of melanoma colonies is usually quite simple, as they often express brownish melanin pigmentation without special staining. However, some colonies are hypomelanotic (Chapter 8). The intensity of brown-black pigmentation of hypomelanotic colonies can be enhanced with DOPA (3,4 dihydroxyphenylalanine). DOPA is the first oxidation product of tyrosine, a precursor to melanin, and is taken up by melanoma cells in culture and undergoes rapid enzymatic oxidation to an indolic quinone or semiquinone-like substance that is deep brown and incorporated into melanin. Overnight incubation with 1 mM DOPA at 37°C can be used, and washing steps are not required. Hypomelanotic colonies from 95% of patients are darkened by this procedure. A more detailed melanin procedure, which combines both the DOPA reaction and a metal reduction reaction with ammoniacal silver (which depends on the reducing capacity of the melanins), is discussed later, in the section on staining of dried slides.

Fixation of Wet Agar Cultures

It is sometimes necessary to preserve soft agar cultures in the semi-solid state at various times in their growth period so that all samples can be evaluated simultaneously. Additionally, some samples need to be subjected to fixation for preparation of permanent slides. Glutaraldehyde may represent an optimal fixative for this purpose, as it is compatible with a very wide array of histochemical, immunologic, and electron microscopic procedures [7]. We have used a 3% solution of glutaraldehyde in Hanks' balanced salt solution for this purpose, and have found that 1.0 ml of this solution will preserve plates indefinitely if the fixed dishes are placed in a humidified plastic box and stored in the refrigerator. Specimens preserved in this fashion are useful for analysis of colony morphology as well as for subsequent enumeration with the automated tumor colony counter (Chapter 15).

PERMANENT SLIDES FOR HISTOLOGIC ANALYSIS

Most morphologic techniques of analysis require availability of fixed tissues that are dehydrated and thereby made amenable to a variety of staining procedures. Until recently, morphologic study of colonies by conventional histological and histochemical techniques has been an arduous task, inasmuch as most staining techniques have required careful picking of individual colonies from the agar under an inverted microscope, followed by deposition on microscope slides. With that approach, careful selection of relatively nondisrupted stained colonies for review is usually required, further limiting morphological study. Additionally, the relationship of the small fraction of cells that form colonies in vitro from normal marrow or spontaneous human tumors to background nonproliferative and "stromal" cells has not been studied, inasmuch as only colonies or clusters have been picked for fixation and staining. Several techniques have previously been described in an attempt to simplify or standardize the picking procedure [8–10], however, these are, in our opinion, suboptimal.

In the course of our reviewing various techniques that might be suitable for preservation and morphological study of whole Petri dishes of marrow and tumor colonies, we considered that the general approach that has long been used for permanent fixation of protein electrophoresis or immunodiffusion in agar on glass slides or plastic films [11] might be adaptable to intact agar layers from in vitro colony assays. This perspective provided the basis for a simple method for fixation and preservation of agar cultures for a wide variety of studies. We recently reported a procedure that we standardized and found to provide excellent routine morphology as well as being amenable to a wide variety of specialized staining techniques [2].

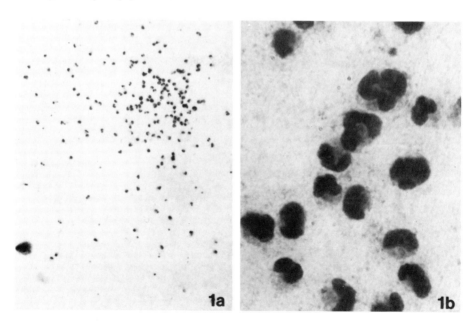

Fig. 1. a. Morphological appearance of normal human bone marrow colony-forming units in culture grown in soft agar. The stain produces a stippled background staining of the agar, which does not interfere with morphology in thin films. Lower left, nonproliferative megakaryocyte (Wright-Giemsa, × 100). b. High-power photomicrograph of human neutrophils in a soft-agar colony grown from the bone marrow (Wright-Giemsa, × 1,000). Reproduced from Cancer Res [2] with permission of the publishers.

Dried Slide Method

The method for preparing dried slides from agar cultures is as follows:

1) With a Pasteur pipette, each Petri dish is carefully filled with Hanks' balanced salt solution and incubated for 15 minutes at room temperature to elute the majority of extracellular proteins from the plating layer. The supernatant solution is then aspirated or decanted from the dish, with care not to disrupt or pour out the colony-containing plating layer. This initial wash step is optional but reduces background staining somewhat.

2) Each dish is then filled with a fixative solution consisting of 3% glutaraldehyde in Hanks' balanced salt solution and is allowed to fix for ten minutes at room temperature. The supernatant fixative is then removed as above.

3) Each dish is then filled with distilled water, agitated, submerged in a tray (eg, a disposable plastic weigh boat) containing at least 50 ml of distilled water, and allowed to incubate for ten minutes to elute the salts and fixative from the plating layer. During this time, the plates can be agitated and decanted gently by hand in order to displace the plating layer from the Petri dish so that it may float freely in the water. Frequently, the 0.3% agar plating layer will slip free from the 0.5% agar feeder layer, which remains attached to the plate. When this occurs, the dish containing the feeder layer can be discarded. Should the plating layer and feeder layer remain attached to one another, gentle spurts of water from a Pasteur pipette can be used to tease the plating layer from the thicker feeder layer. Sometimes, a small rent must be made in the plating layer so that the pipette can be introduced between the two layers. They may then be dissociated from each other with additional gentle spurts of water. The separated plating layer is thin and filmy and can usually be identified easily even when both the plating and feeder layers are floating in the water. After the feeder layer is discarded, the tray is carefully drained, leaving the plating layer in a few drops of water. The thin plating layer can then be poured gently onto a microscope slide. Alternatively, a cleaned microscope slide can be introduced into an undrained tray by hand, and the plating layer (or a piece of it) allowed to spread on the slide, which is then taken gently out of the water. If the entire area of the colony-containing layer is not needed, standard microscope slides (2.5×7.5 cm) can be used, and the few millimeters of excess agar can be pinched off with a finger where it hangs over the edges of the slide. Wider slides can be used when the entire 35 mm diameter layer must be recovered.

4) A prewetted cellulose acetate membrane (as is used for electrophoresis) that has been cut to the dimension of the slide is carefully placed on top of the still-wet agar layer, and any bubbles are gently expressed. The cellulose acetate strip provides for uniform evaporation of the water from the agar, which otherwise dries unevenly onto the slide. Alternatively, low-fiber filter paper can be substituted for cellulose acetate strips; however, filter paper frequently leaves un-

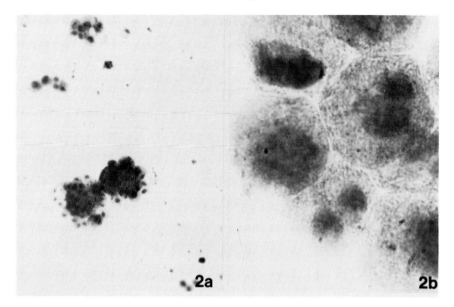

Fig. 2. a. Morphological appearance of human melanoma colonies grown directly from biopsy in soft agar. Different amounts of melanin are expressed in various cells (Papanicolaou stain, × 100). b. High-power photomicrograph of melanoma cells shows excellent preservation of cellular morphology in dried films from soft-agar culture. Cytoplasmic stippling is due to diffuse distribution of fine melanin granules (Papanicolaou stain, × 1,000). Reproduced from Cancer Res [2] with permission of the publishers.

desired fibers adherent to the agar which may obscure morphology. Once the slide has dried (four to 12 hours), the cellulose acetate strip either comes loose spontaneously or can be moistened and gently pulled off the slide. Slides prepared in this fashion can either be stored in the unstained state or be stained immediately with any of a variety of stains.

The most useful generation stain that we have applied is a Papanicolaou stain [12]. While standard staining times are often satisfactory, longer staining times for the nuclear hematoxylin are sometimes required. Romanovsky stains such as Wright-Giemsa often stain the background agar a somewhat bluish-pink but are usable when the plating layer is thin. Figures 1–5 illustrate typical stained dried films of human bone marrow granulocyte colonies and tumor colonies from a variety of human cancers that we have cloned directly in soft agar from biopsy samples. Of particular note are the excellent morphology and staining qualities of hematopoietic and tumor colonies. Agar plates of colony-forming units in culture that have been stained histochemically for peroxidase [6] prior to fixation maintain good histochemical localization through the fixation and drying steps.

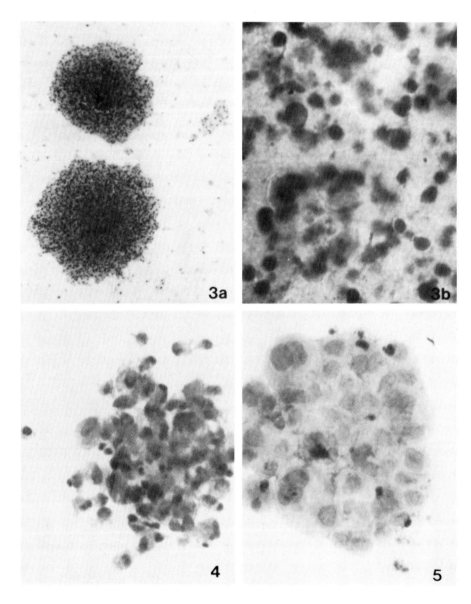

Fig. 3. a. Morphological appearance of human squamous cell carcinoma colonies grown in soft agar from a needle biopsy of the liver from a patient with known liver metastases. The dark central areas and cracking near the center of the right-hand colony are due to flattening of the large spherical colony onto a microscope slide in the agar film (Papanicolaou stain, × 100). b. High-power photomicrograph of squamous cells in a colony depicted in a. The wide intercellular spacing and bridging were characteristic morphological features manifest in squamous carcinoma colonies (× 1,000). Reproduced from Cancer Res [2] with permission of the publishers.

Melanin Staining of Dried Slides

Assessment of melanin content of cells has been evaluated with various methods by Meyskens (Chapter 8), who found the approach used by Mishima [13] most useful. His modification of Mishima's technique is as follows:

The dried slides are fixed in buffered formalin at 4°C for 24 hours, after which they are transferred to a 1mM DOPA solution (pH 7.4) at 37°C for 30 minutes. This is replaced with a fresh DOPA solution for an additional 16 hours, after which the slides are rinsed in distilled water (two minutes), immersed in 10% ammoniated silver nitrate solution (58°C for ten minutes), and rinsed in distilled water (ten minutes). Two-percent gold chloride is then poured onto the slide for 30 seconds, and the slide is transferred to 6% hyposulfite for two minutes and washed in running water for 15 minutes. An example of the DOPA stain applied to a dried slide of "hypomelanotic" melanoma colonies appears in Figure 6.

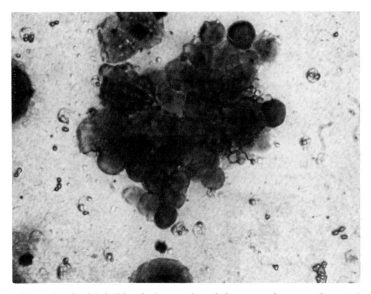

Fig. 6. Example of a dried slide of a hypomelanotic human melanoma colony stained for melanin with the modification of the Mishima technique [13] described in the text (× 400). The unstained colonies were clear and would otherwise be thought not to contain melanin.

Fig. 4. Typical malignant adenocarcinoma colony at 14 days grown in soft agar from a patient with metastatic colonic carcinoma and malignant ascites. Excellent morphology of the cancer cells in the colony as well as a background nonproliferative plasma cell at the left side of the field are preserved in the dried agar film (Papanicolaou stain, × 400). Reproduced from Cancer Res [2] with permission of the publishers.

Fig. 5. Human ovarian cancer colony from a patient with a metastatic papillary cystadenocarcinoma. The colony size was typical for 14-day culture (Papanicolaou stain with hematoxylin exposure prolonged to 12 minutes to enhance nuclear detail, × 400). Reproduced from Cancer Res [2] with permission of the publishers.

Other Stains, Immunofluorescence, and Autoradiography

We have also been successful in applying an enzymatic technique for nonspecific esterase and have also used several other histochemical stains applied directly to the dried slides. For a more general review of other potentially applicable histochemical stains, the reader is urged to consult one of the standard histochemistry texts [eg, 1], as many of the procedures can be directly adapted to the dried slide technique. The PAS and mucin stains are discussed by Hamburger in Chapter 6 of this volume.

We have demonstrated specific cell membrane localization of carcinoembrionic antigen on the surface of cells in carcinoma colonies using the indirect immunofluorescent technique (Fig. 7), as well as cytoplasmic localization of fluorescent anti-immunoglobulin in dried slides of human myeloma colonies. Autoradiographic slides prepared with the high-speed scintillation technique [14] displayed specific localization of DNA-synthesizing cells that had been pulsed with high-specific-activity (^{3}H) thymidine shortly before fixation. Additionally, we have used this autoradiographic technique to demonstrate ^{125}I uptake by thyroid cancer clusters and colonies exposed to radioiodine for one hour prior to fixation (Fig. 8). How-

Fig. 7. Human bladder carcinoma colony grown for seven days, prepared with the dried slide technique and stained with an indirect immunofluorescent method for carcinoembrionic antigen (× 100). Prominent cell surface staining of the TCC cells is evident, while appropriate control slides tested simultaneously were negative. Rabbit anti-CEA was kindly provided to us by Dr. Charles Todd, City of Hope Medical Center, Duarte, California. Fluorescein-conjugated goat anti-rabbit gamma globulin was used as the second antibody.

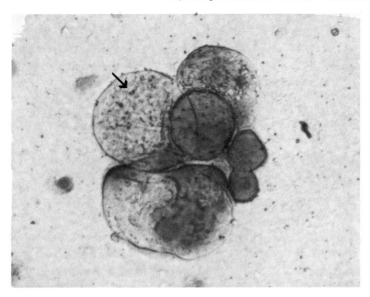

Fig. 8. Thyroid carcinoma at the early cluster stage in vitro. This specimen was prepared for autoradiography using the highspeed scintillation technique [14]. Bovine TSH was added to the dishes for two hours prior to a one-hour incorporation of ^{125}I. Extensive washing of the agar was carried out prior to preparation of dried slides for autoradiography. Prominent grains (reflecting radioiodine incorporation) can be appreciated in the small follicle (see arrow) (Papanicolaou stain, × 1,000).

ever, prominent background grains occur on such autoradiographs unless longer periods of washing are used (generally several days), rather than the abbreviated washing procedure as described earlier in step 3.

The development of this simple technique for fixation and drying of agar cultures onto microscope slides will probably have wide application in experimental studies of hematopoietic and tumor colony formation. Such films can also be dried onto hydrophilic plastic supports or directly onto Petri dishes if there is not a separate feeder layer. As discussed in Chapter 10, Von Hoff has modified this technique and embedded the whole agar layer in paraffin and cut sections in the usual manner with a microtome.

Diagnostic Applications

Although many applications of the dried slide method will probably be directed toward research questions, we recognize that the ability to prepare permanent stain slides of such colonies may have practical applications in clinical diagnosis. For example, in the study by Von Hoff and his colleagues on bronchial washings from lung cancer patients [15] and in our studies of bladder irrigations in patients suspected of having transitional cell carcinoma of the bladder [16; and Chapter 7], samples that were reported to be negative in the routine cytology laboratory have formed typical tumor colonies in vitro. Additionally, as discussed in Chapter 9 by Casper et al, tumor colony growth has been obtained from histologically negative marrows from patients with neuroblastoma. Histologic review of the fixed-stained slides or sections of these colony-containing plates confirmed the colonies to have morphology consistent with these tumor types. In contrast, Ozols, Willson, and Young (Chapter 19) observed ovarian tumor colony formation only from peritoneal washings that were also positive on cytologic examination.

ELECTRON MICROSCOPY OF TUMOR COLONIES

Electron microscopy (E/M) can provide excellent "markers" of the histogenic origin of colonies grown from tumor samples, particularly in instances when light microscopy is considered to be inadequate. For example, we have used transition electron microscopy to identify both microvilli and tight junctions with desmosomes in bladder carcinoma [16] and premelanosomes in melanoma cells. A number of other ultrastructural features such as neurosecretory granules (neuroblastoma) and myofibrils (rhabdomyosarcoma) are additional obvious candidates for study with E/M techniques. E/M studies are never routine and require meticulous preparation, particularly because of the relatively small size of the tissue specimens (the colonies) which must be localized appropriately for embedding and sectioning for transition electron microscopy. Scanning electron microscopy can also provide additional valuable information on colonial morphology.

Transition Electron Microscopy

The procedure used at the University of Arizona is as follows: The agar plates are fixed by gently layering on 2–3 ml of 3% glutaraldehyde in Hanks' balanced salt solution (room temperature) and allowing the plates to stand at 4°C for one hour. The uncovered agar Petri dishes are then warmed in a 58°C oven for one hour to evaporate enough liquid to thicken the agar. Colonies are identified under a dissecting microscope, and small blocks containing one or two colonies are carefully cut out with fine scalpels and placed in 0.1 M cacodylate buffer (pH 7.4) containing 7% sucrose. The blocks are postfixed in 2% osmium tetroxide for one hour at 4°C. After a rinsing in 0.1 M cacodylate buffer, the blocks are dehydrated in a series of increasing concentrations of ethanol and embedded in Epon. Thin sections (600–900 Å) are cut with a diamond knife on a Porter-Blum MT-2 ultramicrotome. Sections are then mounted on uncoated 200-mesh copper grids, stained with aqueous uranyl acetate and lead citrate [17], and examined on a Philips 300 electron microscope. A representative electron micrograph of cells from a bladder carcinoma colony prepared with this technique is shown in Figure 9.

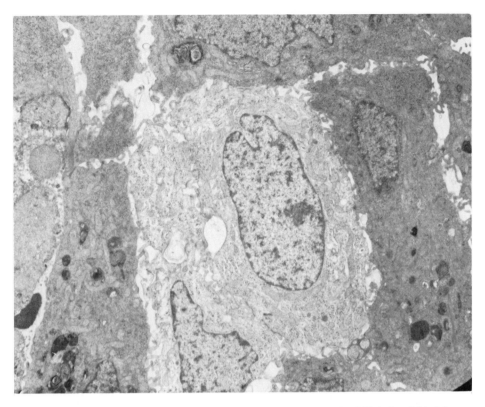

Fig. 9. Medium-power (12,880 ×) transmission electron micrograph of "light" and "dark" cells within a human bladder carcinoma colony grown in soft agar with the assay as described by Stanisic et al in Chapter 7. Common structural features include prominent microvilli apparent on the cell surfaces, cytoplasmic mitochrondria, and a high nuclear cytoplasmic ratio.

A slightly different protocol has been applied by Harris and Von Hoff from the University of Texas at San Antonio [personal communication, 1980] for the preparative stages for both transmission and scanning electron microscopy. Agar layers are first fixed by adding 4% glutaraldehyde in 0.1 M sodium cacodylate buffer, pH 7.4, to the Petri dishes, which are then refrigerated for 24 hours. The dishes are then allowed to come to room temperature, and the fixative is decanted and replaced with 0.1 M sodium cacodylate buffer, pH 7.4, which is changed every 15 minutes for one hour. Samples are then postfixed for 90 minutes at room temperature with 2% osmium tetroxide in 0.1 M sodium cacodylate (pH 7.4). The dishes are then rinsed with distilled water for 60 minutes with changes every 15 minutes, followed by en bloc staining with an aqueous solution of 5% uranyl acetate in the cold overnight. For dehydration, a standard ethanol series is used: 75% (ten minutes), 95% (ten minutes), 100% (15 minutes), 100% (15 minutes); followed by two 15-minute changes of 100% propylene oxide. Samples are then embedded with Spurrs firm mixture in a 1:1 mixture with propylene oxide and held overnight in the cold. The samples are then placed in 100% Spurrs firm mixture at room temperature for three hours and overnight in the cold. Samples are transferred to embedding capsules containing 100% Spurrs plastic, allowed to sit for three hours, followed by hardening for two to three days in a 70°C oven. Thin sections are then cut and examined with a transition electron microscope. An example of cellular detail obtained with this technique appears in Figure 10.

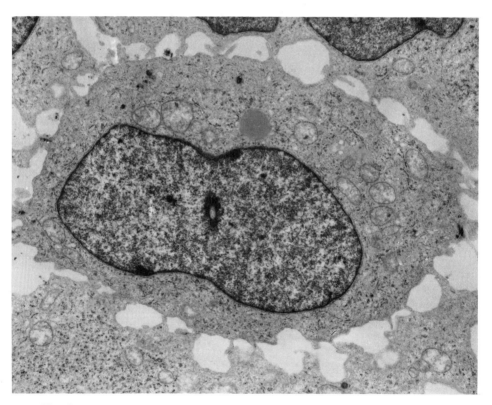

Fig. 10. Transmission electron micrograph of cells in a human colon carcinoma colony that has grown for 14 days in soft agar ($\times$ 17,500). A number of tight junctions formed between cells can be identified. Electron micrograph courtesy of Dr. Gary Harris and Dr. Daniel D. Von Hoff [18].

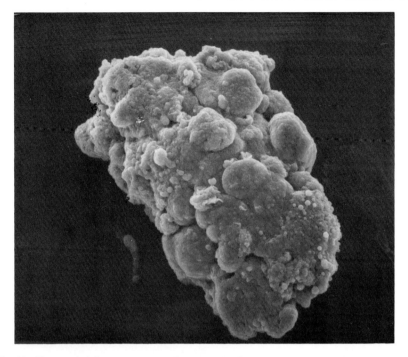

Fig. 11. Human rhabdomyosarcoma colony at eight days of culture studied by scanning electron microscopy (× 610). Excellent morphologic detail of colonial surface features can be obtained. Scanning electron micrograph courtesy of Dr. Gary Harris and Dr. Daniel D. Von Hoff [18].

Scanning Electron Microscopy

Harris and Von Hoff have used the same procedures for fixation, postfixation, and dehydration in preparation for scanning electron microscopy as described above for their transmission electron microscopic studies. Drying of the specimens is achieved with the critical-point drying technique with CO_2, followed by coating with gold/palladium. An example of one of their scanning electron micrographs of a tumor colony appears in Figure 11.

DISCUSSION

The development of simple techniques for morphologic assessment of tumor colonies in agar cultures will undoubtedly have wide application in experimental and clinical studies of tumor colony formation. In this chapter are summarized some of the straightforward morphologic applications that we and others have used in studies of in vitro colony formation. Many more sophisticated analyses should also be possible. For example, in melanoma colony formation, several different colony variants with respect to pigmentation are apparent. A combination of morphologic techniques as described in this chapter, plus the cytogenetic procedures discussed by Dr. Trent in Chapter 14, could be used to establish whether variation in expression of melanin content could be recognized on a chromosomal basis, rather than as a single or polygenic difference in phenotypic expression. Combining relatively simple morphologic techniques such as those that we developed for use with dried slides with other techniques to qualitatively alter growth conditions may provide extremely useful information on proliferation and differentiation of clonogenic neoplastic cells. Use of such techniques to identify markers of differentiation may well represent a major research application in the near future, inasmuch as a variety of agents are now thought to induce differentiation in normal and neoplastic cells.

REFERENCES

1. Metcalf D: "Hemopoietic Colonies." Berlin: Springer-Verlag, 1977.
2. Salmon SE, Buick RN: Preparation of permanent slides of intact soft-agar colony cultures of hematopoietic and tumor stem cells. Cancer Res 39:1133–1136, 1979.
3. Salmon SE, Liu R: Direct "wet" staining of tumor or hematopoietic colonies in agar culture. Br J Cancer 39:779–781, 1979.
4. Schaeffer WI, Friend K: Efficient detection of soft agar grown colonies, using a tetrazolium salt. Cancer Lett 259–262, 1976.
5. Bol S, Van den Engh G, Visser J: A technique for staining haemopoietic colonies in agar cultures. Exp Hematol 5:551–553, 1977.
6. Zucker-Franklin D, Grusky G: The identification of eosinophil colonies in soft agar cultures by differential staining for peroxidase. J Histochem Cytochem 24:1270–1271, 1976.
7. Lillie RD, Fullmer HM: "Histopathologic Technique and Histochemistry," Ed 4. New York: McGraw-Hill, 1976.
8. Dicke KA, Platenburg MGC: Technical manual of the thin agar layer technique. In Van Bekkum D, Dicke K (eds): "In Vitro Culture of Hemopoietic Cells." Rijswijk, The Netherlands: Radiobiological Institute TNO, 1972, pp 466–570.
9. Goube De Laforest P, Riou-Lasma-Vons N, Boizard G: A microcytocentrifugation technique for cytological studies on hemopoietic colonies grown in agar. Exp Hematol 6:361–364, 1978.
10. Testa NG, Lord BI: A technique for the morphological examination of hemopoietic cells grown in agar. Blood 36:586–589, 1970.
11. Weime RJ: "Studies on Agar Gel Electrophoresis." Brussels: Arscia Uitgaven N V, 1959.
12. Papanicolaou GN: "Atlas of Exfoliative Cytology." Cambridge, Massachusetts: Harvard University Press, 1954, p 6.
13. Mishima Y: Modification of combined DOPA-premelanin reaction: New technique for comprehensive demonstration of melanin and tyrosinase sites. J Invest Dermatol 34:355–360, 1960.

14. Durie BGM, Salmon SE: High speed scintillation autoradiography. Science 190:1093–1095, 1975.
    Durie BGM, ibid: Science 195:208, 1977 (includes two corrections to preceding paper)
15. Von Hoff DD, Weisenthal LM, Ihde DC, Mathews MJ, Layard M, Makuch R: Growth of lung cancer colonies from bronchoscopy washings. Cancer, in press, 1980.
16. Buick RN, Stanisic TH, et al: Development of an agar-methylcellulose clonogenic assay for cells in transitional cell carcinoma of the human bladder. Cancer Res 39:5051–5056, 1979.
17. Luft JH: Improvements in epoxy embedding methods. J Biophys Biochem Cytol 9:409–414, 1961.
18. Harris GJ, Zeagler J, Hodek A, Casper J, Von Hoff DD: Ultrastructural analysis of colonies growing in a human tumor cloning system. Submitted, 1980.

13
Cell Kinetic Analysis of Human Tumor Stem Cells

Brian G. M. Durie and Sydney E. Salmon

INTRODUCTION

Current knowledge of human tumor cell kinetics has been derived primarily from studies of the proliferative behavior of entire populations within tumors [1]. While it has long been recognized that only subsets of tumor cells are critical for population renewal within a tumor, relatively few methods of cell cycle or cell kinetic analysis have been perfected and applied to the study of human tumor stem cells as they can be cultivated within an in vitro soft agar culture system. Initial studies of various tumor types with the in vitro culture technique, including the one-hour suicide index of tumor stem cells with tritiated thymidine or hydroxyurea, suggest that cell cycle parameters of the tumor stem cells may be different from those of the general population of tumor cells as assessed by more conventional techniques of cell cycle analysis. It would seem that although thymidine labeling index, pulse labeled mitosis curves, and flow microfluorometry may reflect tumor cell kinetics [2], they do not allow accurate quantitation of the kinetics of the critical self-renewing tumor stem cell or clonogenic population. The objectives of this chapter are 1) to summarize some of our initial kinetic findings obtained with fresh biopsies of human tumors as assessed in the in vitro human tumor stem cell assay, 2) to discuss some of the existing and potential methods that may prove useful for this purpose, and 3) to provide an approach to relating cell kinetics and drug sensitivity.

The relationship between the kinetic status of clonogenic tumor cells and drug sensitivity of various tumor types has long been questioned. However, the approaches used to tackle this important problem have usually been indirect. In view of accumulating evidence that the in vitro tumor stem cell assay is predictive of in vivo clinical drug sensitivity or resistance (Chapters 10 and 18), it would appear most reasonable to carry out a variety of studies of cell kinetics and drug sensitivity testing simultaneously for various tumor types. Such studies should be able to clarify whether, for any given drug, sensitivity or resistance of the clonogenic cells has a predominantly kinetic, biochemical, or other basis.

As discussed in Chapter 2, it is extremely important to appreciate the elusive nature of true stem cells, even in the clonogenic assay system. By definition, the term stem cell usually implies a cell both histologically primitive and with an extensive capacity for self-renewal. Such cells give rise to further stem cells and to other committed populations [3]. The essential property of stem cells is self-renewal. However, depending upon local growth conditions [4], stem cells may or may not proliferate and exhibit self-renewal either in vivo or in vitro. Specific properties of tumor stem cells include 1) production of tumor cells with a limited proliferative potential; 2) a tendency for tumor stem lines to undergo cytogenetic change and produce subclones of new stem lines with properties different from the original stem cell population; and 3) true self-renewal, from a single cell. Therefore, one must be able to discriminate between true stem cell growth, production of tumor cells with limited proliferative potential, and the production of new tumor subclones as well.

METHODS OF KINETIC ANALYSIS

A number of methods are available for cell kinetic analysis and are summarized in Table I. For an overview of the population dynamics of a tumor, it is necessary to integrate the tumor burden stage or total cell number with measurements of the rate of change of cell number on the basis of direct or indirect measurement of DNA content and its rate of synthesis or accumulation within cells. Direct evidence of cell growth has, of course, been documentation of cell division.

TABLE I. Methods of Cell Kinetic Analysis on Human Tumors

A. Entire tumor cell population
   1. Tritiated thymidine as marker of DNA synthesis
      a. Pulse labeled mitoses curve
      b. Labeling index of tumor cells
   2. DNA content analysis
      a. Quantitative Feulgen staining
      b. Flow cytometry using fluorescent stains (eg, propidium iodide, mithramycin)

B. Clonogenic tumor cell population (evaluation of in vitro tumor cell colony growth)
   1. Assessment of plating efficiency
   2. Suicide index: tritiated thymidine, hydroxyurea
   3. Assessment of rate, type and duration of growth (eg, time to form clusters or colonies)
   4. Thymidine labeling index and pulse labeled mitosis curve of developing colonies

Measurement of overall tumor cell growth has required some estimate of cell loss. This latter measurement has proved to be a particularly difficult component of most analyses and is usually computed rather than directly measured. Important additional considerations often include clinical disease activity (eg, remission or relapse), as well as the source of cells used for study — for example, from a primary tumor or metastatic lesion. These same types of considerations are important to apply in the stem cell system.

TRITIATED THYMIDINE LABELING AND COLONY GROWTH

Our initial studies suggest that the tritiated thymidine labeling index or the number of cells in S as determined by flow cytometry or other methodologies are critical determinants of the plating efficiency of stem cells grown in agar culture. In our studies of 97 bone marrow samples from patients with multiple myeloma, there was a highly significant correlation between the tritiated thymidine labeling index of the plated cells and the subsequent likelihood of no growth or growth of clusters, or of colonies [5]. In addition, the labeling index was significantly correlated with the actual number of clusters or colonies grown [5]. Similar data have been found in smaller numbers of patients with both ovarian cancer and malignant melanoma. Of interest, in a recent study by Schlag et al [6], in vitro colony growth of cells from primary and metastatic carcinomas was evaluated. These authors suggested that a consistently greater number of colonies were grown from patients with breast and gastrointestinal carcinoma when cells were obtained from metastatic sites rather than from the primary. However, determinations of absolute cloning efficiencies have yet to be defined from these sites, and the relationship of the host cell numbers and other determinants of clonogenicity (Chapter 11) must be considered. Although there are a number of possible explanations for this phenomenon, other workers have previously observed significantly higher labeling indexes of metastatic tumor cells [7]. This can certainly be the basis for further study. For example, in patients with breast cancer, one could assess such parameters as the tritiated thymidine labeling, estrogen receptor content, plus the kinetics and drug sensitivity of tumor colony-forming units (TCFUs). Since the triatiated thymidine labeling index has been shown to have prognostic importance in several tumor types, including breast cancer, leukemias, lymphomas [8], and multiple myeloma [9], it would appear to be worth evaluating those tumor types in this fashion.

TUMOR STAGE, DISEASE ACTIVITY, AND COLONY GROWTH

Initial results with the tumor colony assay in patients with several tumor types, (including both solid tumors and hematologic malignancy) have indicated significant correlation between the stage of disease and/or disease activity of the disease at the time samples were obtained for culture and the subsequent degree and type of colony growth. An example of this is shown in Table II, describing the incidence of no growth, cluster formation, and colony growth in patients with multiple myeloma at various stages of the disease using the myeloma assay as discussed in Chapter 3. The best colony growth was obtained at the time of pretreatment sampling as well as the early relapse. In contrast, no growth or significant cluster formation was found at the time of remission or late relapse. In both myeloma and other tumor types there appears to be a relationship between the *rate* of colony growth and disease activity. In general, a much more rapid rate of colony formation has been observed with samples from patients at the time of relapse than during other phases of the disease. As also discussed in Chapter 1, and by Von Hoff in Chapter 10, the cloning efficiency also rises late in the stage of disease. Such observations suggest that there is a dynamic interaction between clonogenic tumor cells and the host and treatment factors, which impacts on growth support, growth inhibition, and clonal progression during the patient's clinical course of disease. Unless the patient is cured with treatment, the neoplastic clone and/or its subclones are present, and proliferation proceeds at varying rates.

SUICIDE INDEX AND COLONY GROWTH

Since clonogenic cells may or may not be in cycle at the time of initial plating, the assessment of the proportion of cells in S by the suicide technique is a powerful tool. The standard method for calculating suicide index, for example using tritiated thymidine, is outline in Table III. In our culture studies performed to date, one-hour suicide indices have been calculated for both tritiated thymidine and hydroxyurea using methodology reported previously [10]. Data have been obtained in patients with multiple myeloma, ovarian cancer, and malignant melanoma. In general, the suicide index, particularly with tritiated thymidine, was extremely high (range of 40–80%), indicating a high fraction of stem cells in S phase at the time of initial in vitro incubation or very rapid recruitment of cells into the S phase. This usually corresponded to a high tritiated thymidine labeling index of the tumor cells, although not invariably so. In a minority of patient samples (5–10%), there was negligible suicide. Our initial studies have not identified any particular correlates of this phenomenon, although the majority of such samples also had a very low tritiated thymidine labeling index, plus low suicide, with both hydroxyurea and tritiated thymidine. As shown in Figure 1 for myeloma and ovarian colonies when comparing tritiated thymidine suicide with hydroxyurea suicide, hydroxyurea was found to be much less effective and reliable as an S phase marker of clonogenic tumor cells.

TABLE II. Myeloma Stem Cell Growth From Bone Marrow of Patients: Relation of Disease Status and In Vitro Colony Growth*

Clinical status	Number of patients	Colonies	Clusters	No growth
Pretreatment	19	9 (47%)	7 (37%)	3 (16%)
Remission	17	4 (24%)	9 (53%)	4 (24%)
Relapse	22	14 (64%)	4 (18%)	4 (18%)
Totals	58	27 (47%)	20 (34%)	11 (19%)

*The standard myeloma colony assay described in Chapter 3 was used; all samples were plated at 5×10^5 bone marrow cells per dish.

TABLE III. Suicide Index

$$\text{Suicide index of TCFU (\%)} = \frac{\text{Number of control tumor colonies} - \text{Number of tumor colonies following preincubation with }^3\text{H-thymidine}}{\text{Number of control tumor colonies}} \times 100$$

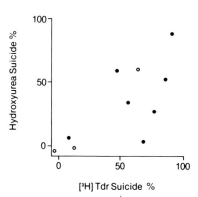

[³H] Tdr and Hydroxyurea Suicide

Fig. 1. Comparison of one-hour suicide indices with hydroxyurea and ^{3}H thymidine on TCFUs from patients with multiple myeloma (●) and ovarian carcinoma (○).

The standard suicide technique uses a one-hour incubation period in vitro prior to cell plating. Although suicide with this one-hour incubation is usually high, there is the potential to measure cell cycle parameters of stem cell colonies using variable time exposure to tritiated thymidine. As summarized in Table IV, tumor cells are set up in continuous exposure to tritiated thymidine for periods of a few minutes to several hours. From the rate of increase in the suicide index with time, both the generation time and length of S phase can be calculated. This method has already been used by Wu to calculate the cell kinetic parameters of clonogenic cells in normal hematopoietic cell samples [11]. The major limitation in the application of this technique to tumor cell samples would seem to be the high suicide we have observed with even a short incubation time. It might, therefore, prove hard to obtain sufficient time points to measure accurately true generation time and duration of S. However, the technique is clearly worth evaluating in detail in the tumor stem cell assay system. Based on the high one-hour suicide results we have obtained, we must consider that tumor cells may be held up at the G_1-S interface (a type of synchrony) or that the cell cycle phase durations of the clonogenic tumor cells may be quite different from the general tumor cell population. If the latter is the case, then the S phase may constitute 80% of the cycle duration of tumor stem cells.

STEM CELL KINETICS AND IN VITRO DRUG SENSITIVITY

Perhaps the most important reason for looking at kinetics of TCFUs is in correlation with the drug sensitivity or resistance of the cultures clonogenic cells. As has been amply discussed in Chapters 10 and 18, there is now excellent evidence that the in vitro drug sensitivity assay system is highly predictive in clinical drug treatment programs. An important consideration is the mechanism of drug sensitivity or resistance in individual cases. The in vitro culture system allows for the possibility of precise discrimination between kinetic, biochemical, and other types of drug resistance or sensitivity. In the studies carried out to date, it would seem that the stem cell kinetics and the intrinsic drug sensitivity are independent prognostic parameters in individual patients. Thus, although there is strong evidence that the pretreatment kinetics an determined by the S phase fraction (measured by tritiated thymidine labeling index) significantly correlate with remission duration and survival, initial cell kill with chemotherapy is determined primarily by intrinsic drug sensitivity or resistance [9] (Chapter 18). Table V summarizes such data for a patient with multiple myeloma. These results suggest that kinetics as expressed by tritiated thymidine labeling index and in vitro drug sensitivity or resistance are independent as prognostic factors. At the time of initial diagnosis, the patient had a myeloma cell labeling index of 7% (high for multiple myeloma), 170 myeloma colonies per 5×10^5 cells plated and was sensitive in vitro to melphalan, Adriamycin, BCNU, and vincristine. The patient

TABLE IV. Measurement of Cell Cycle Parameters of Clonogenic Tumor Cells in Agar Culture Using Variable Exposure to ^{3}H-Thymidine*

1) Continuous exposure to ^{3}(H) thymidine for 20 minutes, one hour, three hours, five hours, seven hours, etc

2) Generation time (Tg) = $\dfrac{\text{time for 100\% killing}}{1 - \text{fraction of cells in S}}$

3) Length of S = Tg × fraction of cells in S

*Based on the model of Wu [11].

TABLE V. Serial In Vitro Studies of ^{3}H-Thymidine Labeling Index, Tumor Colony Formation and Drug Sensitivity of Marrow Plasma Cells in a Patient With Multiple Myeloma

Date (clinical status)	Myeloma cell labeling index	Number of myeloma colonies or clusters	Drug sensitivity results/clinical treatment
September 1975	7%	170	Sensitive to: Melphalan Adriamycin BCNU Vincristine/ Clinical response to melphalan combination therapy
June 1976 (in remission)	< 1%	Occasional clusters	In vitro testing not feasible/ therapy unchanged
March 1977 (in relapse)	11%	90	Resistant to: Melphalan Adriamycin BCNU/clinically treated with vincristine, BCNU, Adriamycin, and prednisone without response
June 1977 (died)	–	–	–

subsequently responded clinically (> 75% tumor regression) to a combination schedule of the agents to which her cells were sensitive in vitro. However, as might be predicted from the high initial labeling index [9], remission duration was short. Throughout the time of remission the in vitro tritiated thymidine labeling index of myeloma cells was extremely low, and soft agar cultures produced only rare myeloma colonies or clusters. At the time of relapse, the labeling index again became high, and significant myeloma colony growth was observed, with 90 colonies per 5×10^5 cells plated. At this point in vitro drug sensitivity testing was repeated and the patient's TCFUs manifested resistance to the drugs that had previously produced substantial cell kill in vitro and resulted in an initial remission.

ADVANTAGES OF SOFT AGAR CULTURE

A number of methods have been proposed for the evaluation of in vitro tumor cell kinetics and drug sensitivity. For example, Livingston et al have proposed a method based upon suppression of tritiated thymidine labeling index by preincubation of cells with appropriate drug [12]. Although it would seem that this technique has utility in patients with high initial labeling and can give rapid results within several days, as opposed to several weeks for the stem cell assay, it has a major disadvantage that it does not directly evaluate the stem cell population and is difficult to count when the labeling index is low. Likewise, a number of workers have used drug suppression of incorporation of tritiated thymidine (by scintillation counting) as an indicator of in vitro drug sensitivity [13]. Although rapid results can again be obtained, nonclonogenic cells and nontumor cells can incorporate the label. Such factors render interpretation of the data difficult. Using flow cytometry, changes in the pattern of DNA histograms following drug exposure can also be assessed. This technique has provided clinically useful information in the hands of several workers [14]. However, to date it has not been possible to identify the clonogenic population or to quantitate the drug effect upon this critical subgroup of tumor cells. Possibly, with the future availability of surface markers to identify stem cells [15], it may prove possible to use this methodology to identify stem cells and to quantitate drug effects upon their DNA content distributions or, possibly, some other related parameters. However, for clinical samples, a significant limitation will remain the requirement for a very large number of cells. Potentially, a way around this will be the use of a sterile pathway through a cell sorter (with use of noncytotoxic dyes), followed by culturing of sorted populations in the soft agar system.

The combined use of flow cytometric techniques using a cell sorter and in vitro soft agar culture could thus prove to be a very powerful tool. Not only could cells be sorted in a sterile fashion, on the basis of surface or DNA fluorescence, but also chromosome content of the cell population could be assessed. Several workers have successfully used flow systems to sort cells on the basis of chromosome content [16]. Cytogenetically different subclones can potentially be sorted, subsequently grown in soft agar, and evaluated with respect to both kinetics and in vitro drug sensitivity. The major advantage of the in vitro culture system is the ability to address this type of biologic question directly.

DIRECT EVALUATION OF SOFT AGAR COLONIES

Although the suicide technique assesses the S phase component of the plated cells, it does not allow quantitation of the rate, quantity, or quality of colony growth. However, it is possible to look directly at colonies on a serial basis. One can count the number of cells contained in clusters or colonies and measure their doubling time as well as the number of mitoses and the degree of differentiation using differentiation markers for each type of tumor sample. The number of mitoses can be directly assessed by overlaying cultures with colchicine prior to sampling, as outlined by Trent (Chapter 14) for cytogenetic analyses. One can also use tritiated thymidine as proposed by Elson and co-workers [17] to detect ongoing DNA synthesis by assessing incorporation of tritiated thymidine by means of autoradiography of individually plucked clusters and colonies. Differentiation can be detected by production of markers found in differentiated cells such as immunoglobulin produced by myeloma cells.

One can potentially extrapolate from serial measurements of cell number, number of mitoses, and number of cells with evident differentiation to derive estimates of the growth rate of clonogenic cells. Measurements obtained to date on tumor colony growth in vitro suggest that the doubling time of the clonogenic cells is quite short in comparison to the observed total tumor population doubling in vivo. For example, ovarian cancer frequently reaches the 32-cell stage within seven days (168 hours) after plating and is consistent with a 28-hour doubling time for the TCFUs. Growth rates for TCFUs in neuroblastoma are often equally rapid, whereas melanoma and myeloma TCFUs appear to double more slowly (about one-half this rate) in vitro. From the clinical standpoint, these differences in growth rates in vitro appear superficially to correspond to relative differences in the aggressiveness of these neoplasms in the untreated state in vivo. Assessment of growth rates with an image analyzer (Chapter 15) may also permit analysis of potential heterogeneity of doubling times of TCFUs within individual culture plates and with differing growth conditions. From the final size of clusters or colonies, as well as their morphology and karyology, one can also obtain information about whether growing populations are tumor cells with limited proliferative potential, for example, forming only small clusters or whether one or more subclones are present, indicated by different cytogenetic markers.

Initial studies in malignant melanoma colonies have been extremely productive, particularly in the assessment of subcloning. As outlined in Chapter 8, single melanoma biopsy specimens frequently grow two types of colonies, pigmented and nonpigmented. Conceivably, these two variant types of colonies may prove to have both different growth kinetics and drug sensitivities. Initial cytogenetic analysis of melanoma, both in colony culture and in short-term cell line culture, indicates a dramatic tendency for frequent subcloning, with abnormal segregation (with associated unequal cell division) being a very common phenomenon [18]. Assessment of plating efficiencies and suicide indices of morphologically and/or cytogenetically different types of colony variants can potentially provide information about subclones with differing proliferative potential.

CONCLUSION

The application of the tumor stem cell assay to the study of human tumor cell kinetics has the potential for major advances in our knowledge of proliferative characteristics of clonogenic human tumor cells. From simultaneous evaluation of in vitro drug sensitivity, in vitro doubling time, and the thymidine suicide index, plus flow cytometry, cytogenetic analysis, and assessment of differentiation markers, substantial insights into the basic biology of tumor cell growth may be attained. As discussed in other chapters, using modifications of the two-layer system, one can assess both local cell-mediated and humoral factors that might influence in vitro kinetics and drug sensitivity. The next few years should see the acquisition and integration of valuable new information that should allow more rational approaches to the treatment of tumors by taking full advantage of information on both kinetics and drug sensitivity of clonogenic human tumor cells.

REFERENCES

1. Steel GG: "Growth Kinetics of Tumors." Oxford: Clarendon Press, 1977.
2. Mendelsohn ML: Principles, relative merits, and limitations of current cytokinetic methods. In Drewinko B, Hymphrey RM (eds): "Growth Kinetics and Biochemical Regulations of Normal and Malignant Cells." Baltimore: Williams & Wilkins, 1977, pp 101–112.
3. Lajtha LG: Stem cell concepts. Differentiation (Springer-Verlag) 14:23–34, 1979.
4. Buick RN, Fry SE, Salmon SE: Effect of host cell interactions on clonogenic carcinoma cells in human malignant effusions. Br J Cancer (in press).
5. Durie BGM, Young LA, Salmon SE: Clinical importance of kinetics in myeloma stem cell culture. Proc Am Assoc Cancer Res Am Soc Clin Oncol 20:431, C-582, 1979.
6. Schlag PM, Schreml W, Vergani G: Comparison of human tumor cloning of primary tumors and metastases (abstr) 2nd Workshop on Human Tumor Cloning Methods, Tucson, Arizona, 1980.
7. Simpson-Herron L, Griswold DP Jr, Corbett TH: Studies of the growth, population kinetics, and host lethality of CD8F mammary adenocarcinoma. Cancer Treat Rep 62(No. 4):519–528, 1978.

8. Livingston RB, Sulkes A, Thirlwell MP, Murphy WK, Hart JS: Cell kinetic parameters: Correlation with clinical response. In Drewinko B, Humphrey RM (eds)' "Growth Kinetics and Biochemical Regulation of Normal and Malignant Cells." Baltimore: Williams & Wilkins, 1977, pp 767–785.
9. Durie BGM, Salmon SE, Moon TE: Pretreatment tumor mass, cell kinetics, and prognosis in multiple myeloma. Blood 55:364–372, 1980.
10. Preisler HD, Shoham D: Comparison of tritiated thymidine labeling and suicide indices in acute myelocytic leukemia. Cancer Res 38:3681–3684, 1979.
11. Wu AM: A method to measure the generation time and length of DNA synthesizing phase of hemopoietic progenitor cells in a heterogenous population. Cell Tissue Kinet (in press).
12. Livingston RB, Titus GA, Heilbrun LK: In vitro effects on DNA synthesis as a predictor of biologic effect from chemotherapy. Cancer Res 40:2209–2212, 1980.
13. Raich PC: Prediction of therapeutic response in acute leukemia. Lancet 1:74–76, 1978.
14. Tobey RA, Crissman HA: Use of flow microfluorometry in detailed analysis of effects of chemical agents on cell cycle progression. Cancer Res 32:2766–2732, 1972.
15. Price GB, Stewart S, Krogsrud RL: Characterization of stem cells and progenitors of hemopoiesis by cell sorting. Blood Cells 5:161–174, 1979.
16. Barlogie B, Latreille J, Fu C-T, Franco J, Meistrich M, Andreeff M: Characterization of hematologic malignancies by flow cytometry. Blood Cells (in press).
17. Elson D, Titus GA, Lam T, Livingston RB, Von Hoff DD: A study of the growth kinetics of human tumor cells in double-layer agar culture (abstr). 2nd Workshop on Human Tumor Cloning Methods, Tucson, Arizona, 1980.
18. Saxe DF, Meyskens FL Jr: Abnormal chromosome segregation in human melanoma. Abstract 366, vol 31, no. 6, p 108A. Proc Am Soc Hum Genet, Minneapolis, 1979. Abstract no. 366, page 108A, vol 31, no 6, Am J Human Genetics, November 1979.

14
Cytogenetic Analysis of Human Tumor Cells Cloned in Agar

Jeffrey M. Trent

In the early 1900's, Theodore Boveri, a classical cytogeneticist, suggested that chromosomal changes were involved in the etiology of cancer [1]. Despite the publication of this precocious hypothesis, the first evidence for a tumor-specific chromosome alteration was not obtained until nearly 50 years later [2]. The consistent finding of a "minute" chromosome (called the Philadelphia, or Ph^1, chromosome) in approximately 85% of patients with chronic myelogenous leukemia (CML) was the initial, and remains the most widely studied and accepted, example of tumor-specific chromosome change. Since the discovery and subsequent identification of the Ph^1 chromosome [3], an increasing number of tumors have been studied with detailed cytogenetic techniques. Utilizing various chromosome-banding techniques, nonrandom chromosome change has been reported in a variety of human cancers [4]. Among the most convincing nonrandom chromosome changes in human tumors are CML, with most commonly a 9/22 translocation [3]; meningioma, with partial or complete loss of a chromosome 22 [5]; Burkitt lymphoma, with a chromosome 14q+,8q− [6]; non-Burkitt lymphomas, with translocations of chromosome 14 [7]; acute promyelocytic leukemia with a chromosome 15/17 translocation [8]; breast, colon, ovary, bladder, and cervical carcinoma, with changes in chromosome 1 regions q23−q32 [9]; endometrial adenocarcinoma, with alterations in group-D chromosomes [10]; and ovarian adenocarcinoma with deletion of part of the long arm of chromosome 6 [11].

The enormous increase since 1960 in the identification of chromosomes or segments of chromosomes nonrandomly involved in various human cancers is the direct result of the development and application of chromosome-banding techniques [12]. Cryptic or minimal internal rearrangements of tumor cell chromosomes, as well as identification of "marker" or grossly rearranged chromosomes that were undetectable by standard staining, are now discernible with chromosome banding. However, it should be noted that the majority of chromosome-banding analyses of human tumors has involved tumors derived from hematopoietic origin. Recent reviews have shown that less than 5% of all published banded chromosomal analyses of human tumors have involved carcinomas

[4, 13]. The reasons for the paucity of published reports of solid tumor karyology reside in the usually low proliferative fraction of solid tumors and the inability to procure chromosomes morphologically suited for chromosome-banding analysis.

With the development by Hamburger and Salmon [14] of a clonogenic assay for colony-forming cells from hematopoietic and solid tumors and its subsequent application to cytogenetic analysis [15, 16], an innovative and perhaps significant new technique for solid tumor karyology was devised. Preliminary evidence for the usefulness of this clonogenic assay in tumor karyology will be presented in the paragraphs that follow.

METHODOLOGY

The method for growth of hematopoietic and solid tumor stem cells is that described by Hamburger and Salmon [14], which is detailed in Appendix 1 of this text. The methodologic application of this technique to cytogenetics is described in Appendix 3. Additionally, procedures for the chromosome banding of colony cell chromosomes are also described in Appendix 3.

APPLICATIONS TO TUMOR KARYOLOGY

The use of short-term agar culture in the cytogenetic analysis of human tumors has been reported only rarely [17]. However, the techniques employed required hand picking of individual tumor colonies, followed by manual dissociation prior to cytogenetic analysis. This laborious addition to the already time-consuming procedures for cytogenetic analysis, coupled with often inherently low mitotic indices, is, perhaps, responsible for the scarcity of cytogenetic analyses of clonogenic progenitor/stem cells.

We have demonstrated that it is possible to perform detailed cytogenetic analysis in a variety of hematologic and solid tumors cloned in agar (Table I). In addition, substantial analytic enhancement of countable mitotic figures from tumor cells cloned in agar has been observed (Table II). Due in part to the selective in vitro circumstances for clonogenic tumor cell growth and in part to the inhibition of normal fibroblastic proliferation by the agar, substantial enrichment in the number of tumor cell mitoses is achieved in agar as compared to direct or liquid techniques.

Figures 1 and 2 illustrate a second and important adjunct to the increases observed in tumor mitoses — the procurement of chromosomes that are morphologically suitable for chromosome banding. Samples taken from the colony stage of tumor growth (Figs. 1 and 2) provide insights into the location and number of mitoses within generating colonies. Samples taken from the cluster stage of colony growth (Fig. 1) provide chromosomes suitable for a variety of chromosome-banding techniques. G-, C-, and NOR-banding of tumor stem cells in human

TABLE I. Cytogenetics of Human Tumors Cloned Directly in Agar Culture

Tumor types successfully analyzed (evidence of neoplastic origin[a])	Modal chromosome assessment successful/total (% successful)	Banding analysis successful/total (% successful)
Carcinomas		
Bladder (M, C)	4/8 (50%)	1/1
Breast (M, C)	2/2 (100%)	1/1
Kidney (M, C)	1/1	NA†
Lung (M, C)	2/2 (100%)	NA
Ovary (M, C)	15/22 (68%)	8/9 (89%)
Uterus (M, C) (cervix and corpus)	2/3 (66%)	1/1
Sarcoma and other malignancies		
Diffuse lymphoma (M, B, C)	1/3 (33%)	NA
Melanoma (M, B, C)	5/10 (50%)	1/3 (33%)
Multiple myeloma (M, B, C)	3/8 (38%)	NA
Neuroblastoma (M, B, C)	2/3 (66%)	1/1
Totals	37/62 (60%)	13/16 (81%)

[a]Morphology (M), biomarker (B), cytogenetic (C); † = not attempted. Tumors also successfully cultured, but not yet studied cytogenetically; carcinomas: adrenal (M), colon (M, B), pancreas (M), prostate (M), thyroid (M, B), upper airways (M); other malignancies include chronic lymphocytic leukemia (M, B), Ewing tumor (M), fibrosarcoma (M), glioblastoma (M), liposarcoma (M), nodular lymphoma (M, B, C), rhabdomyosarcoma (M).
Source: Reproduced from Br J Cancer [15] with permission of the publishers.

TABLE II. Mitotic Index* of Human Tumor Cells by Direct, Liquid, and Agar Culture Techniques

Sample number	Tumor type	Direct	Liquid	Agar
1	Adenocarcinoma, breast	0.0022%	0.0010%	0.17%
2	Adenocarcinoma, ovary	0.0070%	–	1.70%
3	Adenocarcinoma, ovary	–	0.0030%	0.60%
4	Adenocarcinoma, ovary	0.0032%	–	0.28%
5	Adenocarcinoma, breast	0.0022%	0.0041%	0.04%

*Mitotic index is calculated as

$$\frac{\text{Total number of mitotic figures counted}}{\text{Total number of cells counted}}$$

Source: Reproduced from Br J Cancer [15] with permission of the publishers.

hematopoietic and solid tumors have been successfully performed (Table I, Fig. 3). These techniques, which ordinarily are not performed on solid tumor samples because of technical difficulties [4], are beginning to provide useful information concerning tumor-related chromosome change.

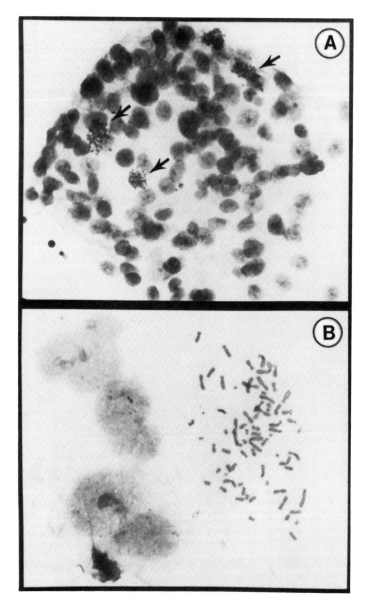

Fig. 1. A) Human ovarian tumor colony harvested for chromosome analysis. Arrows indicate mitotic figures (× 1,070, day 10). B) Cytogenetic sampling from the cluster stage of colony growth. Chromosomes harvested at this stage are amenable to a variety of chromosome banding techniques (× 2,600, day 3). Reproduced from Cancer Genet Cytogenet [16] with permission of the publishers.

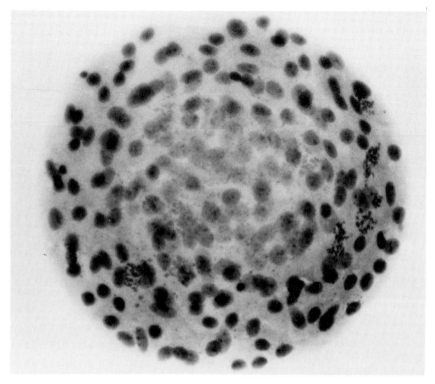

Fig. 2. Human bladder carcinoma colony grown in soft agar. Mitotic figures are peripherally distributed around a necrobiotic center (Giemsa, ×1,417). Reproduced from Br J Cancer [15] with permission of the publishers.

An example of the utility of this cloning system for tumor karyology has been our study over three years of 22 cases of ovarian adenocarcinoma. Successful assessment of the chromosome range (33–200) and modal chromosome number (37–72) of 15 of 22 samples (68%), as well as successful banding of nine of 11 (82%) cases attempted, has provided substantial information for the karyotypic profile of these ovarian cancers. Among the large variety of chromosome aberrations observed in these tumors were centric fusion, centric fission, rings, nonreciprocal translocations, and the generation of a variety of "marker" chromosomes. In agreement with the studies of Atkin and Pickstall [18], substantial numeric and structural alterations were observed in chromosome 1. Interestingly, as recently reported for endometrial adenocarcinoma [10], regions 1q23–32 in ovarian tumor cells were often involved in nonreciprocal translocations but not the often observed triplications of this segment, as seen in other tumors [19]. With the agar colony system, a significant and previously unrecognized nonrandom

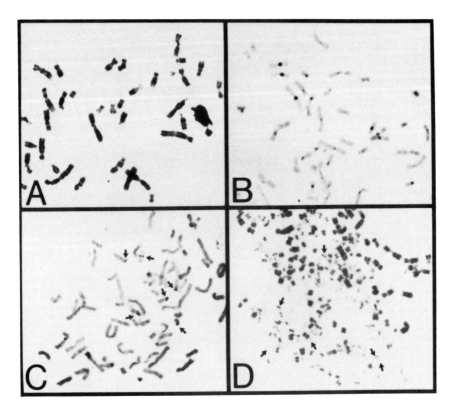

Fig. 3. Banded chromosomes from tumor cells of ovarian origin. A) G-banded metaphase displaying a variety of complex chromosome changes. B) C-banded metaphase displaying heavily stained regions of constitutive heterochromatin. C) N-banding showing silver staining of nucleolus organizer regions. Metacentric as well as the normal acrocentric silver staining is visible (arrows). D) Multiple copies of double minute bodies (arrows) in addition to numerous tumor cell chromsomes, stained by Giemsa. Reproduced from Br J Cancer [15] with permission of the publishers.

chromosome change was observed in cells from ovarian carcinoma — deletion of part of the long arm of chromosome 6 (6q—) (Fig. 4). The finding of a chromosome 6q— in four of five patients studied with detailed G-, C-, and NOR-banding techniques is further supported by several additional cases in the recent literature (Table III).

It is important that the finding of the 6q— chromosome in ovarian adenocarcinoma has been observed in cells taken from both colony and direct samples. Furthermore, the finding of an isochromosome for the long arm of chromosome 1 (i1q) in two patients studied with the colony technique (Fig. 4) has previously

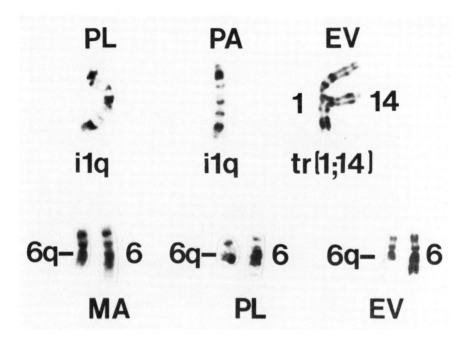

Fig. 4. Examples of chromosome 1 alterations from patients PL (case 2, Table III), PA (case 4), and EV (case 1) are presented. Isochromosomes for the long arm of chromosome 1 (i1q) observed in cases 2 and 4 are examples of chromosome change that has previously been reported in ovarian tumors [16]. Triradial formation between chromosomes 1 and 14 [tr(1:14)] was among the most unusual chromosome alterations observed in these tumors. Additionally, inversions of chromosome 1 were common in tumor cells in this study. Examples of deletion of 6q from patients MA (case 3), PL (case 2), and EV (case 1) are also presented. Breakpoints of 6q in these cases were all within two band regions (q15–21:). The remaining patient with deletion of 6q (case 4, not illustrated) was also shown to have a breakpoint in this band region (6q15:).

been reported as a common marker in ovarian cancer, as well as other solid tumors [20, 9]. Thus, our initial experience, correlated with previously published work, suggests that the colony assay may indeed provide a valuable new tool for cancer-related cytogenetic analysis.

In addition to the detailed studies on ovarian carcinoma, preliminary evidence for the application of this procedure to other tumor types is accumulating. Saxe and Meyskens [21] have studied seven cases of malignant melanoma grown in soft agar. Chromosome numbers ranged from 26 to 146, with no discernible modal number found in any tumor. The study of these colony cells has pro-

TABLE III. Summary of Cytogenetic Evidence for 6q Deletion in Ovarian Carcinoma

Case	Ovarian tumor source	No. of mitoses analyzed by banding	Deletion of 6q	Breakpoints	Reference
1	Solid/ascites	96	Present	q15	Trent & Salmon [11]
2	Ascites	37	Present	q15–16	Trent & Salmon [11]
3	Solid	31	Present	q21	Trent & Salmon [11]
4	Solid	95	Present	q15	Trent & Salmon [11]
5	Ascites	20	Absent	–	Trent & Salmon [11]
6	Solid	10	Present	q–	Not reported[a]
7	Solid	10	Present	q–	Not reported[a]
8	Solid	10	Present	q–	Not reported[a]
9	Ascites	50	Present	q12	Tiepolo & Zuffardi [33]
10	Ascites	NR	?Present?	q16	Kakati et al [20]
11	Ascites	NR	Absent	–	Kakati et al [20]
12	Cell line	50	Present	q25	Woods et al [34]
13	Cell line	60	Present	q21	Woods et al [34]
14	Cell line	50	Present	q1?	Woods et al [34]
15	Cell line	NR	Absent	–	Woods et al [34]
16	Cell line	14	Absent	–	Freedman et al [35]

[a]Personal communication, Dr. C. Kusyk, M.D., Anderson Hospital and Tumor Institute, Houston, TX.

vided evidence for the mechanism of the observed aneuploidy. Radical aberrations in chromosome segregation resulting in multiple chromosome bridges and random chromosome gain or loss result in markedly differing karyotypes among cells of a single tumor colony. This has been suggested by the authors as one possible mechanism for the heterogeneous and often ineffectual response to chemotherapy commonly seen in this tumor. Additionally, preliminary success in obtaining multiple mitoses in neuroblastoma [22; and Chapter 9], breast, and bladder cancer [23; and Chapter 7] has further supported the widespread application of this technique to solid tumor karyology.

FUTURE CONSIDERATIONS

Clonality of Human Tumors

Evidence suggesting that a variety of neoplasms have a clonal origin has been conclusively provided by cytogenetic [24] and isoenzyme analysis [25]. With data provided by these biomarkers, models of "clonal evolution" or "tumor progression" have been proposed [26] (Fig. 5). Briefly, tumors that show diploid or near-diploid mitoses and appear monoclonal by cytogenetic and isoenzyme analysis

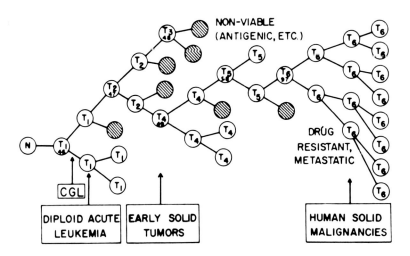

Fig. 5. Model of clonal evolution in neoplasia. Carcinogen-induced change in progenitor normal cell (N) produces a diploid tumor cell (T_1, 46 chromosomes) with growth advantage permitting clonal expansion to begin. Genetic instability of T_1 cells leads to production of variants (illustrated by changes in chromosome number, T_2 to T_6). Most variants die, due to metabolic or immunologic disadvantage (hatched circles); occasionally one has an additional selective advantage (for example, T_2, 47 chromosomes), and its progeny become the predominant subpopulation until an even more favorable variant appears (for example, T_4). The stepwise sequence in each tumor differs (being partially determined by environmental pressures on selection), and results in a different, aneuploid karyotype in each fully developed malignancy (T_6). Biological characteristics of tumor progression (for example, morphological and metabolic loss of differentiation, invasion and metastasis, resistance to therapy) parallel the stages of genetic evolution. Human tumors with minimal chromosome change (diploid acute leukemia, chronic granulocytic leukemia) are considered to be early in clonal evolution; human solid cancers, typically highly aneuploid, are viewed as late in the developmental process.
Figure reproduced from Science [26] with permission of the publishers.

(eg, chronic granulocytic leukemia) are considered to have been sampled early in clonal evolution. In contrast, most solid tumors studied have appeared multiclonal, displaying marked deviation from the diploid state. Thus, the complex pattern of chromosome change commonly seen in solid tumors is thought to occur as a result of sampling late in clonal evolution. Within this framework, Nowell [26] has suggested that for solid tumors, a correlation exists between the state of clonal evolution and the corresponding response of the tumor to clinical therapy. Thus, he infers that the further into tumor progression (evidenced by marked clonal heterogeneity), the more variable or reduced the response to standard treatment. The cytogenetic analysis of individual tumor samples with our bioassay system may reveal a portion of the clonal evolution present within solid tumors. With this basic information, coupled with results from our in vitro drug-sensitivity assay as well as in vivo clinical response, it may be possible to test the afore-mentioned theory in a clinically relevant setting for a variety of cancers. We intend to compare the cytogenetic findings prior to administration of chemotherapy with the in vivo response of the patient to chemotherapy. Coupled with this study will be a correlation of karyotypic change in vitro and in vivo after exposure to various chemotherapeutic agents. By this method it is hoped that evidence for chromosomal alteration in response to drug resistance, as well as potentially clinically prognostic information, will be accrued. In similar studies of the cytogenetic profiles and subsequent response to chemotherapy in 90 patients with acute nonlymphocytic leukemia, evidence for the relevance of chromosome analysis as an index to patient prognosis has been observed [27]. Specifically, patients exhibiting marked aneuploidy in all the cells observed at the time of presentation of disease carried a poorer prognosis than those patients with mixed or entirely diploid mitoses. There seems reason, then, to suggest that our studies of the clonality of human solid tumors may provide useful clinical information similar to that obtained for hematopoietic neoplasms.

POSSIBLE ROLES OF DOUBLE MINUTES AND HOMOGENEOUSLY STAINING REGIONS IN HUMAN CANCERS

The usefulness of detailed cytogenetic analysis in elucidating mechanisms for acquired drug resistance has recently been dramatically illustrated in both murine and human tumor cells resistant to the antifolate methotrexate (MTX) [28]. Briefly, tumor cell lines were exposed to increasing concentrations of MTX with cells selected displaying resistance to very high concentrations of this drug. When these cells were analyzed by G-banding, extended nonstaining regions were observed on several chromosomes. Upon biochemical characterization, these extended DNA segments were demonstrated to contain multiple copies of the gene dihydrofolate reductase (dHRF), the enzyme necessary to overcome the MTX block. This represented the first example of somatic gene amplication in a mammalian system. Importantly, the cytologic profile of the extended G-negative regions (now called homogeneous staining regions; HSRs) visible using chromosome banding were found to vary in size in direct proportion to the concentration of MTX. Apparently, when MTX was removed, copies of the amplified genes for dHFR were excised from chromosomal DNA. The decrease in size or the disappearance of the HSR could then be correlated with the appearance of double minute chromatin bodies (dms). Dms are cytologically identifiable phenomena that may be associated with the functional adaptability of human tumors. These small paired chromatin bodies (Fig. 3D) have been observed in MTX-resistant cell lines as well as in direct samples and cell lines derived from several human tumors, including neuroblastoma, breast, colon, and ovary, among many others [29].

It is important that dms have not been observed in samples from normal tissues [30]. Double minute chromatin bodies have been shown to contain DNA but lack centromeres necessary for normal chromosome segregation [29]. However, often large numbers of dms (>200 copies/cell) are maintained in a high percentage of tumor cells. Because of their persistence in tumor populations and their absence in normal tissues, dms have been suggested to play an adaptive role in tumor growth. Although hard evidence for the functional expression of dm-DNA is not yet available, such a functional role for dms in tumor adaptation has served as the basis for a theory of "preadaptive amplification" responsible for the function and occurrence of dms. Briefly, I suggest that segments of DNA, perhaps segments directly involved in cell proliferation, are "preferentially amplified" following malignant transformation. These amplified regions could then be excised from chromosomal DNA to form the pattern of dms seen in various tumor samples. The suggestion for a preadaptive role for dms is based upon the observation of multiple dms in direct samples from clinically untreated tumor samples. Thus, dms are not envisioned as resulting as a direct response to a selective environment (postadaptive change), but rather the selective environment (eg, chemotherapy) may direct the survival of cells *previously* mutated and maintaining the ability to proliferate. Examples of preadaptive mutation are well established in genetic literature in prokaryotic [31] as well as in lower eukaryotic systems [32]. With the experimental rationale for tumor cell lines with and without dms that has been used in bacterial systems [32], preliminary testing of this hypothesis is under way.

It may soon be possible to correlate the occurrence of either HSRs or dms with clinical response to therapy. Monitoring the action of various chemotherapeutic agents in producing or possibly restricting clonal heterogeneity as well as in HSR and dm formation may become an important prognostic tool in clinical medicine.

SUMMARY

Cytogenetic analysis of human tumors is rapidly coming of age as a useful tool in the diagnosis and prognosis of cancer. With the use of chromosome-banding analysis, an increasing display of tumors containing cytologically recognizable nonrandom chromosome change is accumulating. Additionally, with avenues of assessing drug resistance and sensitivity, current cytogenetic analyses are yielding clinically relevant information. Finally, the blending of cytogenetics with biochemistry and cell biology is yielding an impressive array of research tools for the subcellular study of cancer. DNA hybridization, gene cloning, autoradiography of various drug and hormonal chromosomal binding sites, gene mapping, and numerous other related techniques are providing fresh insights into the genetic basis of cancer. It seems clear that during the next decade substantial progress in understanding chromosome structure and function in normal cells as well as tumor cells will occur. Technical advances in the methodology for growing and harvesting tumor cells for detailed cytogenetic analysis will undoubtedly be important in this effort. Application of the described soft agar colony technique may provide a useful tool for study of the genetics of human solid tumors.

REFERENCES

1. Boveri T: Beitrog Zum Studium dis Chromatins in der Epithelzellen der Carcinome. Beitr Pathol 14:249, 1912.
2. Nowell PC, Hungerford DA: A minute chromosome in human chronic granulocytic leukemia. Science 132:1497, 1960.
3. Rowley JD: A new consistent chromosomal abnormality in chronic myelogeneous leukemia identified by quinacrine fluorescence and Giemsa staining. Nature 243:290, 1973.
4. Mitelman F, Levan G: Clustering of aberrations to specific chromosomes in human neoplasms. Hereditas 89:207, 1978.
5. Mark J, Levan G, Mitelman F: Karyotype patterns in human meningioma. Heriditas 75:213, 1973.
6. Manlov G, Manlova Y: Marker band in one chromosome 14 from Burkitt lymphomas. Nature 237:33, 1972.
7. Fukuhara S, Rowley JD: Chromosome 14 translocations in non-Burkitt lymphomas. Int J Cancer 22:14, 1978.
8. Golomb HM, Rowley JD, Vardiman J: Partial deletion of long arm of chromosome 17: A specific abnormality in acute promyelocytic leukemia? Arch Inserv Med 136:825, 1976.
9. Kovacs G: Abnormalities of chromosome 1 in hematological malignancies. Lancet 1:554, 1978.
10. Trent JM, Davis JR: D-group chromosome abnormalities in endometrial cancer and hyperplasia. Lancet 2:8138, 1979.
11. Trent JM, Salmon SE: Karyotypic analysis of human ovarian carcinoma cells cloned in agar. Am J Hum Genet 31:113A, 1979 (Abst).
12. Caspersson T, Zech L, Johansson C, Modest ES: Identification of human chromosomes by DNA-binding fluorescent agents. Chromosoma 30:215, 1970.
13. Kakati S, Sandberg AA: Chromosomes in solid tumors. Virch Arch B Cell Pathol 29:129, 1978.
14. Hamburger AW, Salmon SE: Primary bioassay of human tumor stem cells. Science 197:461, 1977.
15. Trent JM, Salmon SE: Human tumor karyology: Marked analytic enhancement via short-term agar culture. Br J Cancer (in press).
16. Trent JM, Salmon SE: Potential applications of a human tumor stem cell bioassay to the cytogenetic assessment of human cancer. Cancer Genet Cytogenet 1:291–296, 1980.

17. Moore MAS, Metcalf D: Cytogenetic analysis of human acute and chronic myeloid leukemia cells cloned in agar culture. Int J Cancer 11:143, 1973.
18. Atkin NB, Pickstall VJ: Chromosome 1 in 14 ovarian cancers. Hum Genet 38:25, 1977.
19. Rowley JD: Abnormalities of chromosome no. 1: Significance in malignant transformation. Virch Arch B Cell Pathol 29:139, 1978.
20. Kakati S, Hayata I, Oshimura M, Sandberg AA: Chromosomes and causation of human cancer and leukemia. X. Banding patterns in cancerous effusions. Cancer 36:1729, 1975.
21. Saxe D, Meyskens FL: Abnormal chromosomal segregation in human malignant melanoma (in preparation).
22. Von Hoff DD, Casper J, Bradley E, Trent JM, Hadoch A, Reichert C, Makuch R, Altman A: Direct cloning of human neuroblastoma cells in agar: A potential assay for diagnosis, response and prognosis. Cancer Res (in press).
23. Buick RN, Stanisic TH, Fry SE, Salmon SE, Trent JM: Development of an agar/methylcellulose clonogenic assay for progenitor cells in transitional cell carcinoma of the human bladder. Cancer Res 39:5051, 1979.
24. Nowell PC: Preleukemia: Cytogenetic clues in some confusing disorders. Am J Pathol 89:459, 1977.
25. Fialkow P: The origin and development of human tumors studied with cell markers. N Engl J Med 291:26, 1974.
26. Nowell PC: Clonal evolution of tumor cell populations. Science 194:23, 1976.
27. Golomb HM, Vardiman J, Rowley JD, Testa J, Mintz J: Correlation of clinical findings with quinacrine-banded chromosomes in 90 adults with acute nonlymphocytic leukemia. N Engl J Med 299:613, 1978.
28. Alt F, Kellens R, Bertino J, Schimke R: Selective multiplication of dihydrofolate reductase genes in methotrexate-resistant variants of cultured murine cells. J Biol Chem 253:1357, 1978.
29. Barker PE, Hsu TC: Double minutes in human carcinoma cell lines with special reference to breast tumors. J Natl Cancer Inst 62:257, 1979.
30. Quinn LA, Moore GE, Woods LK, Morgan RT, Semple T: Occurrence of double minute chromosomes (dm) in fresh and cultured human tumor cells and normal cells. Proc Am Assoc Cancer Res 20:26, 1979 (abstr).
31. Cavalli-Sforza LL, Lederberg J: Isolation of pre-adaptive mutants in bacteria by sib selection. Genetics 41:367, 1956.
32. Luria SE, Delbruck M: Mutations of bacteria from virus sensitivity to virus resistance. Genetics 28:491, 1943.
33. Tiepolo L, Zuffardi O: Identification of normal and abnormal chromosomes in tumor cells. Cytogenet Cell Genet 12:8, 1973.
34. Woods LK, Morgan RT, Quinn LA, Moore GE, Semple T, Stedman KE: Comparison of four new cell lines from patients with adenocarcinoma of the ovary. Cancer Res 39:4449, 1979.
35. Freedman RS, Phil E, Kusyk C, Gallager HS, Rutledge F: Characterization of an ovarian carcinoma cell line. Cancer 42:2352, 1978.

15
Use of an Image Analysis System to Count Colonies in Stem Cell Assays of Human Tumors

Bernhardt E. Kressner, Roger R. A. Morton, Alexander E. Martens, Sydney E. Salmon, Daniel D. Von Hoff, and Barbara Soehnlen

As is well demonstrated elsewhere in this text, the clonogenic assay for tumor colony-forming cells has applicability to a broad scope of human tumors and has proved valuable in studies of biology, clinical course, and chemosensitivity of human cancers. The development of this promising new area of clinical research, however, has precipitated a substantial new laboratory problem; namely, the need for automation in counting tumor colonies. This need was not fully apparent until it became clear that the clonogenic assays predicted clinical and biological features of human cancers. In the initial studies, careful qualitative and quantitative evaluations of tumor clusters and colonies in soft agar were conducted by the clinical research laboratory staff of two of the authors (S.E.S., D.D.V.H.). As their studies proceeded, we recognized that there was a major need for a precise automated instrument for selective counting of tumor colonies and therefore initiated a join developmental project with Bausch & Lomb Incorporated on the application of image analysis to this task.

Our experience with visual counting of colonies growing in this assay has defined a number of problems:

1) Visual counting of experiments is time-consuming and therefore very expensive in terms of professional time. It is not uncommon to spend nine hours to count a 100-plate experiment.

2) We have shown (Fig. 1) that in the size range of interest the number of colonies changes very rapidly with size. This means that an error of only ±10% in the visual estimate of colony size can result in +90% to −45% counting error. Since in visual counts small differences in size are very difficult to determine accurately, this problem can contribute to very substantial counting errors.

3) Investigator fatigue becomes a significant factor in counting a series of large assays on one or more specimens, and has been shown to lead to nonreproducibility.

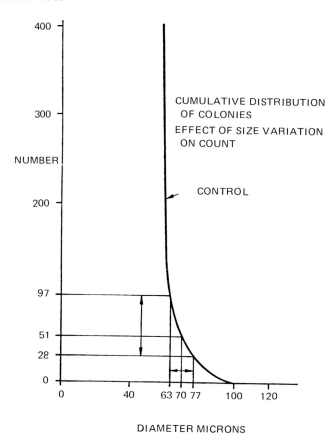

Fig. 1. Plot of cumulative number of colonies versus colony diameter in microns to show extreme sensitivity of counts to changes in size limits.

4) There is variability in counting from one investigator to another, in part because of differences in criteria with respect to cluster and colony size, and additionally because of fatigue, as mentioned above. Errors of 100% and more have been often observed when two people count the same plate.

5) A large amount of time is required to train an individual to count colonies visually with an acceptable accuracy.

6) Standardized definitions, such as size, cut-off criteria and information on the rate of development of clusters and colonies, are highly desirable and virtually require automation if they are to be carried out on more than just a few plates.

Whereas we expected that automated counting equipment could potentially increase counting speed and provide reproducible criteria, we also anticipated substantial difficulties with regard to the instrument's ability to discriminate between the colonies and extraneous objects. In fact, our concerns were substantiated during our initial attempts to use a relatively simple image-analysis system (The Bausch & Lomb Omnicon Alpha™) in this application.

This instrument's analytical capabilities for differentiating between colonies and other objects were not adequate to produce reliable results. Although the Alpha performed quite well on plates with a large number of colonies (compared to the number of artifacts) it failed to analyze accurately plates with only a few colonies — ie, those with low fractional survival of TCFUs, which are of primary importance in indicating sensitivity to drugs.

We therefore initiated a series of studies using the more powerful, software-based Bausch & Lomb Omnicon™ FAS-II image analysis system with the objective of matching or exceeding human capabilities for correctly identifying tumor clusters and colonies while rejecting artifacts. This objective was attained, and efforts toward reducing counting time and providing additional automated features were initiated.

MATERIAL

Tumor biopsies were prepared and cultured with the method of Hamburger and Salmon [1–3], which is detailed in Chapters 3–10. The standard agar concentrations were used, but many plates (particularly for solid tumors) did not require conditioned media. In addition to plates cultivated at 200,000–500,000 cells/dish, and various reference specimens, additional cell concentrations ($1 \times 10^4 - 1 \times 10^6$) were prepared to test the linearity of the automated equipment in relation to visual counting. Representative cultures from a wide variety of tumor types were investigated (eg, breast, melanoma, ovary, lung, neuroblastoma, and myeloma) from fresh biopsies as well as from human tumor cell lines (myeloid-leukemia HL60, and myeloma 8226). Plates with varying degrees of red cell contamination were also studied, as were plates with a variety of extraneous objects in the agar layers. Colony counts on triplicate samples were performed by two of us (B.S., D.D.V.H.) immediately prior to counting with the FAS-II. Some specimens were also fixed with 3% glutaraldehyde in phosphate-buffered saline so that they could be preserved in a refrigerated humidified chamber for varying periods prior to automatic counting.

EQUIPMENT

The equipment used in these experiments consisted of a microscope equipped with a fast automatic stage and a precision television scanner, the Omnicon FAS-II unit, and a printing terminal. The FAS-II [4], which is a versatile, general-purpose image-analysis system, was modified, as described below, to ensure optimum performance in this application. Also, dedicated software, including special algorithms for object discrimination, was developed.

Both regular and inverted stage microscopes have been used in this application. The inverted stage offers greater convenience in loading and removing the Petri dishes, and also has the very desirable feature of permitting the dish cover to be left in place during counting because the image is formed through the bottom of the dish. Thus the Petri dish contents are continually protected from dust, the drying of the agar is retarded, and the possibility of contamination by biologically active agents (bacteria, mold spores, viruses) is greatly reduced.

The initial machine-counting of the assay plates has been attempted using the regular Omnicon automatic microscope stage designed primarily for operation at high magnification. That stage proved to be unacceptably slow, and a new fast stage was developed specifically for colony counting. This stage can move at speeds of up to 35 mm per second, and its introduction made the sample scanning time a small fraction of the total analysis time.

In its present configuration the stage is equipped with a shallow well designed to accept and locate accurately the round 35 mm plates. This arrangement is designed for manual placement and removal of Petri dishes. Several accessories for handling groups of dishes will be added in the future.

We found experimentally that the optimum magnification, as a compromise among depth of field, resolution, and scanning speed, is 2.5 ×. At this magnification, the size of each field corresponding to a single TV frame, on the agar surface is 3.3 × 4.4 mm. The scanning sequence is as follows:

1) From the fixed start position the stage moves the center of the dish directly under the objective to allow for focusing adjustments, if required.

2) The stage moves the inner edge of the meniscus to the optical axis to start the analysis.

3) The TV scanner scans this field to acquire the image for analysis.

4) After completion of the image processing, the stage moves to the next field in the scanning pattern. This pattern of 35 fields has been devised to provide maximum coverage (86% of the useful agar surface area limited by meniscus) of the plate in minimum time, with larger area coverage optionally available.

The area for analysis is confined to the flat portion of the agar surface, because the meniscus at the Petri dish walls may cause optical distortion. The overall coverage with this scanning pattern is 51% of the total Petri dish area. Initially, when comparing visual with machine counts to account for the difference in the areas counted, we applied a constant factor to the machine results because we had assumed that the visual counts came from the total area of the plate. As it turned out, investigators doing visual counts also stay within the level area of the agar. For this reason, the constant was dispensed with (Fig. 2).

In order to understand why the FAS-II has been successful in simulating human judgment to perform these analyses, it is necessary to discuss some fundamental concepts. The agar is imaged by the microscope objective onto the photosensitive

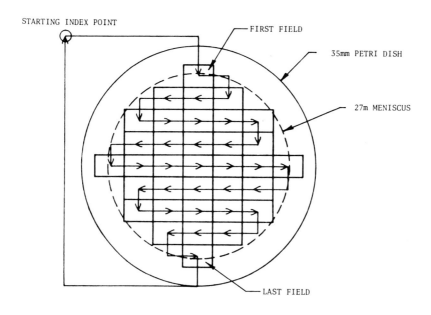

Fig. 2. Diagram of scanning pattern for stage motion in Omnicon stem cell analysis.

surface of a vidicon tube in the scanner, which converts the optical image into electrical video signals. The precision scanner has been designed specifically for image analysis, as distinct from scanners used in closed-circuit or broadcast television. As such, the scanner has a much greater scanning linearity, lower noise, and higher stability to ensure faithful transformation of images into equivalent video signals. These are suitable for extraction of quantitative size, shape, and density information in the FAS-II.

FAS-II is a system in which a computer plays a major role. Nearly all capabilities and functions of this instrument are defined in terms of software programs, which facilitates the adaptation of the FAS-II to a wide range of diverse applications. Consequently, the major tasks in adapting this instrument to colony counting in stem cell assays, in addition to the design of a fast stage and some modifications in the detection circuitry, were the formulation of the colony selection and measurement algorithms and the software realization of these algorithms.

Besides the computer, the FAS-II includes the precision scanner discussed above, detection circuitry, and special signal processing and interface circuits that allow the computer to accept the video information. The interaction between the system and its operator occurs by means of a keyboard and a television screen which, in addition to the image of the specimen, also displays a variety of messages, queries, and prompters to guide the operator. A printing terminal is also included to generate hard-copy output of results in a desired format.

METHODS

Our initial attempts to use the less powerful image analyzer (Omnicon Alpha) did not yield acceptable results because the instrument was unable to discriminate satisfactorily between the colonies and extraneous objects, such as tissue debris, agar imperfections, bubbles, and occasional colonies of mold. Some of these artifacts are always present in the culture plates despite efforts to prevent them. The limitations of the instrument were most obvious in drug-survival curves.

When visual counts are performed, the individual doing them identifies, on the basis of training and previous experience, which objects in the Petri dish are to be counted or not counted. The decision whether to count or not to count an object is based on this object's appearance – ie, size, shape, optical density, color, and surface texture. Our studies have revealed, however, that shape and optical density are the primary discriminating factors. The person performing a visual count uses a three-step procedure:

1) Decides to accept or reject an object, depending on whether it is a colony or not.

2) If it is a colony, judges whether its size (diameter or maximum chord) exceeds the predetermined minimum size to qualify it for acceptance.

3) If these conditions are satisfied, adds one more count to the running total, and proceeds to the next object.

The techniques for selective automatic counting of stem-cell colonies developed for the FAS-II emulate this procedure.

1) The image from a field on the agar surface is *detected*. This means that the optical density of all objects in that field is analyzed and only those objects that fall into a predetermined optical density range are accepted for further investigation. The detection rejects some of the extraneous objects, while at the same time converting the video signals representing the remaining objects into a digital form for further processing.

Agar substrates are optically nonhomogeneous in transmitted light, so that the optical density of the background against which the objects on the agar surface are detected varies considerably. To deal with this problem and the inherent shading in the optics special circuitry to track the background has been developed to ensure uniform detection across the entire field.

2) The system measures the individual areas of all detected objects, rejecting all that are outside a specified range. This is a fast preselection process that results in a considerable overall time saving by reducing the number of objects to be subjected to the more time-consuming shape measurements.

3) Dimensionless shape parameters (related to elongation and boundary "roughness") are then measured on the remaining objects. These shape parameters and their numerical criteria have been chosen after considerable experimental effort and it appears that they are applicable to a wide range of cancer types. The instrument performs the final selection of the colonies based on the numerical values of these parameters. The four steps of this process are illustrated in Figure 3.

4) After the final selection, the remaining objects now accepted as clusters or colonies are measured. We used two types of measurements: equivalent circular diameter and maximum horizontal chord. The former is derived by measuring the areas and computing diameters of circles corresponding to those areas; the latter is the longest intercept in the direction of the scan. Both of these measurements are reported.

The foregoing four-step procedure is termed "Selective Counting."

RESULTS

From the many experiments and investigations performed over a nine-month period, several findings stand out.

1) Conventional regression analyses on repeated FAS-II versus FAS-II counts showed a four- to eight-fold improvement in stability or reproducibility when compared with visual versus visual counts (Table I).

2) Correlation coefficients relating machine versus visual counts were typically greater than 0.95, with regression coefficients ranging from 0.84 to 0.94. Comparison data from one of the most recent experiments are also listed in Table I. We found, however, that correlation coefficients can be a misleading measure of error, due to the strong influence of the spread or range of counts in a given experiment. Thus individual counts having a certain displacement from a regression line, but which were grouped close to the control counts, produced regression coefficients that differed considerably from those derived from counts that had the same apparent displacement but were grouped farther away from the control value.

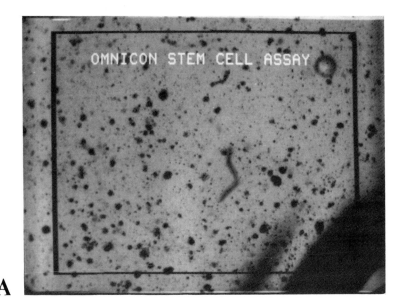

Fig. 3. Four monitor photographs showing the detection and selection process (the white "blips" on panels C and D are tags indicating the selected features): (A) A normal image with colonies, cracks, agar wrinkles, and a bubble; (B) a detected image of the same view;

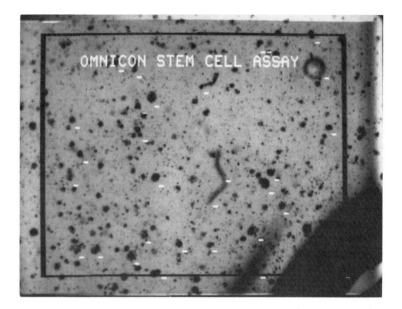

(C) tagged features remaining after size selection only; (D) tagged features remaining after size and shape selection. These are now only the colonies of interest.

TABLE I. Tabulation of Two Types of Errors in Comparisons of Machine and Visual Counts

Comparison[a]	Regression coefficient (r^2)	% RSM error
Original counts		
Visual vs visual	0.84	23
FAS-II vs FAS-II	0.98	11
Visual vs FAS-II	–	20
% Survival data		
Visual vs visual	0.91	17
FAS-II vs FAS-II	0.98	8
Visual vs FAS-II	0.94	13

[a]Comparison of visual vs visual and FAS-II vs FAS-II refers to blinded repeat studies with each approach. The analysis has been carried out separately on both original counts and computations of percent survival in relation to controls in drug-treated experiments.

To deal with this problem, a "% RMS error" function was derived for characterizing various comparisons in this application (Fig. 4). Among those advantages, this measure of error represents the overall accuracy of data matching to a greater degree and with more stability than do regression coefficients. Our goal, of course, has been to develop a system to the point where machine counts would match visual counts at least as well as visual counts can match other visual counts. On the basis of either type of error analysis, we have achieved that goal, as is shown quantitatively by the examples in Table I.

More importantly, the ability of the FAS-II to predict the percent survival of TCFU after drug exposure (standard assays as in Chapter 18) closely matches the results of visual (manual) counting. This is shown for three separate studies in Table II.

3) Part of the reason for the relatively poor performance of human counting is shown in Figures 1 and 5. Contrary to the most desired functional relationships for limits in scientific work, those in which a limit value is chosen on a "plateau," the nature of cumulative counting in stem cell assays requires a judgment to be made on the "edge of a cliff!" The problem is of course compounded for an investigator who is required to estimate diameters using a reticle or other eyepiece scale. For example, in Figure 1 a 10% error in estimated diameter at the 70 micrometer level can cause cumulative count values having relative errors of −45% to +90%. It is apparent, therefore, that only an accurate, stable detection system that provides a constant size limit can ensure the required reproducibility, even if other colony recognition factors were not involved. Figure 5 shows a comparison of cumulative counts from a control dish and two drug-treated dishes.

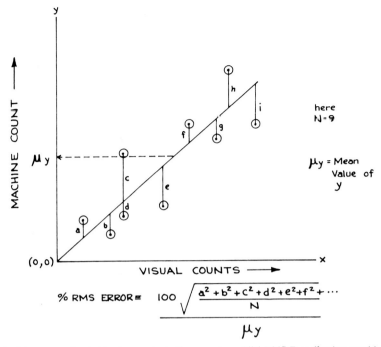

Fig. 4. Diagrammatic plot to demonstrate the meaning of "% RMS Error" using machine versus visual counts.

TABLE II. Survival of TCFU After Drug Exposure: Visual Counts vs FAS-II Counts*

Tumor type	Drug	Visual count (% survival)	FAS-II count (% survival)
Squamous carcinoma of the lung	Methotrexate	85	87
	Adriamycin	82	85
	Chlorambucil	80	75
	Dihydroxyanthracenedione	52	48
Adenocarcinoma of the lung	Methotrexate	100	100
	Adriamycin	10	15
	Dihydroxyanthracenedione	30	20
Ovarian carcinoma	Methotrexate	100	100
	Adriamycin	100	100
	Chlorambucil	100	100
	Cis-platinum	42	30
	Hexamethylmelamine	100	100

*A comparison of percent survival predicted by both visual and FAS-II analyses for three separate experiments.

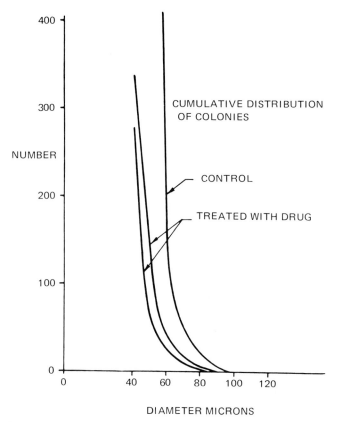

Fig. 5. Plot of cumulative number of colonies versus colony diameter for a control plate and two separate drug experiments.

4) The need for automatic recognition of colonies is demonstrated in Figure 6, which shows the results of counting a typical stem-cell control plate of an ovarian cancer. Without automatic recognition the number of objects (colonies plus extraneous objects or background) is much greater than the actual number of colonies counted using automatic recognition to select the colonies of interest, while excluding from the result all other objects.

Since the Omnicon Alpha lacks adequate ability to differentiate objects based on their shape, the RMS error, as contrasted to that of the FAS-II, was 38% for similar experiments.

5) The average time for the FAS-II to perform an analysis is about three minutes per dish without the use of special programming or the incorporation of high-speed hardware options. With a high-speed system configuration and software, this time will be reduced to less than one minute per dish.

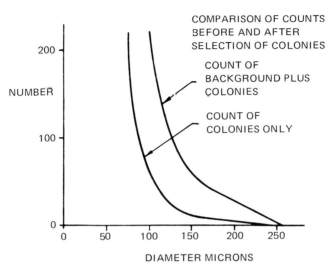

Fig. 6. Plot of cumulative number of colonies versus colony diameter from the same dish with and without use of selection criteria.

DISCUSSION

The extensive experimental evidence obtained to date by personnel from three independent laboratories (University of Arizona, University of Texas, and Bausch & Lomb) indicates that automatic selective colony counting using image analysis technology is not only feasible but also produces results that are in several important ways (speed, repeatability, uniformity of criteria) superior to visual counts.

The development of new instrument capabilities for performing quantitative biological observations on human tumor-colony formation could have major impact on fundamental cancer research. Such capabilities could provide an essential function of selective colony counting for drug screening and routine clinical testing of drug sensitivity on cultured biopsy specimens. Because analytical procedures with the FAS-II system are nondestructive and do not require fixation or staining of the cultures, serial observations can be readily made on multiple samples. Investigations carried out in this fashion could clarify the growth kinetics of clonogenic cells that give rise to clusters and colonies in the presence or absence of cytotoxic drugs or hormone exposure. Such data are of significant importance, as we lack information about what proportion of cells that give rise to clusters are also the progenitors of colonies, and whether spontaneous human tumors contain a hierarchy of progenitors with differing degrees of proliferative and differentiating capability. By examining serial histograms of the size distributions, as well as other features of clusters and colonies, this question may be clarified substantially.

At a more practical level, scanning of the dishes shortly after initial plating would permit enumeration of cellular clumps that might be the result of sample preparation, and would permit, if required, subsequent use of these data for a type of "background subtraction." Alternatively, for routine purposes, several control plates can be fixed in glutaraldehyde (Chapter 12) shortly after they have been plated and stored until final counting to provide a simultaneous evaluation of sample background. Such considerations may be of only limited importance, however, as the common artifacts present in such plates are easily recognized as such by the FAS-II and rejected automatically during the selective counting of colonies.

While initially we had hoped that the less sophisticated Omnicon Alpha would prove adequate to quantify tumor colony formation based merely on size and optical density criteria, these two functions proved to be inadequate because they fail to ensure adequate rejection of artifacts in the plates (bone marrow spicules, clumps, agar holes, or bubbles, etc). Consequently, additional criteria relating to shape, as well as other descriptors, available with the image analysis and detection of the FAS-II proved essential for dealing with some of the most important specimens. For example, in drug assay experiments, wherein reduction of colony-forming units to less than 10% of control was achieved with a drug, the failure to reject extraneous objects had an adverse impact on the accuracy of the data. This is particularly important in the tumor stem cell assay system, since often fewer than 100 colonies are present on the control plates. Thus the usual simple video colony counters that cannot discriminate between colonies and artifacts fail to compute accurately the most important experimental results. Video colony counting might, for instance, report 40–50 surviving colonies on a plate, whereas in reality there were fewer than ten. Since the reduction in survival of colony-forming cells to 40–50% of control is not indicative of clinical response or substantial drug activity (Chapters 16–18), this limitation of simpler instruments drastically reduces their usefulness. Plating of increasing numbers of cells in serial dishes in control experiments did not detect this limitation, and it was only through extensive empirical studies with clinical samples from actual drug assays that we were able to conclude that the image-analysis capabilities of the FAS-II were essential to the enumeration process.

The tumor stem cell assay system is undergoing transition from a research procedure to a clinical procedure. In that context we are considering various options that can be retrofitted to the FAS-II to increase its speed and specimen throughput capability. For example, one logical way to optimize the scanning procedure is to introduce square or rectangular Petri dishes or sample wells having an area similar to that of 35 mm round Petri dishes. The flat area of a 31 X 31 mm square dish could be completely covered by thirty-six 3.3 X 4.4 mm fields. The compatibility of such Petri dish shape with the tumor stem cell assay would have to be, of course, experimentally confirmed. Another alternative is the use of multi-chamber plates or groups of plates that could be readily handled with a large mechanical stage especially designed for that purpose. Further refinements in the reporting of assay results are also in progress.

The widespread application of automated tumor colony counting may well provide the basis for major advances in new drug screening as well as in clinical oncology. In both circumstances it is essential that standardized criteria for tumor colony counting be available that have identical meaning from laboratory to laboratory. As is underscored by the percent error in manual counts from experienced investigators (as shown by our data), even greater error could be anticipated in manual counting from laboratory to laboratory, since the criteria for manually counting colonies are not easy to teach, and the process of serial counting is subject to substantial problems of fatigue.

We view our efforts as having been successful in solving the major problems of automated tumor colony counting for the stem cell assay system and look forward to the application of this new technology to basic and clinical cancer research and patient management.

REFERENCES

1. Salmon SE, Hamburger AW, Soehnlen B, Durie BGM, Alberts DS, Moon TE: Quantitation of differential sensitivity of human-tumor stem cells to anti-cancer drugs. N Engl J Med 298:1321–1327, 1978.
2. Hamburger AW, Salmon SE: Primary bioassay of human tumor stem cells. Science 197: 461–463, 1977.
3. Hamburger AW, Salmon SE, Kim MB, Trent JM, Soehnlen BJ, Alberts DS, Schmidt HJ: Direct cloning of human ovarian carcinoma cells in agar. Cancer Res 38:3438–3444, 1978.
4. McCarthy CJ, Stevens RE: Dimensions in image analysis. Am Lab, vol 2, 1978.

IV. Measurement of Drug Sensitivity and Clinical Correlations

16
In Vitro Drug Assay: Pharmacologic Considerations

David S. Alberts, H.-S. George Chen, and Sydney E. Salmon

The design and analysis of a relevant in vitro human tumor stem cell assay system for chemotherapeutic agents requires knowledge of the kinetics of the anticancer drug's spontaneous, in vitro degradation rate, in vivo activation state, and plasma disappearance kinetics. Information on spontaneous degradation is necessary to calculate the effective concentration·time product of the active drug in direct contact with the tumor cells in vitro, and pharmacokinetic data are necessary to predict the clinically relevant drug concentration·time products to be used in vitro. Data on the biochemical pharmacology and on cellular kinetics may also be required for optimal design of in vitro exposure conditions for antimetabolities. This chapter will provide both practical pharmacokinetic data and rationale for the selection of anticancer drug concentration·time products to be used in the in vitro drug assay systems. In addition to this pharmacokinetic analysis relating to correlations with drug sensitivity or resistance, we have also provided brief discussion of the effects of varying dose exposure time in vitro of cycle-active drugs using the tumor stem cell assay as a model system.

RELATION OF PHARMACOKINETICS TO ASSAY DESIGN AND ANALYSIS

Explanation of Terminology

There are three important pharmacokinetic parameters used throughout this analysis of pharmacology relevant to in vitro sensitivity studies of tumor colony-forming units (TCFUs):

1) the plasma concentration·time product (CXT)
2) the peak plasma drug concentration
3) the plasma terminal phase drug half-life ($t_{1/2}$)

These are standard pharmacokinetic terms, and methods for their calculation have been clearly described [1].

The plasma CXT of an anticancer drug is probably the most important pharmacokinetic parameter in determining drug efficacy and toxicity. In this analysis, when we speak of an anticancer drug, we are referring to the active form of the agent. There is ample evidence from in vitro cell culture and sensitive mouse tumor model systems to suggest that an anticancer drug's CXT is directly related to the magnitude of its cellular lethality [2]. Thus, as the drug's plasma CXT increases, its toxicity also increases, whether it be to TCFUs or normal renewal systems. The CXT of an anticancer drug can be calculated by fitting a multi-exponential equation to the measured plasma disappearance curve after drug administration and then integrating the area under the curve. An alternative method of measurement is to calculate the areas of the triangles and rectangles that fit under the drug's plasma disappearance curve (ie, the "trapezoidal rule").

The second important pharmacokinetic parameter is the peak plasma concentration of the agent. Studies in mouse tumor model systems have shown that the anticancer activity of alkylating agents and anthracyclines is directly related to dose [3, 4] and, thus, peak plasma concentration. The peak concentration of the agent in plasma is usually determined by direct measurement. For certain agents (eg, cyclophosphamide) the relevant moiety is a biotransformed active product, and not the administered parent compound.

Finally, the drug's terminal phase plasma half-life ($t_{1/2}$) is important because it is one determinant of the drug's CXT and dosing schedules and reflects the dynamics of the drug's elimination, whether by active metabolism, chemical degradation, or urinary and/or fecal excretion, etc. The $t_{1/2}$ can be determined by fitting a multiexponential equation to the plasma disappearance curve for the drug. The $t_{1/2}$ (terminal) is the time required to halve the concentration of the drug during its terminal elimination phase. The $t_{1/2}$ per se has not been used with respect to the in vitro assay design or analysis, but it is useful for establishing optimal in vivo dosing schedules for clinical trials, which are required to validate predictions from in vitro assay of drug sensitivity.

Important Pharmacokinetic Parameters

We have tabulated (Appendix 4) the important pharmacokinetic parameters of 36 of the most commonly used anticancer drugs so that these may be used in the design of rational in vitro drug assay and clinical trials. Included in these tables for each drug are data concerning dosing schedules and routes of administration, peak plasma concentrations, plasma CXTs, terminal phase $t_{1/2}$s, urinary excretion (%), and specific journal references. Pharmacokinetic parameters for these anticancer drugs were either obtained from the published literature or computed from reported results, which we analyzed by the methods described in the paragraphs that follow.

If there was enough pharmacokinetic information for an accurate determination of parameter values, the plasma or serum concentrations were then expressed as

$$C = \sum_{i=1}^{N} A_i \exp(-B_i t) \qquad (1)$$

where C is the drug concentration at time t after drug administration, A_i and B_i are constants, and N ranges from one to three. The additional constraint of $\sum_{i=1}^{N} A_i = 0$, for $N \geq 2$ has been posed for Eq 1, if the drug is administered by oral, intraperitoneal, or intrapleural routes.

The area under the plasma disappearance curve (CXT) and the terminal phase plasma $t_{1/2}$ for the anticancer drug were calculated according to

$$CXT = \sum_{i=1}^{N} \frac{A_i}{B_i} \qquad (2)$$

and

$$t_{1/2} = \frac{\ln 2}{B_N} \qquad (3)$$

The parameters A_i and B_i were either directly obtained from the literature or derived by curve fitting of the reported data using NONLIN [5]. In several cases only the total body clearance was reported. The drug's plasma CXT was then calculated by

$$CXT = \frac{\text{Total dose administered}}{\text{Total body clearance}} \qquad (4)$$

If the number of data points was not sufficient to allow the above procedures to be completed, the terminal phase plasma $t_{1/2}$ was then estimated by graphic methods or by linear regression of the data points in the terminal phase. The CXT was then calculated using the trapezoidal rule plus the last plasma concentration point multiplied by 1.443 $t_{1/2}$.

Our summary of the pharmacokinetics of standard anticancer drugs as well as current phase I–III investigational agents is presented in Appendix 4. For several standard drugs (ie, Adriamycin, cytosine arabinoside, methotrexate) the results from model simulation are also included for comparison. Throughout these tables, it has been assumed that a drug's pharmacokinetics are linear with dose. Thus it has been assumed that for a different drug dosage, peak plasma concentrations and plasma CXT could be obtained using linear scaling.

Pharmacokinetic Principles Used in the Design of the In Vitro Drug Assay

Three principles have been used to design the concentrations and exposure times of a specific anticancer drug to be placed into the in vitro tumor stem cell assay system. First, we have tried to use at least three different drug concentrations for a minimum of 1 hour of incubation. The 1-hour exposure time was established on practical grounds, including 1) the need to standardize time exposure for assays involving eight to ten different drugs, 2) considerations that short-term exposures (eg, 5–10 minutes) might not be adequate with respect to time for uptake of all drugs, and 3) pharmacokinetic data suggesting that significant cellular exposure to most drugs is greatest during the first hour after administration. Three concentration points were used to ensure an adequate representation of the dose-response curve for any given drug against a specific patient's tumor. Once a relatively large body of information is available from assays in which sensitivity is manifest, consideration can be given to using two drug doses plus a control.

Second, for new drugs a wide range of drug concentrations are used, usually covering two or more logs, and ranging from doses considerably below the clinically achievable CXT to doses far above those conventionally achievable. This is necessary to assure that the tumor stem cells are given a maximum opportunity to manifest a dose response to the drug. As discussed in Chapter 22, this is particularly important in the area of new drug development, as pharmacologically achievable concentrations may well not have been defined. Additionally, drugs requiring hepatic bioactivation to exert cytotoxic effect (eg, cyclophosphamide) can be used in vitro if they are treated with a liver microsomal system (Chapter 22). Perhaps unique to our approach has been the routine use of very low drug concentrations. As mentioned in Chapter 1, most published studies of in vitro drug sensitivity of fresh human tumor cells to anticancer drugs (with assays not involving clonogenicity) have used unrealistically high dosages often with long exposure times that are not clinically achievable. Furthermore, in our opinion, even clonogenic assays employing only one drug dose level have only limited value. For example, it is valuable to know whether a tumor is completely resistant to a given drug in vitro. Even if it does not show sensitivity to the drug at low concentrations and CXTs, inhibition of its growth at higher concentrations could lead to the design of a high-dose drug trial or suggest the use of the drug by local administration. For example, if a tumor stem cell assay predicted that a high melphalan CXT was needed to inhibit ovarian tumor colony formation to 10% of control, intraperitoneal administration of melphalan (eg, 2–3 mg/kg) might prove useful for treatment of malignant ascites, even though high concentrations are not achievable at that site after IV administration. In fact, we have had success employing precisely this approach in relation to the drug assay for a patient with malignant ascites and a sensitivity index of 5.3 (intermediate sensitivity) for melphalan. Independently, investigators in the Division of Cancer Treatment, National Cancer Institute, have explored a similar approach with methotrexate, Adriamycin, and 5-fluorouracil "belly baths" in ovarian cancer [6–8; and Chapter 19]. Part of their rationale for these studies has been derived from use of the ovarian tumor colony assay for exploring effects of high doses of drugs.

Third, under normal conditions we select the lowest in vitro drug concentration to represent only 5–10% of the achievable peak plasma concentration and plasma CXT in vivo for any given drug. This is done for three reasons. The drug concentration entering tumor cells in vivo may only be a small fraction of that which is measured in the plasma. Also the CXT of a given anticancer drug may have marked variation from patient to patient. For example, drugs like 5-fluorouracil and melphalan, when taken orally, may have extremely variable absorption, leading to a wide range of plasma CXTs [9, 10]. Variations in the rate of elimination of various drugs (eg, as a function of hepatic or renal function) from patient to patient is also a well-known phenomenon that can influence the plasma CXT. Finally, as discussed in detail in Chapters 17 and 18, correlations of in vitro tumor stem cell sensitivity with clinical response have indicated that if inhibition of tumor colony is not achieved at low drug concentrations, then in vivo drug resistance to systemically administered drugs is almost a certainty.

In Vitro Tumor Stem Cell Dose-Response Curves and the Concept of the Drug "Sensitivity Index"

Shown in Figure 1 are linear survival-concentration curves on ovarian cancer stem cells from one series of patients whose cells were exposed in vitro to vinblastine. On the vertical axis is the percent survival of the TCFUs, and increasing vinblastine concentration (from 0.05 to 10 μg/ml) is depicted along the horizontal axis. Vinblastine is incubated with the tumor cells for 1 hour at 37°C prior to plating. Note the marked heterogeneity of response with some patients' tumor cells relatively sensitive and others much more resistant to vinblastine exposure. Several of the curves show a relatively steep decline in surviving TCFUs in response to low drug concentrations, followed by a later plateau in responsiveness. This tiphasic-type dose-response curve may have two explanations. First, there may be at least two populations of tumor stem cells, one more sensitive than the other to drug exposure. Second, the steep portion of the dose-response curve may represent a cell population in the proliferative phase of the cell cycle, whereas the plateau portion of the curve may represent "resting" clonogenic cells, which would be kinetically resistant to vinblastine. The latter consideration is clearly of importance for "cycle active" drugs such as vinblastine, the effects of which are largely limited to cycling tumor stem cells. In fact, for cycle active drugs, such as vinblastine or cytosine arabinoside, it would appear important to routinely relate the fractional lethality from the steep portion of the survival curve to the ^{3}H-thymidine suicide index. Priesler [11] has recently reported on precisely such an approach in the response of normal myeloid progenitors (eg, CFU-C) to cyto-

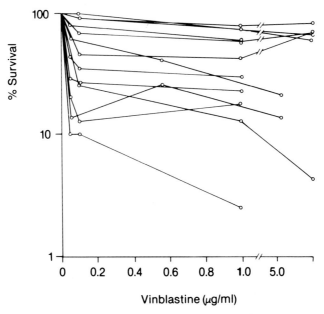

Fig. 1. In vitro survival of ovarian TCFUs following exposure to increasing concentrations of vinblastine for 1 hour at 37°C. Vinblastine concentrations have been extrapolated between 1 and 5 μg/ml. Each curve corresponds to the results for an individual patient whose tumor biopsy was assayed.

sine arabinoside. He found an excellent correlation between cytosine arabinoside lethality and the ^{3}H-thymidine suicide index. However, with tumor cells, such a study might also disclose a dichotomous relationship in which, even though the ^{3}H-thymidine suicide index is high, the TCFUs may still be inherently resistant to the anticancer drug at pharmacologically achievable concentrations.

We quantitate a sensitivity index for these in vitro drug assays by measuring the area under linear survival-drug concentration curves (1-hour exposure) out to an upper concentration limit that is defined by clinically achievable dosage exposures. Details on the calculation of the sensitivity index are included in Chapter 17. The index can be expressed in area units and in relation to a cut-off or boundary concentration established from experimental observations for each drug. For example, Figure 2 shows two examples of drug sensitivity in multiple myeloma using the 0.1 μg dose as the upper boundary for measurement of "area under the curve." The shaded area under the melphalan dose-response curve for one patient represents the sensitivity index of his myeloma stem cells to the drug. This patient, whose sensitivity index was 3.11 units, went into a clinical complete

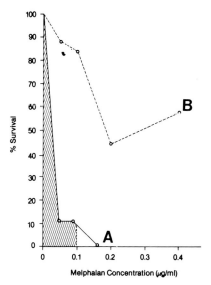

Fig. 2. In vitro survival of myeloma TCFUs following exposure to increasing concentrations of melphalan for 1 hour at 37°C. The shaded area under the melphalan dose-response curve for patient A's TCFUs yields a sensitivity index of 3.1 area units (see explanation in Chapter 17). Patient A (o——o) went into a clinical complete remission following treatment with melphalan, whereas patient B's TCFUs (o- - - o) failed to respond to melphalan in vitro and his myeloma showed progressive growth in vivo.

remission of myeloma following melphalan treatment, whereas the second myeloma paitent represented by the dotted line failed to respond to melphalan, and had a sensitivity index of 8.9 units.

Relation of Drug Sensitivity Index and In Vivo Plasma Disappearance Kinetics

When we compared the in vitro drug concentrations and CXTs associated with the "sensitive" range of sensitivity index areas to the in vivo peak plasma concentrations and CXTs for a series of drugs we discovered that the "cut-off" concentrations (an empirically established boundary for calculation of sensitivity) were only 5–10% of those that were clinically achievable. For example, for melphalan, the cut-off concentration for a sensitivity index of five would be a one-hour exposure to 0.1 $\mu g/ml$ of the drug. In vivo this represents less than 10% of the achievable melphalan peak plasma concentration or CXT which average 2.6 $\mu g/ml$ and 2.5 $\mu g \cdot hr/ml$, respectively after an intravenous bolus dose of 0.6 mg/kg (see Appendix 4).

TABLE I. In Vitro-In Vivo Pharmacokinetic Correlations

	Adriamycin IV	BCNU IV	Bleomycin IV
In vivo			
Route and dosage	60 mg/m^2	95 mg/m^2	15 units/m^2
Peak plasma concentration (μg/ml)	1.0	1.97	2.81
CXT (μg/hr/ml)	2.76	1.02	4.99
Reference number (Appendix 4)	[7]	[15]	[19]
In vitro			
Dose range studied (μg/ml)	0.01–1.0	0.1–3.0	0.01–1.0
Cut-off concentration (μg/ml)	0.2	0.2	0.1
In vitro/in vivo ratios (%)			
Peak concentration ratio (%)	20.0	10.0	3.6
CXT ratio (%)	7.3	19.6	2.0

As shown in Table I a similar relationship between the in vitro cut-off concentration and in vivo peak plasma concentration and CXT has been found for several other commonly used drugs. For example, our in vitro cut-off concentration for Adriamycin is 0.2 μg/ml for 1 hour. Its peak plasma concentration and CXT after a 60-mg/m^2 intravenous bolus dose are 1.0 μg/ml and 2.8 μg·hr/ml, respectively. Expressed as a percentage the ratios of in vitro to in vivo peak concentration and CXT for Adriamycin are 20.0% and 7.3%, respectively. Similar data apply to cell cycle-specific drugs like methotrexate, whose peak concentration was 4.4% and CXT ratio was 2.2% after a 50-mg/m^2 intravenous bolus dose.

Duration of In Vitro Drug Exposures: One Hour vs Continuous Contact

In an effort to determine whether lethality to TCFUs caused by a specific anticancer drug was schedule dependent, we carried out comparative studies of different time course exposures for tumor cells to the same drug concentration. One-hour exposure in liquid culture prior to plating was compared to continuous contact of the TCFUs with the drug. We recognize that prolonged liquid culture could lead to selective alteration of cell populations attributable to factors other than drug sensitivity per se. Therefore, we carried out our studies of continuous exposure of cells to anticancer drugs by including stable drugs in the agar gel matrix. While it would be of definite interest to study varying time periods of continuous contact of drugs in agar, it is now impractical to compare drug incubation in liquid culture prior to plating with incorporation of a drug into the agar for the entire period of culture. These studies have focused on antimetabolites and vinca alkaloids, which are thought to be stable in vitro and "cycle active." For methotrexate at 5 μg/ml there appeared to be little difference in colony growth inhibition whether cells were exposed for 1 hour prior to plating or for up to 14 days;

	Melphalan		Methotrexate	Cis-platinum
	IV	PO	IV	IV
	0.6 mg/kg	0.6 mg/kg	50 mg/m²	100 mg/m²
	2.64	0.28	4.58	2.49
	(1.06–4.94)	(0.07–0.63)		
	2.47	0.88	8.90	1.94
	(0.87–5.50)	(0.15–2.50)		
	[36]	[37]	[40]	[42]
	0.01–1.0	0.01–1.0	0.05–5.0	0.01–1.0
	0.1	0.1	0.2	0.2
	3.8	35.7	4.4	8.0
	4.0	11.4	2.2	10.3

TABLE II. Effect of Duration of In Vitro Vinblastine Exposure to Drug-Induced Lethality

		Sensitivity index[a]	
Patient	Tumor type	Standard 1-hour exposure	Prolonged exposure[b]
1	Ovarian	16.1	8.7
2	Ovarian	9.2	5.0
3	Breast	17.1	10.8
4	Endometrial	14.8	16.3
5	Endometrial	10.8	12.9

[a]Area under the in vitro tumor stem cell dose-survival curve to an upper cut-off concentration limit of 0.2 μg/ml.

[b]Tumor cells were exposed to the drug continuously in the agar in vitro for up to 14 days. Patients 4 and 5 are of interest inasmuch as the lethality observed with continuous contact is clearly no greater than that achieved with a 1-hour exposure. In the latter case (5) continuous contact with the cytotoxic drug was associated with increased survival of tumor colony-forming units.

however, it is possible that various nutrients within the culture media may have reversed methotrexate's antitumor activity during continuous in vitro exposure in agar. Results of continuous contact in agar with vinblastine in three of five patients (1, 2, and 3, Table II, and Fig. 3) show an increased lethality as compared to a 1-hour exposure prior to plating, but a plateau of resistant colony-forming tumor stem cells. These experiments provide clear evidence of inherent resistance to vinblastine, inasmuch as the in vitro CXT in continuous contact experiments is so often 300-fold that of the 1-hour experiments. Hence resistance to vinblas-

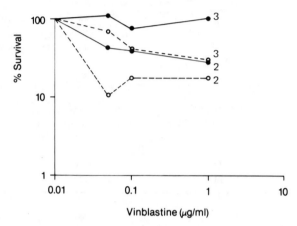

Fig. 3. In vitro survival of TCFUs from patients 2 and 3 from Table II following exposure to increasing concentrations of vinblastine for 1 hour priot to plating in agar (—•—) vs continuous contact in agar for up to 14 days (– – – ○ – – –).

tine in these experiments cannot be explained on a kinetic basis, as colonies do indeed form in the plates. Recent interest in vinblastine infusions for breast cancer [12] might increase the relevance of continuous contact experiments such as those reported here.

ACKNOWLEDGMENT

We wish to thank Dr. Joseph G. Gross for his helpful suggestions.

REFERENCES

1. Gibaldi M, Perrier D: One-compartment model (Chapter I); Multicompartment models (Chapter II). In: "Pharmacokinetics." New York: Marcel Dekker, 1975, pp 1–96.
2. Alberts DS, van Daalen Wetters T: The effect of phenobarbital on cyclophosphamide antitumor activity. Cancer Res 36:2785–2789, 1976.
3. Valeriote FA, Tolen SJ: Survival of hematopoietic and lymphoma colony-forming cells in vivo following the administration of a variety of alkylating agents. Cancer Res 32: 470–476, 1972.

4. Alberts DS, van Daalen Wetters T: Rubidazone vs Adriamycin: An evaluation of their differential toxicity in the spleen colony assay system. Br J Cancer 34:64–68, 1976.
5. Metzler CM: NONLIN: A computer program for parameter estimation in nonlinear situations. Technical Report 7292/69/7292/005. Kalamazoo, Michigan: Upjohn Company, 1969.
6. Dedrick RL, Myers CE, Bungay PM, DeVita VT Jr: Pharmacokinetic rationale for peritoneal drug administration in the treatment of ovarian cancer. Cancer Treat Rep 62: 1–11, 1978.
7. Ozols RF, Willson JKV, Grotzinger KR, Young RC: Cloning of human ovarian cells in soft agar from malignant effusions and peritoneal washings. Cancer Res 40:2743–2747.
8. Speyer JL, Collins JM, Dedrick RL, Brennan MF, Londer H, DeVita VT Jr, Myers CE: Phase I and pharmacological studies of intraperitoneal (IP) 5-fluorouracil (5-FU). Proc Am Soc Clin Oncol, Am Assoc Cancer Res 20:C–251, 1979.
9. Cohen JL, Irwin LE, Marshall GJ, Darvey H, Bateman JR: Clinical pharmacology of oral and intravenous 5-fluorouracil (NSC-19893). Cancer Chemother Rep 58:723–731, 1974.
10. Alberts DS, Chang SY, Chen HSG, Evans TL, Moon TE: Pharmacokinetics of oral melphalan in man. Clin Pharmacol Exp Ther 26:737–745, 1979.
11. Preisler HD, Shaham D: Comparison of tritiated thymidine labeling and suicide indices in acute myelocytic leukemia. Cancer Res 38:3681–3684, 1978.
12. Yap HY, Blumenschein GT, Hotobagyi GN, Tashima CK, Loo TL: Continuous 5-day infusion vinblastine (VBL) in the treatment of refractory advanced breast cancer. Proc Am Soc Clin Oncol, Am Assoc Cancer Res 20:C–179, 1979.

17
Quantitative and Statistical Analysis of the Association Between In Vitro and In Vivo Studies

Thomas E. Moon

The prediction of clinical or in vivo response to cancer therapy and the corresponding determination of optimum patient treatment through the use of in vitro assays has been the goal of many investigators. The in vitro human tumor stem cell assay has the theoretical advantage over other assays by evaluating clonal growth of the patient's own tumor. If a high positive correlation exists between the in vitro and the in vivo results, then the in vitro assay could be effectively used to select more effective treatments for each patient and have a prediction of their likelihood of success or failure. Also, the assay could be used as a preclinical screen to identify new agents of potential clinical value. Important to these efforts is the development of techniques to quantitate accurately results of the in vitro assay, predict in vivo outcome, and evaluate the consistency of this association method. This chapter will focus on quantitative analyses and statistical techniques that we have found useful in developing a quantifiable assay system with a firm foundation of established pharmacological principles. Aspects related to the design of preclinical studies that use the stem cell assay system to screen and identify potentially effective new agents will also be discussed.

QUANTIFICATION OF THE IN VITRO STUDY

The appearance of the in vitro concentration-percent survival curves led to their quantitative description. For purposes of quantitative analysis, as shown in Figure 1, the percentage of surviving tumor colony forming cells is plotted on a linear scale along the vertical axis, and drug concentration is plotted on the horizontal axis. Such plots are frequently prepared for semilogarithmic display; however, as will be discussed, simple linear plots have value for getting an overall display of

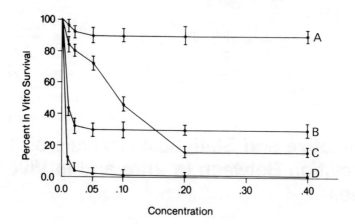

Fig. 1. In vitro concentration-percent survival curves.

drug-induced lethality when there is marked clonal heterogeneity. Percent survival is calculated by dividing the number of surviving tumor colony-forming cells in the treated sample by the number of colonies in the untreated (control) sample. Each series of connected lines represents a separate patient's data for cell survival at varying drug concentrations in vitro. It can be observed that for each patient, several clinically achievable concentrations are employed. Three replications are carried out at each concentration and indicated by the two standard error confidence band about the corresponding mean survival concentration point.

If the in vitro survival curve rapidly drops toward the horizontal axis, as seen by curve D, a high percentage of in vitro drug-induced lethality is obtained at low drug concentrations (ie, in vitro sensitivity). Assuming close correlation between the required concentration-time product for the in vitro lethality and that which is required in vivo, would lead to the prediction of high in vivo effectiveness (ie, the patient would be clinically responsive to the treatment). Conversely, a minimal decrease in the survival curve, as shown by curve A, corresponds to a small percentage of lethality at any concentration used in the in vitro assay and predicts in vivo resistance. An intermediate drop or a plateau of the in vitro survival curve, as seen in curves B and C, indicates a moderate lethality followed possibly by a reduction in lethality beyond a certain concentration of the drug. Conceptually, this suggests the presence of subpopulations of tumor stem cells that are kinetically or biochemically resistant to the agent tested.

Because the number of in vitro concentration-percent survival points obtained for each patient is frequently six or fewer, the ability to develop and validate a general mathematical model describing the various shapes of the in vitro survival curves is limited. Based upon pharmacokinetic and tumor kinetic methods, several potential mathematical models could be considered. These would include first-order kinetic or other models that incorporate the rate of the initial decrease of the in vitro survival curve, followed by the existence of a change in the rate or appearance of a plateau. However, the experimental data did not approximate exponential cell kill in most instances. Thus, the calculations of the duration of the shoulder and the initial slope (D_0^{-1}) from semi-log plots, as is common in radiobiology and model tumor systems, appear to distort the actual data too severely. The complexity of the in vitro human stem cell survival curves is presumably due to clonal heterogeneity with respect to drug sensitivity.

The use of the area under linear in vitro concentration-percent survival curves does furnish a conceptually and computationally simple approach to describe and quantify how fast and how far the survival curve drops with increasing concentrations. The sensitivity index for the in vitro concentration-percent survival curve is thus defined as the area under the survival curve between 0 and an upper cut-off concentration limit. As discussed by Dr. Alberts in Chapter 16, the upper cut-off concentration is defined by clinically achievable concentration-time products for the particular drugs studied. Figure 2 shows that the sensitivity index would be much smaller for curve B than for curve A.

The sensitivity index (or area under the curve) is most readily calculated by the trapezoidal method (Fig. 2). The trapezoidal method adds the areas of the rectangles and triangles that can be placed entirely under the in vitro survival curve between the 0 and the upper cut-off concentration. Recalling that

$$\text{Area of rectangle} = \text{base} \times \text{height}$$

$$\text{Area of triangle} = \tfrac{1}{2}\,\text{base} \times \text{height}$$

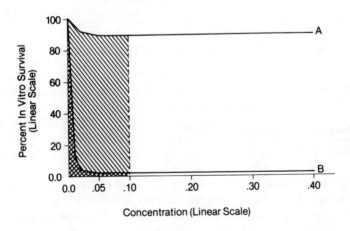

Fig. 2. Calculation of sensitivity index using the trapezoidal method.

then the sensitivity index for curve B, Figure 2, is

$$S = \frac{1}{2}(0.01 - 0.0) \times (100 - 12) + (0.01 - 0.0) \times (12 - 0)$$
$$+ \frac{1}{2}(0.02 - 0.01) \times (12 - 4) + (0.02 - 0.01) \times (4 - 0)$$
$$+ \frac{1}{2}(0.05 - 0.02) \times (4 - 1) + (0.05 - 0.02) \times (1 - 0)$$
$$+ (0.10 - 0.05) \times (1 - 0)$$
$$= 0.765$$

The sensitivity index for curve A is 9.15.

QUANTIFICATION OF THE IN VIVO STUDY

In contrast to the in vitro study, the in vivo study generally does not employ several different treatment concentrations, but rather the treatment is administered to the patient at a standard dosage schedule. The patient is then observed for in vivo or clinical tumor response, with standard clinical criteria used to quantify tumor response.

In vivo tumor response is commonly classified by the percentage of tumor reduction, as compared to the pretreatment tumor size (eg, partial or complete response) or simply as either response (eg, at least 50% tumor reduction) or no response (less than 50% reduction). The results of the in vivo study will be quantified for this chapter by using the above simple dichotomous classification of response vs no response and correspondingly coded as either 1 (response) or 0 (no response). Survival duration and response duration are other commonly used indicators of in vivo response and can be analyzed with similar methods, as discussed below.

ASSOCIATION BETWEEN IN VIVO AND IN VITRO STUDIES

Having individually quantified the in vitro and the corresponding in vivo studies for each patient, the next step was to evaluate the degree of association between sensitivity index and observed in vivo results.

Simple Prediction of Response

The initial attempt was to draw, as shown in Figure 3A for myeloma and ovarian cancer patients, a separate line plot for each drug showing the sensitivity index for each patient and the corresponding in vivo result. In vivo response was determined independently by the clinician, who did not know the results of the in vitro study. A plus sign indicates an in vivo response and a closed circle indicates the lack of in vivo response. A distinct pattern was observed. Patients with an in vitro sensitivity index to melphalan of less than four units frequently obtained an

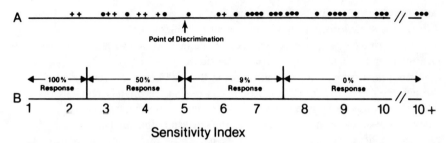

Fig. 3. Sensitivity index versus in vivo response (+ = in vivo response; ● = lack of in vivo response).

in vivo response, whereas those with a sensitivity index greater than six units consistently did not achieve an in vivo response. It was thus strongly suggested by this and similar figures for other treatments that a small value for the sensitivity index did correlate with subsequent in vivo response and a large value for the sensitivity index correlated with the lack of in vivo response. The definition of a point of discrimination that could be used to classify each patient's sensitivity index into two disjoint prognostic regions was needed. Patients whose sensitivity index was less than this point of discrimination would be predicted to have an in vivo response, whereas patients whose sensitivity index or area under the curve was greater than the point would be predicted not to have an in vivo response.

The point of discrimination was determined by statistical discriminant analysis [1]. As shown in Figure 3A, the point of discrimination for melphalan-treated patients was initially estimated to be approximately 5 area under the in vitro survival curve units. The point of discrimination differed for different drugs. Also, other patient characteristics such as tumor type and prior therapy, which might affect the point of discrimination were not initially considered.

The degree of association between the sensitivity index and the observed in vivo response was evaluated by the Fisher exact test [1] ($P < 0.001$).

It can be observed from Figure 3A that some patients with a sensitivity index less than the point of discrimination did not obtain an in vivo response, whereas some patients who did have an in vivo response also had a sensitivity index greater than that point. Thus, the consistency of the above simple prediction of response method simply to classify every patient as either in vivo resonse or no response is a viable approach, but clearly it has limitations.

Prediction Gradient of Response

The next approach was developed to extend the above simple method by dividing the sensitivity index scale into several parts and observe the corresponding in vivo response rate in each subsequent part. Figure 3B shows that there is a gradient of in vivo response ranging from 100% for a sensitivity index of less than 2.5

units to 0% in vivo response for a sensitivity index greater than 7.5 units. Figure 4 shows the same information plotted on a two-dimensional graph with percentage of in vivo response along the vertical axis. A gradient of in vivo response with a decreasing proportion versus an increasing sensitivity index is observed.

A smooth curve was drawn that begins at the 100% response rate, passing through the top of the frequency bars in Figure 4 and ultimately reaching the 0.0% of response for large sensitivity index values. This suggests an explicit relationship between sensitivity index and the corresponding in vivo response. Because of the observed sigmoid shape of the curve, the analogy of the in vitro-in vivo studies with other dose-response models and their corresponding statistical analysis, the logistic regression model was used to quantify explicitly the relationship between the in vitro-in vivo studies. The use of such a quantitative relationship to predict, prior to actual treatment, a probability of in vivo response based upon the corresponding sensitivity index was also considered.

Using the combined in vitro-in vivo melphalan-treated patient data shown in Figure 3 as a training set, the parameter of a logistic regression model was statistically estimated [2]. The curve representing the model is shown in Figure 5. The actual sensitivity index in vivo response for each patient is also indicated.

The excellent fit of the training set by the logistic regression model is due in part to the retrospective comparison of the data. The most effective way to evaluate the logistic model or any model is by a prospective evaluation based upon independent patients who are not considered in the training set. The evaluation of this model to predict in vivo response is in progress.

One can use this model to predict patient in vivo response, prior to treatment, based upon their observed sensitivity index. For example, a patient with multiple myeloma who is tested using melphalan in the in vitro human tumor stem cell as-

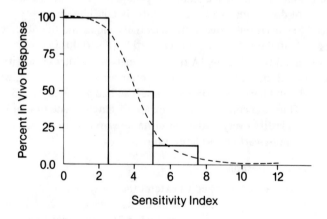

Fig. 4. Two-dimensional plot of sensitivity index versus in vivo response.

say and determined to have a sensitivity index of 2.0 units would be predicted to have a 0.96 probability of attaining an in vivo response if treated with melphalan at clinically achievable doses. In contrast, a patient whose sensitivity index was 6.0 units would have a predicted probability of less than 0.10 of attaining a clinical response.

Because of the desirability to furnish the clinician with a somewhat simpler method of predicting a patient's in vivo response, the sensitivity index scale can be separated into disjoint regions based upon the probability of in vivo response. One method that has been found to be of value is to determine three disjoint regions, as indicated in Figure 5 and defined as follows:

In vitro sensitivity index units for mephalan	Predicted in vivo response
<3.0	Sensitive (response)
3.0–5.3	Intermediate or indeterminate
>5.3	Resistant (no response)

The definition of three disjoint regions for which a patient may be predicted to be in vivo sensitive, resistant, or intermediate enables the clinician to identify treatments that have very different predicted probabilities of furnishing the pa-

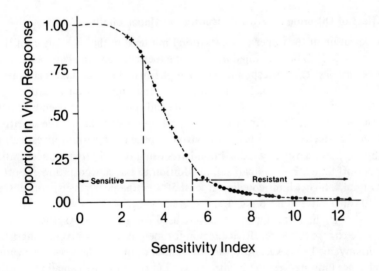

Fig. 5. Calculated logistic regression model of sensitivity index versus in vivo response for melphalan-treated myeloma patients (+ = in vivo response; ● = lack of in vivo response).

tient with an in vivo response. Also, this method allows for the comparison of different therapeutic agents and subsequent use of those agents that have the highest predicted probability of furnishing the patient an in vivo response to treatment.

Table I shows the disjoint sensitivity index boundaries for five standard cancer chemotherapeutic agents. For each of the agents, we have a sufficiently large training set to develop estimates of separate logistic models. Other agents and combinations are being evaluated.

Incorporation of Other Prognostic Factors

The existence of an interim in vitro sensitivity region related to the prediction of in vivo response (Fig. 5) suggests that there may be other factors that are related to in vivo response. The identification of patient characteristics that are associated with clinical response or other measures of prognosis have been greatly facilitated by the statistical publications of D. R. Cox [2] and the development and distribution of corresponding computer programs. Thus, identification of commonly observed patient characteristics such as prior therapy, histology, stage, and others can readily be incorporated into the above logistic model in an explicit manner. For example, such factors clearly are important in high-risk or rapidly fatal neoplasms. In acute myeloid leukemia, it is important to select an effective treatment (as might be predicted from in vitro sensitivity), but the patient's age, infection status, degree of thrombocytopenia, and availability of supportive care are also important in the predictive equation.

The Effect of Differing In Vitro Concentration Upper Limits

The selection of the upper concentration limit used in the calculation of the sensitivity index has been defined by clinical experience and in vivo plasma disappearance kinetics. The corresponding in vivo pharmacokinetics data relates to plasma levels and not to drug levels found inside the tumor. A quantitative relationship between drug levels in human plasma and tumor has not been well defined. Thus, the effect of different in vitro concentration upper limits on the sensitivity index and its effectiveness to predict in vivo response does warrant investigation.

Table II shows sensitivity index boundaries using two different concentration upper cut-off limits. For comparison, in addition to the concentration cut-off previously used, a higher limit equaling at least 50% of the peak plasma concentration was used for the five indicated drugs.

Even though different upper limits were used for the five drugs, there was relatively little change in the sensitivity region for melphalan, bleomycin, vincristine, and Adriamycin. There was, however, substantial change in the resistant region. Such changes indicate that the in vitro survival curve drops to a small proportion

TABLE I. Drug-Sensitivity Boundaries

Drug	Sensitivity index		
	Sensitive	Intermediate	Resistant
Adriamycin (29)[a]	<2.2	2.2– 7.3	> 7.3
Bleomycin (15)	<2.4	2.4– 3.8	> 3.8
Melphalan (33)	<3.0	3.0– 5.3	> 5.3
Platinum (14)	<8.7	8.7–11.4	>11.4
Vincristine (8)	<4.7	4.7– 6.4	> 6.4

[a]The numbers in parentheses indicate the sample size of the training set used to estimate the logistic regression model.

TABLE II. Drug-Sensitivity Boundaries Using Different Concentration Upper Limits

Drug/upper limit		Sensitive	Intermediate	Resistant
Adriamycin	0.2	< 2.2(1/2)[a]	2.2– 7.3(4/7)	> 7.3(1/20)
	0.5	< 2.3(1/2)	2.3– 9.6(3/5)	> 9.6(2/22)
Bleomycin	0.1	< 2.4(1/1)	2.4– 3.8(1/2)	> 3.8(0/12)
	2.0	< 6.5(1/1)	6.5–32.5(0/1)	>32.5(1/13)
Melphalan	0.1	< 3.0(3/4)	3.0– 5.3(4/7)	> 5.3(1/22)
	0.5	< 3.3(2/3)	3.3–15.0(3/8)	>15.0(3/22)
Platinum	0.2	< 8.7(2/2)	8.7–11.4(1/3)	>11.4(0/9)
	1.25	<51.5(1/1)	51.5–72.0(2/4)	>72.0(0/9)
Vincristine	0.1	< 4.7(3/3)	4.7– 6.4(2/3)	> 6.4(0/2)
	0.2	< 5.5(3/4)	5.5–10.3(2/3)	>10.3(0/1)

[a]The numbers in parentheses indicate the proportion of patients that had an observed in vivo response. Changes in the upper limit were reflected by different estimated logistic models and different numbers of patients classified in each sensitivity region.

of colony survival for relatively low drug doses for the majority of patients who subsequently are found to be sensitive in vivo. Patients who are found to be resistant in vivo have a large area under their in vitro survival curve (sensitivity index). This area substantially increases when the concentration upper limit is increased.

The effect of a plateau (or substantial change in slope) in the in vitro survival curve on sensitivity index is clearly indicated by the platinum boundaries (Table II). Regardless of the treatment concentration and proportion survival at which the in vitro survival curve plateaus, the sensitivity index will increase as the upper concentration limit is increased. The increase will be substantial if the ultimate proportion survival is greater than 10% and the upper cut-off concentration is significantly enlarged. Since all patients who receive platinum had either minimal decrease or a plateau in their in vitro survival curve, the increase in the upper cut-off concentration did markedly change all three sensitivity index boundaries.

For each of the five drugs shown in Table II, a very similar proportion of patients is observed in each sensitivity region regardless of which upper concentration was used. The proportions indicate the in vivo response rate and were predicted by the corresponding logistic model. Changes in the concentration upper limit were reflected by different estimated logistic models, corresponding changes in sensitivity index boundaries, and different numbers in each region.

This information suggests that the actual upper concentration limit is relatively unimportant in the calculation of the sensitivity index. The upper limit must be sufficiently large, however, to be able to obtain a good estimate of the initial drop or rate of change as well as the point of plateau (if one occurs in an in vitro survival curve).

DESIGN OF DRUG SCREENING STUDIES THAT USE THE HUMAN TUMOR STEM CELL ASSAY

In addition to the clinical applications described above, the application of the human tumor stem cell assay to new drug screening offers considerable promise (Chapter 22). To use potentials of the clonogenic assay system in preclinical stud-

TABLE III. Sample Size Related to the Identification of New Drugs

Hypothetical probability that a patient will be sensitive to a new drug	Probability the assay will identify at least one in vitro tumor sensitive to a new drug									
	Number of tumors of same histology									
	1	2	3	4	5	6	7	8	9	10
0.1	0.1	0.19	0.27	0.34	0.41	0.47	0.52	0.57	0.61	0.65
0.2	0.2	0.36	0.50	0.59	0.67	0.74	0.79	0.83	0.87	0.89
0.3	0.3	0.51	0.66	0.76	0.83	0.88	0.92			
0.4	0.4	0.64	0.78	0.87	0.92					
0.5	0.5	0.75	0.88	0.94						
0.6	0.6	0.84	0.94							
0.7	0.7	0.91								
0.8	0.8	0.96								
0.9	0.9									

ies to screen and identify new and effective agents, sample size and related study design aspects need careful consideration. As shown in Table III, more than one patient with the same histology should be tested in order to evaluate the activity of new agents.

If only one patient with the same histology is tested in the human tumor stem cell assay system, then as many as one-fourth to one-half of the new agents that would produce at least a 50% clinical response rate would be missed. A sample size of three different patients with the same histology would insure the assay system a 50% chance of not missing a new drug that would produce a 20% clinical response, a 66% chance of not missing a new drug with a 30% clinical response rate, and at least an 88% chance of not missing a new drug that produces at least a 50% clinical response. Also, a sample size of eight patients with the same histology would insure that four out of five new drugs with a 20% clinical response rate would be identified by the human tumor stem cell assay system.

Patient characteristics such as histology and prior therapy need be carefully evaluated as to their effect on the ability of the human tumor stem cell assay system to identify new agents correctly. For example, if the hypothetical probabilities that a patient will be sensitive to a new drug depend upon the patient's prior therapy and histology, as shown in the following table

	Prior therapy	
	No	Yes
Responsive histology	0.5	0.1
Less responsive histology	0.1	0.01

then, 88% of such new drugs would be correctly identified by the assay system when three patients with no prior therapy and responsive histology types are assayed. However, only 3% of the same drugs would be correctly identified if all three patients had prior therapy and less responsive histology as indicated by the hypothetical table above.

Finally, the probability that a new drug will be effective in vitro may be different from the probability that a patient will be sensitive in vivo (ie, the in vitro sensitivity rate estimated from the drug screening test may be different from the in vivo response rate).

In fact, if the false-positive rates for the in vitro drug screening assay and the in vitro test used to predict in vivo response are approximately 0, then the probability that a new drug will be effective in vitro will be less than the probability that a patient will be sensitive in vivo (ie, the in vitro sensitivity rate may underestimate the actual clinical response rate). The results are obtained as follows:

$$P(DE) = P(DE|CS)P(CS|TS)P(TS) + P(DE|CS)P(CS|\overline{TS})P(\overline{TS})$$
$$+ P(DE|\overline{CS})P(\overline{CS}|TS)P(TS) + P(DE|\overline{CS})P(\overline{CS}|\overline{TS})P(\overline{TS})$$
$$= P(TS)[\{P(CS|TS) - P(CS|\overline{TS})\} \cdot \{P(DE|CS) - P(DE|\overline{CS})\}]$$
$$+ P(CS|\overline{TS})\{P(DE|CS) - P(DE|\overline{CS})\} + P(DE|\overline{CS})$$
$$\leqslant P(TS) + P(CS|\overline{TS}) + P(DE|\overline{CS})$$

where

$P(DE)$ = probability a new Drug is Effective by reducing in vitro colony growth (ie, in vitro sensitivity)

$P(DE|CS)$ = probability a new Drug is Effective when tumor Colonies are Sensitive (ie, sensitivity of the in vitro drug screening test)

$P(DE|\overline{CS})$ = probability a new Drug is Effective when tumor Colonies are not Sensitive (ie, false-positive rate of drug screening test)

$P(CS|TS)$ = probability tumor Colonies are Sensitive when the corresponding patient's Tumor is Sensitive (ie, sensitivity of the in vitro test to predict in vivo response, in vitro-in vivo correlation)

$P(CS|\overline{TS})$ = probability tumor Colonies are Sensitive when the patient's tumor is not sensitive (ie, false-positive rate of in vitro test to predict in vivo response)

$P(TS)$ = probability a patient's Tumor is Sensitive to a new drug

$P(\overline{TS})$ = probability a patient's Tumor is not Sensitive to a new drug

It can be noted that $P(DE) = P(TS)$ only if

$$P(CS|TS) = 1, P(CS|\overline{TS}) = 0$$

and

$$P(DE|CS) = 1, P(DE|\overline{CS}) = 0$$

(ie, the in vitro drug screening test always correctly identifies a new drug as effective or not effective and in vivo response to a drug is perfectly predicted by the in vitro assay).

DISCUSSION

Various statistical methods can be applied to quantify and evaluate the relationship between in vitro and in vivo studies. Beginning with a rather simple procedure based upon the patient's in vitro sensitivity index, or area under the linear survival concentration curve, a discrimination and subsequent classification procedure was obtained to predict the patient's in vivo sensitivity. This procedure was then extended by the use of a logistic regression model to obtain an explicit quantitative relationship between the predicted probability of in vivo response and the corresponding in vitro sensitivity index. The characterization of this quantitative relationship was specialized to the definition of three distinct regions based upon the in vitro sensitivity index. Subsequent prospective evaluation of this logistic regression model with possible inclusion of other patient characteristics is in progress. The final evaluation will indicate the tumor types and the therapies to which the in vitro human tumor cloning system is most applicable.

REFERENCES

1. Armitage P: "Statistical Methods in Medical Research." New York: John Wiley, 1971, pp 375–384; 135–138.
2. Cox DR: "Analysis of Binary Data." London: Methuen, 1970, pp 43–54.

18
Clinical Correlations of In Vitro Drug Sensitivity

Sydney E. Salmon, David S. Alberts, Frank L. Meyskens, Jr., Brian G.M. Durie, Stephen E. Jones, Barbara Soehnlen, Laurie Young, H.-S. George Chen, and Thomas E. Moon

A major impetus for the initiation of investigations of in vitro clonogenicity of human tumor cells has been the objective of developing a practical, quantitative, and biologically relevant test with which in vivo sensitivity of human cancers could be reliably predicted. This major goal remains substantive, despite the fact that the treatment of cancer with chemotherapy has improved considerably within the past decade. Nonetheless, selection of effective therapy for individual patients remains somewhat of a trial-and-error procedure. Predictive techniques (similar to the culture and sensitivity assays used for the management of microbial infections) have not been available. Numerous clinical observations have shown a wide range of responsiveness to particular treatments, among patients with cancers of identical histopathologic types. In our own experience, such individual differences in responsiveness can be particularly well appreciated in the monoclonal neoplasm multiple myeloma, for which tumor burden and response can be quantitated [1–3].

As is well delineated in this text, of the various cells comprising a malignant tumor, the key replicative units appeared to be the small fraction of clonogenic tumor cells or tumor stem cells. As discussed in Chapters 1 and 2, prior studies with animal tumor systems have shown that the tumor stem cells responsible for population renewal of a tumor can be assayed in vivo or in vitro with assays for tumor colony-forming units (TCFUs) [4–6]. Such assays of TCFUs appear to be more reliable measures of reproductive death tests for cytotoxicity such as dye exclusion or radionuclide incorporation [7]. Clonogenic tumor cells appear to be central to the metastatic process in vivo, as they retain the capability to form secondary colonies at distant sites in the body (assuming that they can gain access to the circulation and find "fertile soil" for colonization [8–10]). Measurement of survival of TCFUs from mouse myeloma after exposure to drugs showed that differential sensitivity was present between different myeloma cell lines and that results of the quantitative in vitro colony assay were predictive of in vivo results [6]. We have focused on the same objective with human cancers, since this type of potentially simple assay would make predictive cancer chemotherapy feasible. We, therefore, applied the in vitro assay system discussed in Sections I and II of this text to measurement of drug sensitivity of TCFUs from biopsy specimens from a variety of human cancers. A flow chart summarizing the standard in vitro assay procedure is shown in Figure 1. In June 1978, we published our first report on the use of the human tumor colony assay system for quantitation of drug sensitivity in 18 patients with myeloma or ovarian cancer [11]. That report provided preliminary evidence that the assay system might prove useful for prediction of clinical response as well as playing a role in new drug development. In this chapter, updated experience (until January 1980) correlating drug sensitivity of TCFUs and clinical response of a relatively large number of patients with multiple myeloma, ovarian carcinoma, metastatic melanoma, and other neoplasms is summarized. For purposes of generalization and comparison, summary data from independent but analogous studies by Von Hoff (Chapter 10) have also been included. This body of information now provides strong support for the hypothesis that the in vitro sensitivity or resistance of TCFUs to cytotoxic agents is predictive of clinical response as well as documenting the acquistion of drug resistance by TCFUs subsequent to treatment in vivo. Additional biological information on tumor heterogeneity and drug resistance has also been obtained from our studies.

MATERIAL AND METHODS

Patient Studies

Clinical staging and response in myeloma were measured as previously described by our group [1–3]. The International System for clinical staging of patients with ovarian carcinoma was used [12]. Patients were treated in our institution every three to four weeks with short courses of chemotherapy. Only patients who had clinical trials that were evaluable for response, as well as having sufficient in vitro data, were included in the clinical correlations. Thus, there is substantially more data available on in vitro drug-induced lethality than could be applied directly to therapeutic trials in vivo. Previously, untreated patients with multiple myeloma received bifunctional oral alkylating agents (mustard type). These agents consisted of either melphalan, 40 mg/m^2, or combinations of melphalan, 28 mg/m^2, and cyclophosphamide, 500 mg/m^2. Our standard protocol for patients in relapse from oral alkylating agents was with carmustine (BCNU), 30 mg/m^2, combined with Adriamycin, 30 mg/m^2 [13]. Patients with ovarian carcinoma who had not received prior chemotherapy received Adriamycin, 40 mg/m^2, combined with cyclophosphamide, 800 mg/m^2 [14]. Patients with

TUMOR STEM CELL ASSAY

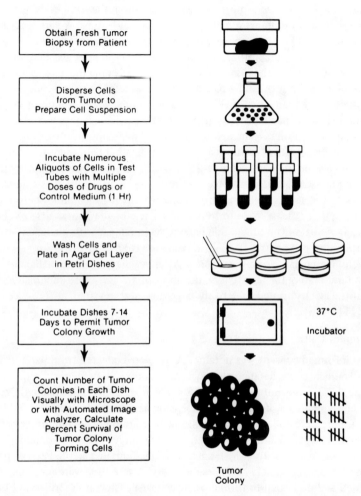

Fig. 1. Flow diagram of steps in drug-sensitivity assay.

metastatic melanoma received any of a number of drugs, including dacarbazine, actinomycin-D, cis-platinum, melphalan, and other agents administered with standard dose schedules. Patients in relapse with ovarian cancer, melanoma, and other neoplasms had their therapy selected either empirically or, when possible, with therapy initiated on the basis of evidence of in vitro sensitivity. Except for this latter category, physicians responsible for treatment were not informed of the results of the in vitro studies until the clinical trial had been completed and evaluated. Southwest Oncology Group criteria of response were used. For myeloma, a complete response required a 75% reduction in the total body tumor mass as well as definite improvement in other clinical featues. A partial response required 50% to 74% reduction in tumor mass and improvement in other clinical features. For ovarian carcinoma, melanoma, and other neoplasms, a complete response was defined as the disappearance of all clinical manifestations of disease for more than one month, and a partial response as a 50% or greater decrease in the product of the diameters of all measurable lesions. Mixed responses were noted (in melanoma particularly) when at least a partial response was observed in some lesions while others remained unchanged. While "improvement" was also recorded for those patients with ovarian cancer or other neoplasms having 25–49% regression in all measurable lesions, such improvement was not considered to represent a sufficient objective response to be classified as clinical sensitivity to the agents used. In vitro studies were performed either before chemotherapy or at least three to four weeks after the previous course of chemotherapy. Some patients with ovarian carcinoma in relapse with malignant effusions were entered on prospective trials correlating in vivo and in vitro response to bleomycin, melphalan, and other agents. After evacuation of the malignant effusion (and plating for sensitivity), the appropriate agent was administered by the intracavitary route in relatively large dose and repeated at four-week intervals [15–16].

Collection of Cells

After informed consent was obtained, cells were collected from patients using preparative techniques detailed elsewhere in this text. Bone marrow cells (from myeloma, lymphoma, and other neoplasms) were collected into a heparinized syringe and prepared as a single cell suspension as described in Chapter 3. Malignant effusions were collected into heparinized vacuum bottles and processed as described in Chapter 6 and Appendix 1. Tumor nodules that were obtained surgically were placed with a small amount of medium in a sterile Petri dish or flask and transported quickly to the laboratory, where they were mechanically dissociated and processed to form a single cell suspension as described in Chapters 4, 6, 7, 8, and Appendix 1. Viability of the single cell suspension was assessed with trypan blue exclusion.

In Vitro Exposure of Tumor Cells to Drugs

Stock solutions of intravenous formulations of melphalan, carmustine, adriamycin, dacarbazine, vinblastine, methotrexate, cis-platinum, actinomycin-D, 5-fluorouracil, and other agents were prepared in sterile buffered saline or water and stored at $-70°C$ in aliquots sufficient for individual assays. Subsequent dilutions were made in medium for cell incubation. As discussed in Appendix 1, tumor cell suspensions were transferred to tubes and adjusted to a final concentration of 1.0×10^6 cells per ml in the presence of the appropriate drug dilution or a control medium. Each drug was tested at a minimum of three dose levels, including low concentrations that we calculated (from pharmacokinetic data [Chapter 16 and Appendix 4]) to be achievable pharmacologically in vivo. The final concentration ranges of the various agents are discussed in Chapter 16 and 17 and range from a low of 0.01 to 0.5 µg for most agents but included dose ranges up to 3–5 µg per ml for BCNU (carmustine), methotrexate, and 5-fluorouracil. The bifunctional alkylating agent, cyclophosphamide, is inactive in vitro in this system, and therefore melphalan was used as the index bifunctional alkylating agent in the in vitro studies and was assumed to have similar effects to other mustard-like alkylating agents, including cyclophosphamide. Future studies will investigate the use of a rat liver microsomal homogenate ("S-9 mix" [17–18]) for investigation of cyclophosphamide and other agents requiring bioactivation. Cells were incubated with drugs for one hour at 37°C in McCoy's 5A medium (initially we used Hank's balanced salts solution but found it to be suboptimal due to variable pH). The cells were then centrifuged at 150g for ten minutes, washed twice in medium, and prepared for culture. Recently, we have begun to explore continuous contact of the drugs incorporated into the agar with our much larger experience with one hour exposure. However, up until this time, we have not correlated clinical sensitivity with the continuous contact experiments.

Culture Assay for TCFUs

The culture system used for drug sensitivity was carried out in accord with the standard procedures described in Appendix 1. In brief, cells to be tested were suspended in 0.3% agar in enriched CMRL 1066 medium supplemented with 15% horse serum to yield the final concentration in a range of 2 to 5×10^5 cells/ml. Freshly prepared 2-mercaptoethanol was added immediately before triplicate plating of the cells. One ml of the mixture was pipetted into a 35 ml plastic Petri dish over a 1 ml agar feeder layer. Conditioned medium was required for myeloma and lymphoma growth (Chapters 3 and 4) but appeared optional and was generally not used for other tumor types. Cultures were incubated at 37°C in 6% CO_2 in humidified air. Colonies (collections of more than 30 cells) appeared in ten to 21 days and plates were counted for drug effects after 7–14 days, with use of an inverted phase microscope. Recently, we have begun to test an automatic image analysis system (Chapter 15) for drug sensitivity assay counting.

Direct staining of the plates for peroxidase (Chapter 12) was used to identify occasional contaminating granulocyte-macrophage colonies present in the bone marrow cultures. At least 30 tumor colonies per plate were required in the control plates to assure an adequate range for measurement of drug effect. Drug effects were scored only for the reduction of the number of colonies and not for any changes in the morphology of the surviving TCFUs. Control studies for development of the drug assay were carried out with the 8226 human myeloma cell line [19] and with human ovarian and breast carcinoma and melanoma cells obtained from patient biopsies and cryopreserved in 10% DMSO with 20% serum and medium. In experiments with melphalan, a one hour exposure was optimal, with increasing lethality with increasing doses. Repeated drug assays initiated on different days with either the myeloma cell line or the cryopreserved fresh cells yielded similar survival curves for a variety of drugs tested under these circumstances.

Data Analysis

The laboratory technologist entered and stored all assay data on a Wang 2300 laboratory computer disc file. Cloning efficiencies were calculated from the total number of cells plated and not corrected for the proportion of nontumor cells in the sample. The standard error of the mean for individual data points averaged 5% of the mean. For graphic analysis, on each patient's cells the mean of triplicate TCFU percent of control (or survival) observations versus drug concentration was plotted on both linear and semilog scales. Points identified as 1% survival were, in each case, no colony formation at the drug dose tested at the standard plating concentration. Because of practical limits on the number of cultures that could be set up on fresh biopsies, higher plating concentrations were not routinely used to define drug dose effects at TCFU survival levels below 1% of control.

A quantitative "sensitivity index" of TCFUs to drugs was determined graphically from the linear survival by drug concentration curves using the techniques detailed in Chapters 16 and 17. The upper boundary concentration for the drugs used are summarized in Chapters 16 and 17. For each drug, areas under the survival curve are ordered, limits of sensitivity selected (on the basis of a preliminary training set of experiments), and patients classified as sensitive if the corresponding area was less than or equal to the selected limit and resistant if above it. Current sensitivity limits for Adriamycin, bleomycin, melphalan, cis-platinum, and vincristine are summarized in Table I of Chapter 17 and are based on reasonable-sized "training sets" to establish sensitivity zones. For purposes of classification and correlation of the in vivo trials, patients who fall in the intermediate zone in vitro (Chaper 17, Table I) are classed as "sensitive" in vitro so that a tabular analysis can be dichotomously set up as "sensitive" or "resistant." Current limits for sensitivity (area under the curve) for other agents studied (and upper cut-off concentrations in $\mu g/ml$) were BCNU < 10 (0.2 $\mu g/ml$), dacarbazine < 8 (0.2 $\mu g/ml$), actinomycin < 1 (0.1 $\mu g/ml$), AMSA < 10 (0.2 $\mu g/ml$), 5-fluorouracil < 7.5 (0.2 $\mu g/ml$), methotrexate < 6.5 (0.2 $\mu g/ml$), vinblastine, and vindesine < 15.0 (0.2 $\mu g/ml$).

Comparison of serial in vitro drug sensitivity studies on individual patients (for purposes of defining stability of the test in relation to treatment administered) were carried out. We assumed that a 50% change in the sensitivity index (area under the curve) was indicative of a significant change in the in vitro results in such serial studies. Although a 50% increase or decrease in the area under the TCFU survival concentration curve may appear liberal, we wish to avoid confusing a modest degree of spontaneous variability (due to laboratory or biological factors) with a substantial change in sensitivity of TCFUs that might result from in vivo drug exposure. Statistical association of the in vitro and in vivo results was obtained with the chi-square test.

Proof of the neoplastic nature of the colonies derived from TCFUs was routinely obtained using the dry slide technique (Chapter 12) as well as with other related marker techniques discussed elsewhere in this text.

RESULTS AND DISCUSSION

Drug Assays

Clinical correlations on drug assays conducted at The University of Arizona are summarized for a series of patients with myeloma, melanoma, ovarian cancer, and miscellaneous neoplasms who had more than 30 colonies in control plates and had drug testing performed at multiple dose levels, including at least two in the pharmacologically achievable dosage range (Chapter 16 and Appendix 4). Several survival curves were obtained to any specific drug within a given disease category, and they varied substantially from tumor type to tumor type. For example, in Figure 2 A, B, and C shows the in vitro survival curves with BCNU for TCFUs from myeloma, melanoma, and ovarian cancer. It is clear from these curves that at the same dosages, myeloma TCFUs generally showed a steep reduction in survival at the lowest dosages tested using the 0.2 μg dosage as the "upper boundary" concentration for defining the area under the linear survival concentration curve. In contrast, using the same criteria, TCFUs from patients with melanoma and ovarian carcinoma were mostly quite resistant. Clinical studies with nitrosoureas have shown them to be of substantial value in multiple myeloma [20] but of little or no value in melanoma and ovarian carcinoma [21]. Figure 3 depicts our experience with actinomycin-D in a series of patients with metastatic melanoma. Marked heterogeneity in response to this cycle nonspecific drug is apparent. Whereas some patients' clonogenic cells were resistant over the dosage range tested, others showed mixed populations of sensitive and resistant cells or predominantly sensitive cells. In such assays we interpret those curves with initial steep reduction in survival at a relatively low dose (eg, to 50% survival) followed by no further reduction in survival with increasing concentration as reflecting the presence of a mixture of sensitive and resistant TCFUs within the sample. Of interest, a number of the melanoma patients studied with actinomycin-D in vitro and depicted on Figure 3 were prospectively treated with this agent in vivo as well. Only the patient whose curve showed the greatest sensitivity (to the far left in Fig. 3) responded to actinomycin-D with a partial remission

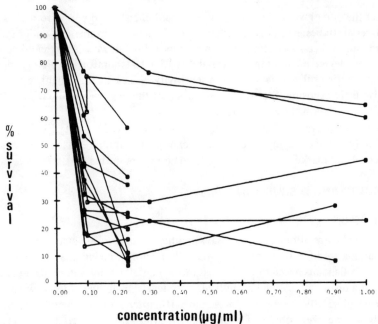

2a

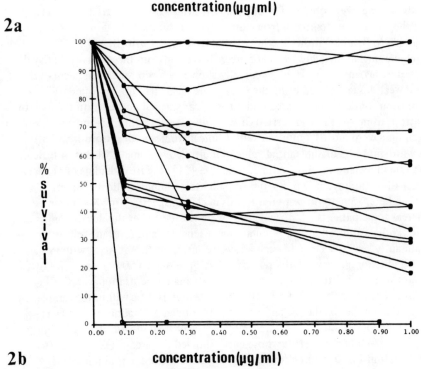

2b

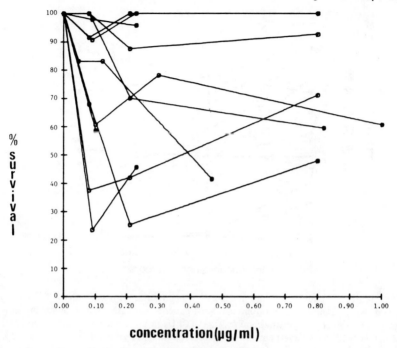

2c

Fig. 2. Linear survival-concentration curves for one hour in vitro exposure of TCFUs to BCNU. (a) Multiple myeloma (15 patients), (b) metastatic melanoma (16 patients), (c) ovarian carcinoma (11 patients). Plots of this type are used for assessing in vitro sensitivity. The boundary upper concentration used for calculating the area under the curve is 0.2 ug/ml Based on current experience, an area of > 5 units indicates drug resistance. Most myeloma patients' TCFUs are clearly more sensitive to BCNU than melanoma or ovarian cases, although TCFUs from one patient with melanoma were quite sensitive to this drug and the patient achieved a mixed response when treated with it clinically.

(greater than 50% shrinkage of a large neck mass). A similar series of curves was observed with cis-platinum in melanoma and again only the most sensitive patient (a woman with a vulvar melanoma) responded and achieved a partial remission (of seven months' duration). In melanoma, we clinically observe mixed responses to several agents for which the in vitro assay indicated sensitivity. This phenomenon has thus far not been observed in other neoplasms. As suggested by Meyskens in Chapter 8, clonal heterogeneity may represent a bigger problem in melanoma and require the simultaneous testing of several lesions, when these are available, to assure that the sample tested in vitro is representative of a population of melanoma stem cells present in vivo. Such studies are under way in our laboratory.

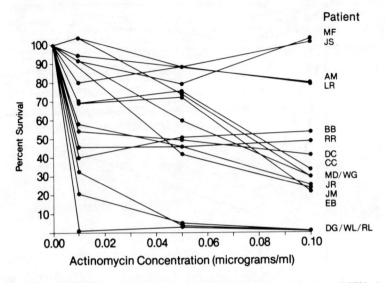

Fig. 3. Linear survival-concentration curves for one hour in vitro exposure of TCFUs from patients with melanoma to Actinomycin. Most patients' cells are resistant and many show clonal heterogeneity of sensitive and resistant TCFUs to this cycle nonspecific agent. Only the patient with the steepest curve with a small fractional survival at low dose achieved a mixed response to this agent clinically. A number of other patients whose TCFUs are depicted on this plot failed to respond to this agent.

Studies with bleomycin in ovarian carcinoma can also be cited as an example of an agent to which drug resistance of TCFUs is frequently manifest. Nonetheless, TCFUs from a small fraction of patients proved to be quite sensitive to this agent. Representative survival concentration curves with bleomycin are illustrated in Figure 4. In all, 41 in vitro studies have been carried out with bleomycin in ovarian carcinoma. The majority of the studies were in patients who had received prior treatment and were studied in relapse. Only two patients' TCFUs expressed exquisite sensitivity in vitro, and both of them had received extensive prior therapy. Four previously treated and five previously untreated patients' TCFUs manifest intermediate sensitivity, and the remaining 30 patients' cells were resistant. Although bleomycin proved to be an effective agent in vivo for a few ovarian carcinoma patients whose TCFUs manifested either sensitive or intermediate curves, in most it had no effect. Analysis of in vitro resistance for five cytotoxic drugs which we tested on cells from 162 patients is summarized in Table I. Not all patients' TCFUs were assessed with all of these agents, although there are sufficient data available to note differences in the percentages of patients whose cells manifest in vitro resistance to one or another of the drugs

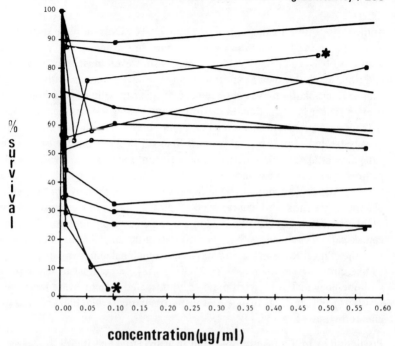

Fig. 4. Linear plot of survival of ovarian TCFUs from 12 patients after a one-hour in vitro exposure to bleomycin. The patient (*) whose cells manifested the most profound reduction in survival (to 2.5%) achieved an excellent partial remission. At the time of relapse, her TCFUs expressed a resistant curve (*).

TABLE I. Percentage of Patients Studied Whose TCFUs Manifest In Vitro Resistance to Five Anticancer Drugs

| | | | Ovarian cancer | | |
| | Myeloma | Melanoma | no prior therapy | prior therapy | Miscellaneous |
Drug (number of assays)	(n = 25) (%)	(n = 33) (%)	(n = 17) (%)	(n = 55) (%)	(n = 32) (%)
1. Adriamycin (116)	<u>63</u>	<u>64</u>	<u>59</u>	<u>82</u>	<u>68</u>
2. Melphalan (107)	<u>57</u>	82	<u>86</u>	<u>86</u>	71
3. Cis-platinum (93)	56	56	38	<u>77</u>	56
4. Vinblastine (65)	*	*	50	<u>73</u>	91
5. BCNU (55)	<u>39</u>	86	*	<u>86</u>	86
Total number of patients, 162					

Underlined percentages ≥ 14 patient biopsies tested in tumor category; * < 7 patient biopsies tested in tumor category, data not shown.

in the various tumor categories. For example, TCFUs from patients with ovarian carcinoma or melanoma more frequently manifest drug resistance to melphalan than was observed in the myeloma category; however, most ovarian patients had had prior alkylating agent exposure. Similar results were seen with BCNU. Resitance to cis-platinum was observed less frequently with TCFUs from untreated ovarian cancer patients than in the other tumor categories studied. In vitro resistance to Adriamycin was more commonly observed with TCFUs from previously treated ovarian cancer patients than with those from the other tumor categories studied. This is perhaps not surprising, because our own clinical protocols for ovarian carcinoma over the past six years have included combinations of Adriamycin and an alkylating agent as initial therapy for ovarian carcinoma patients who had received no prior therapy. In vitro resistance of TCFUs to melphalan was observed in a surprisingly high fraction of ovarian cancer patients studied, perhaps corroborating the single-agent data suggesting that the clinical response rate to single-agent melphalan in ovarian carcinoma can be quite low (in the range of 25–30%) [22]. In contrast to our experience with melphalan, TCFUs from previously untreated patients were less frequently resistant to other drugs than were those from patients who had received prior chemotherapy (Table I).

Prediction of In Vivo Response From In Vitro Drug Sensitivity

As of January 1980, in vitro drug sensitivity tests were successfully performed on fresh biopsy samples from 180 patients with various neoplasms. Most patients who were studied were heavily pretreated and were referred when in relapse from standard agents. More than 1,100 in vitro sensitivity assays (each with three drug concentrations plated in triplicate) were carried out in this patient population [average 6.1 drugs/patient (range 3–14 drugs)]; however, a substantial proportion of the in vitro assays were on patients for whom clinical trials were not feasible or available for evaluation. A total of 193 correlative clinical trials of in vitro and in vivo sensitivity or resistance, which involved 96 patients, were evaluable as of January 1, 1980. Many of these patients could be analyzed for one retrospective clinical correlation and one or more prospective evaluation in relation to clinical trials carried out after the in vitro assay. Each correlation was based on a single clinical trial with a single agent or two-drug combination consisting of agents studied for that patient in vitro. The overall results of these studies are summarized in Table II. The predominant tumor categories included were myeloma, melanoma, and ovarian carcinoma, with an additional miscellaneous category for patients with diffuse histiocytic lymphoma, oat cell carcinoma of the lung, hypernephroma, or sarcoma.

Despite the fact that many of the patients were heavily pretreated, a total of 54 correlations could be made when the in vitro assay indicated drug sensitivity. In 34 of the 54 in vitro assays the patients also had a clinical response to treatment

TABLE II. Correlations of In Vitro and In Vivo Sensitivity of Multiple Myeloma, Metastatic Melanoma, Ovarian Carcinoma, and Other Neoplasms to Anticancer Drugs (Cumulative experience at the University of Arizona: January 1980)

Tumor	Number of patients	Number of clinical trials	Tumor sensitive both in vitro and in vivo[a] (S/S)	Tumor sensitive in vitro and resistant in vivo (S/R)	Tumor resistant in vitro and sensitive in vivo (R/S)	Tumor resistant both in vitro and in vivo[b] (R/R)
Multiple myeloma	20	48	12	6	1	29
Metastatic melanoma	26	37	5[c]	8	3	21
Ovarian carcinoma	40	95	13	8	1	73
Miscellaneous	10	13	4	0	0	9
Totals	96	193	34	22	5	132

[a]True positive rate $[(S/S)/(S/S + S/R)] \times 100 = 61\%$.
[b]True negative rate $[(R/R)/(R/R + R/S)] \times 100 = 96\%$.
[c]Includes two mixed responses.

with the same agent or agents (Table II). Thus, the true positive rate for the in vitro assay is 63% for this patient population. Drugs uncommonly used for certain tumors were sometimes identified and proved clinically effective. Examples include cis-platinum or actinomycin for melanoma and bleomycin or vinblastine for ovarian cancer. In myeloma, in vitro sensitivity was recognized commonly to melphalan, BCNU, and Adriamycin, and to similar agents in lymphoma and oat cell carcinoma of the lung. All clinical responses were at least partial, except for two of the four melanoma patients who had mixed responses (one to BCNU-dacarbazine and the second to m-AMSA). This suggests that more clonal heterogeneity of metastases might be present in melanoma and necessitate multiple biopsies for assay when feasible. Patients who achieved clinical responses with the agents to which they showed sensitivity in vitro uniformly manifested exquisite in vitro sensitivity of their TCFUs. Thus, the concentration-time product or CXT (Chapter 16) associated with an in vitro sensitivity index that falls in the sensitive zone was usually only 5–10% of the pharmacologically achievable CXT or peak concentration achievable in vivo (Chapter 16 and Appendix 4). Our current definition of in vitro sensitivity is an operational one, and the exact limit for any given drug, while already exquisitely sensitive, may require revision as further experience is gained with various tumors. Based on our findings, we disagree with Bateman et al [23], who remarked that our distinction between resistance and sensitivity in vitro was "somewhat arbitrary." Our distinction is no less arbitrary than the definitions of antibiotic sensitivity for bacteria, wherein the minimal inhibitory concentration of the antibiotic is, in fact, of crucial and central importance. Clinical dosages of antibiotics for infectious disease are far greater than the minimal inhibitory concentration; the same is true for effective cytotoxic treatment of cancer. The reason for using area-under-the-curve analysis was distated by the shape of the curves obtained, which were not amenable to simple exponential dose extrapolations. Furthermore, whereas Bateman et al argue that the achievable blood levels are more relevant for prediction, our studies show that lower boundary concentrations than achievable blood levels are more suitable because those patients who are predicted to respond are identified at a very small fraction of the achievable blood levels for all drugs tested. Even in the studies of Ozols and co-workers (Chapter 19), wherein higher doses were studied, the CXT delivered intraperitoneally to the patient was often tenfold greater than the highest dose tested in vitro. Remarkably similar considerations apply to antimicrobial therapy. When the ratio of the peak serum concentration to the minimum bactericidal concentration is 100:1, and therapeutic concentrations can be maintained, clinical results have been "astounding" [24].

Figure 5 depicts two proposed explanations for the requirement of exquisite in vitro sensitivity with the tumor stem assay for prediction of clinical response with systemic drug administration. Both of these explanations may prove relevant. First, the intratumoral drug concentrations achieved in vivo may be far lower than those measurable in the plasma (from which we determine pharmacokinetics in vivo). Second, it should be apparent from the low cloning efficiencies for most of the tumors studied and the survival curves depicted in the chapter, that the in vitro assay is limited to a 1–2 log reduction in survival of TCFUs. However, clinical response may require a 3 or 4 log reduction in survival of tumor stem cells and hence require a CXT of the drug in vivo that is at least ten times that which can be measured in vitro with this assay. If, in the future, cloning efficiencies can be in-

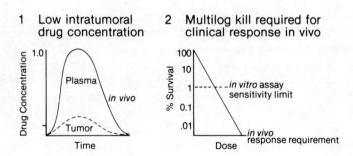

Fig. 5. Proposed explanations for requirements of exquisite in vitro sensitivity for clinical response.

creased to 1%, then it would be feasible to measure a 4 log reduction in TCFU survival in vitro. This is, in fact, now achieved only with the clonogenic assays for acute myelogenous leukemia (Chapter 5) and with some pediatric neoplasms and occasional malignant effusions. One further comment concerning the detection of in vitro sensitivity with this assay system is that the 63% true positive rate compares favorably with several other assays designed to predict drug sensitivity. One is the estrogen receptor test [25] for breast cancer, wherein a high level of estrogen receptor is associated with a 60% likelihood of responding to hormonal treatment. A second example would be the conventional bacterial sensitivity test, which also may have a 60–70% true positive rate for selection of various antibiotics that prove to be clinically useful for specific bacterial strains in individual patients.

In vitro drug resistance was quite commonly observed with the tumor stem cell assay and had excellent correlation with clinical drug resistance as well. A total of 139 correlations were obtained with clinical trials wherein in vitro resistance was observed (Table II). In 134 of the 139 trials, the patients also failed to respond when drug resistance was predicted in vitro. Thus, the true negative rate was 97%, indicating that the tumor stem cell assay has extraordinary power to predict which drug would only cause toxicity and indicate that such agents could be deleted from clinical trial.

Combined Experience With Predictive Testing

It is of interest to compare and combine our clinical experience at the University of Arizona with that obtained independently by Dr. Von Hoff and his associates in San Antonio, as discussed in Chapter 10. The combined summary of our experience with in vitro and clinical drug sensitivity is depicted in Table III. In 316 clinical trials in 197 patients studied at the two institutions, the assay was

TABLE III. Correlations of In Vitro and In Vivo Sensitivity With the Human Tumor Stem Cell Assay: Cumulative Experience at Two Institutions

Investigative group	Number of patients	Number of clinical trials	Tumor sensitive both in vitro and in vivo[a] (S/S)	Tumor sensitive in vitro and resistant in vivo (S/R)	Tumor resistant in vitro and sensitive in vivo (R/S)	Tumor resistant both in vitro and in vivo[b] (R/R)
Tucson	96	193	34	22	5	132
San Antonio	101	123	15	6	2	100
Totals	197	316	49	28	7	232

[a]True positive rate [(S/S) / (S/S + S/R)] × 100 = 64%.
[b]True negative rate [(R/R) / (R/R + R/S)] × 100 = 97%.

64% accurate (95% confidence limits, 53–74%) at predicting which drugs would induce at least partial remission in the heavily pretreated and refractory tumor patients that we treated. The test was 97% accurate (95% confidence) in predicting which drugs would not be useful clinically. Overall, the association of in vitro and in vivo comparisons was clearly established ($P \leq 10^{-9}$). As can be calculated from Table III, in vitro sensitivity was observed in 29% of the laboratory studies on the patients who underwent clinical trials in Tucson, but only in 17% of those patients who underwent the correlative clinical trials in San Antonio. Several factors likely explain this difference. First, while the Tucson group studied primarily three tumor types, the San Antonio group studied a much wider group of tumors, including more tumor types for which there are no effective anticancer drugs. Second, while the average number of drugs studied per specimen in San Antonio was four to five and in Tucson slightly more than six, there was a significant minority of patients studied in Tucson who had ten to 12 drugs studied in vitro. Thus, the probability that an active compound would be identified in Tucson was higher because a greater number of potentially active compounds was tested. The smaller number of drugs per specimen in San Antonio was not by intent, but was limited by the small number of cells that could be gotten into suspension from many of the solid tumor samples. For clinical purposes, it is therefore clearly worth maximizing the number of cells for plating as many standard and new drugs as feasible with the in vitro assay.

Clinical Correlation of Serial In Vitro Studies: Acquisition of Drug Resistance

A total of 45 serial in vitro studies were carried out in 12 patients who had two or more in vitro assays (Table IV). There were 25 instances wherein the patient either received no therapy or unrelated treatment between the two or more in vitro assays, while 20 instances were available wherein the patient had been clinically treated with a specific drug tested in vitro. As discussed in Methods, in

TABLE IV. Correlations of 45 Serial In Vitro Studies in 12 Patients

Clinical information	Number of instances	Area under the survival-concentration curve		
		Decrease (> 50%)	No change (< 50%)	Increase (> 50%)
No therapy given or unrelated therapy	25	0	22	3
Specific drug given that was tested in vitro	20	0	11	9[a]

[a]Acquisition of resistance (increase in the area under the curve) was more common when the patient received the specific drug tested in vitro ($P = 0.03$).

order to determine whether treatment influenced the sensitivity index (area under the curve), we defined a significant change in the area under the curve as being either a 50% increase (decreasing sensitivity) or a 50% decrease (increasing sensitivity). As is evident from Table III, no patient showed a decreased area under the curve after in vivo treatment whether they had received drugs related or unrelated to those studied in vitro. While three of 25 patients who had received either no treatment or unrelated treatment had an increased area under the curve, a very impressive nine of 20 patients who had received the specific drug tested in vitro had an increased area under the curve, reflecting decreasing sensitivity to the drug tested in vivo ($P = 0.03$). This was rather dramatically evident in patients whose curves shifted from being quite sensitive prior to treatment to extremely resistant when in relapse after treatment. This is well demonstrated in Figure 4 in the studies of bleomycin in ovarian cancer wherein the most sensitive patient tested in vitro with this agent responded and then showed virtually complete resistance over the dose range tested at the time of relapse. Qualitatively similar examples of markedly increased resistance have been observed in all the major tumor categories studied with various drugs. In some instances, the patients were initially sensitive to the drug and showed evidence of resistance at the time of relapse, whereas for others, the initial curves would meet criteria that we would call resistant. However, after the patient received the drug in question, the area under the curve increased even more, indicating development of increasing drug resistance of the TCFUs. Based on our experience with serial testing, the general pattern in relapsing patients appears to be one of progressive acquisition of increasing drug resistance to single agents with which the patient has been treated. Thus, the acquisition of drug resistance is a common phenomenon that can be directly detected and quantitated in vitro with this assay. This well may represent a form of "clonal progression," as discussed by Trent in Chapter 14.

The bioassay technique that we have studied offers the potential of facilitating drug selection for cancer chemotherapy. However, we must recognize that this procedure is analogous to those available for study of sensitivity of bacteria to antibiotics. Optimally, drugs that are active in vitro may also be active in vivo, and a minimum cytotoxic concentration can be predicted. Additionally, inactive drugs can be rejected. As discussed by Alberts in Chapter 16, prospective correlations with clinically achievable blood levels and concentration-time products of the various anticancer drugs will likely continue to have major importance. It is important to point out that our in vivo definition of sensitivity is limited to showing objective tumor regression which meets the clinical criteria of a "partial remission." While some patients for whom we saw in vitro sensitivity achieved complete remissions after clinical treatment, we have not limited ourselves to that strict a definition of sensitivity in vivo, and we do not yet know, in fact, whether it will be predictable with this assay. From the clinical standpoint, a partial remission represents a significant biological effect often associated with palliation of the patient's symptoms, but does not imply an improved survival time. In contrast, a complete remission is often associated with improved survival time as well as disappearance of all symptoms and signs of cancer. In fact, achieving a complete remission is an essential prerequisite for that fraction of patients who are cured with chemotherapy. As data accumulate, it is important to measure the overall impact of this assay on patient survival. We have already obtained preliminary evidence that in multiple myeloma, in vitro drug sensitivity (even prior to treatment) does predict for improved survival over patients whose TCFUs manifest in vitro resistance [26]. It well may prove that biological parameters in addition to drug sensitivity per se will be required to optimally predict complete remission and survival. These could include other in vitro measurements such as the doubling time of TCFUs, the tritiated thymidine suicide index (Chapter 13), and self-renewal capacity (Chapter 2), as well as various clinical features of disease such as tumor burden, age, etc. Use of a variety of factors in this fashion is discussed by Dr. Moon in Chapter 17, wherein he indicates that appropriate models for handling such multivariate data are already in use in our analyses. We plan to test various combinations of in vitro and clinical data in these models in the future.

Development of "Cross-Resistance" Tables

An additional area of interest for analysis of data from the assay includes evaluation of cross resistance. This type of analysis has previously been performed by Schabel and associates [27] for various murine tumors and now seems achievable with the tumor stem cell assay. Table V represents a preliminary cross-resistance table for a series of pairs of standard and new drugs in relation to major tumor categories studied. Such analyses may aid in more rapidly identifying common patterns of resistance or sensitivity to a variety of agents for given tumor types. Information of this type could potentially be translated into useful protocols for combination chemotherapy. For example, the combinations of vincristine, BCNU, Adriamycin, and melphalan detected by the analysis of the myeloma in vitro data correspond to known clinically active combinations; some protocols now include all of these agents with evidence of excellent clinical activity [28]. Combinations of actinomycin, or AMSA and vindesine have yet to be tested clinically in melanoma, but would seen reasonable in view of the data of Table V.

Potential for In Vitro Studies of Combination Chemotherapy

While our major focus to date has been on single agent studies in vitro, it appears feasible to consider carrying out some combination studies as well. Planning the proper combinations and doses to be evaluated is difficult because of the large number of possibilities available. Nonetheless, it is not unreasonable to select fixed dose combinations with intermediate or low concentrations of the various drugs to be tested. Additive effects of the combination could be defined as identifiable when TCFU survival of the drug combination is equivalent

TABLE V. Cross-Resistance Table of Selected In Vitro Sensitivity Tests With Myeloma, Melanoma, and Ovarian Carcinoma TCFUs*

	Number of samples tested for both drugs	S_1-S_2 (%)	S_1-R_2 (%)	R_1-S_2 (%)	R_1-R_2 (%)
Multiple myeloma					
1) melphalan vs 2) Adriamycin	(16)	31	13	6	50
1) melphalan vs 2) BCNU	(15)	33	13	27	27
1) vincristine vs 2) melphalan	(9)	44	33	22	0
1) vincristine vs 2) BCNU	(9)	78	0	11	11
1) Adriamycin vs 2) BCNU	(18)	28	33	11	28
Melanoma					
1) actinomycin vs 2) AMSA	(18)	17	5	17	61
1) actinomycin vs 2) BCNU	(15)	7	26	7	60
1) actinomycin vs 2) vindesine	(10)	30	0	20	50
1) AMSA vs 2) vindesine	(11)	46	9	9	36
Ovarian carcinoma (no prior therapy)					
1) melphalan vs 2) Adriamycin	(14)	14	0	14	72
1) melphalan vs 2) cis-platinum	(11)	10	0	45	45
1) bleomycin vs 2) Adriamycin	(10)	30	20	10	40
1) Adriamycin vs 2) cis-platinum	(13)	23	8	38	31

*S_1-S_2 = Percentage of samples sensitive to both drugs listed; S_1-R_2 = Percentage of samples sensitive to the first drug and resistant to the second drug listed; R_1-S_2 = Percentage of samples resistant to the first drug and sensitive to the second drug listed; R_1-R_2 = Percentage of samples resistant to both drugs listed.

to the product of the individual survival percentages for the single agents. Synergistic effects could be defined as being present when more than an additive effect is observed. Brown et al have analyzed the mathematics of such experiments previously [28]. Figure 6 depicts an evaluation of single agents and combinations of agents tested in vitro on TCFUs from a patient with ovarian carcinoma in relapse from Adriamycin, cis-platinum, and cyclophosphamide. All drugs were tested in vitro in triplicate at a dose of 0.1 ug/ml for one hour whether studied individually or in combination. Adequate growth was obtained and assay and points were clear-cut with SEM ± 5%. Agents the patient had previously received had little or no effect on her TCFUs whether they were used alone or in combination. In contrast, a modest reduction in survival was observed with vinblastine (55%) and bleomycin (61%) (Fig. 6). In vitro survival with these two drugs in

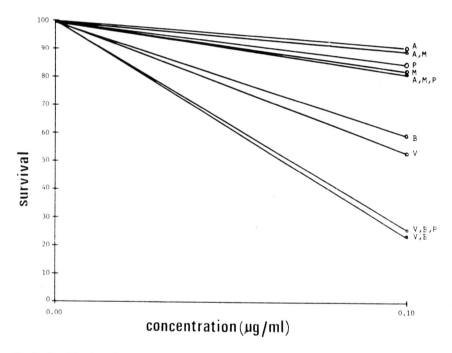

Fig. 6. Combination chemotherapy in vitro. In this study, TCFUs from a patient with ovarian carcinoma in relapse from Adriamycin (A), alkylating agent (M) and cis-platinum (P) were tested against these agents as well as vinblastine (V) and bleomycin (B) as single agents and in two and three drug combinations in vitro. All drugs were used at the 0.1 μg/ml dose for one hour in vitro, even when in combination. Resistance to A, M, and P were manifest when tested singly or in combination. Moderate reductions in survival of TCFUs were induced with V (55%) and B (61%) when used singly, and an additive effect of the two drugs, V + B in combination (25% survival). Addition of P to V + B resulted in no additional potentiation (27% survival).

combination was 25%, and, therefore, was consistent with an additive effect of combination chemotherapy. Addition of an inactive drug (cis-platinum) to vinblastine and bleomycin had no effect on the survival response to these agents in combination. Results such as those depicted in Figure 6 are encouraging and suggest that the definition of drugs to be used in combination chemotherapy could be made more rational. Furthermore, an effect not clearly identified with the single agents in vitro was observed. A virtually identical combination experimental result was found with a second ovarian cancer patient whose TCFUs also manifest intermediate in vitro sensitivity to vinblastine and bleomycin. In this second case, direct addition would predict 16% survival of the combination, and the experimental result was 15% survival. However, extensive experience with such in vitro comparative combination studies (plus in vivo correlations) will be necessary before the utility and relevance of such study designs can be adequately assessed.

Concluding Remarks

In summarizing the overall experience using the human tumor stem cell assay for a variety of cancers, it appears that clinical correlations with the system seem extremely encouraging. The assay now appears to be very accurate for predicting what drugs will be ineffective against an individual patient's tumor. Application of this finding in clinical oncology should reduce morbidity and eliminate costs associated with drug administration to patients who could be confidently predicted not to respond to those agents clinically. Additionally, the assay is very promising for predicting what drug will work against an individual patient's tumor. As mentioned previously, the assay appears to be at least as accurate for detecting sensitivity as the estrogen receptor testing procedure, which has markedly changed the approach for treatment of breast cancer over the past five years. The tumor stem cell assay has the advantage of a broad spectrum of applicability to a wide variety of tumor types and anticancer drugs. It will not surprise us if major efforts are initiated to transform this assay (which is now labor-intensive) into a somewhat simpler and more practical test for routine clinical measurement of drug sensitivity for a wide variety of human cancers.

REFERENCES

1. Salmon SE, Smith BA: Immunoglobulin synthesis and total body tumor cell number in IgG multiple myeloma. J Clin Invest 49:1114–1121, 1970.
2. Durie BGM, Salmon SE: A clinical staging system for multiple myeloma: Correlation of measured myeloma cell mass with presenting clinical features, response to treatment, and survival. Cancer 36:842–852, 1975.
3. Salmon SE, Wampler SE: Multiple myeloma: Quantitative staging and assessment of response with a programmable pocket calculator. Blood 49:379–389, 1977.
4. Bruce WR, Meeker BE, Valeriote FA: Comparison of the sensitivity of normal hematopoietic and transplanted lymphoma colony-forming cells to chemotherapeutic agents administered in vivo. J Natl Cancer Inst 37:233–245, 1966.

5. Park CH, Bergsagel DE, McCulloch EA: Mouse myeloma tumor stem cells: A primary cell culture assay. J Natl Cancer Inst 46:411-422, 1971.
6. Ogawa M, Bergsagel DE, McCulloch EA: Chemotherapy of mouse myeloma: Quantitative cell culture predictive of response in vivo. Blood 41:7–15, 1973.
7. Roper PR, Drewinko B: Cell survival following treatment with antitumor drugs. Cancer Res 39:1428–1429, 1979.
8. Hamburger AW, Salmon SE: Primary bioassay of human tumor stem cells. Science 197:461–463, 1977.
9. Hamburger AW, Salmon SE: Primary bioassay of human myeloma stem cells. J Clin Invest 60:846–854, 1977.
10. Salmon SE: Human tumor stem cells and adjuvant therapy of cancer. In Jones SE, Salmon SE (eds): "Adjuvant Therapy of Cancer II." New York: Grune & Stratton, 1979, pp 27–36.
11. Salmon SE, Hamburger AW, Soehnlen BJ, Durie BGM, Alberts DS, Moon TE: Quantitation of differential sensitivity of human tumor stem cells to anticancer drugs. N Engl J Med 298:1321–1327, 1978.
12. Day TG Jr, Smith JP: Diagnosis and staging of ovarian carcinoma. Semin Oncol 2:217–222, 1975.
13. Alberts DS, Durie BGM, Salmon SE: Doxorubicin/BCNU chemotherapy for multiple myeloma in relapse. Lancet 1:926–928, 1976.
14. Lloyd RE, Jones SE, Salmon SE, et al: Combination chemotherapy with Adriamycin (NSC-123127) and cyclophosphamide (NSC-26271) for solid tumors: A phase II trial. Cancer Treat Rep 60:77–83, 1976.
15. Alberts DS, Chen H-SG, Mayersohn M, et al: Bleomycin pharmacokinetics in man: II intracavitary administration. Cancer Chemother Pharmacol 2:127–132, 1979.
16. Alberts DS, Chen H-SG, Chang SY, et al: The disposition of intraperitoneal bleomycin, melphalan and vinblastine in cancer patients. In Mathé G (ed): "Recent Results in Cancer Research." Berlin: Springer-Verlag, in press.
17. Kovach JS, Ames MM, Powis G, Lieber MM: Use of a liver microsome system in testing drug sensitivity of tumor cells in soft agar. Proc Am Assoc Cancer Res, Am Soc Clin Oncol 21:257(abst. 1030), 1980.
of Arizona Cancer Center, Tucson, 1980.
18. Ames BN, McCann J, Yamasaki E: Methods for detecting carcinogens and mutagens with the Salmonella/mammalian-microsome mutagenicity tests. Mutat Res 31:347–364, 1975.
19. Matsuoka Y, Moore GE, Yagi Y, et al: Production of free light chains of immunoglobulin by a hematopoietic cell line derived from a patient with multiple myeloma. Proc Soc Exp Biol Med 125:1246–1250, 1967.
20. Salmon SE: Nitrosoureas in multiple myeloma. Cancer Treat Rep 60:789–794, 1976.
21. Stanhope RC, Smith JP, Rutledge F: Second trial drugs in ovarian cancer. Gynecol Oncol 5:52–58, 1977.
22. Omura GA, Blessing JA, Buchsbaum AJ, et al: A randomized trial of melphalan (M) vs melphalan plus hexamethylmelamine (M + H) vs Adriamycin plus cyclophosphamide (A + C) in advanced ovarian adenocarcinoma. Proc Am Assoc Cancer Res Am Soc Clin Oncol 20:C–279, 1979.
23. Bateman AE, Peckham MJ, Steel GG: Assays of drug sensitivity for cells from human tumours: In vitro and in vivo tests on a xenografted tumor. Br J Cancer 40:81–88, 1979.
24. Lerner AM: Peak concentration in serum to minimum bactericidal concentration Is a 100:1 ratio needed? Ann Int Med 140:753–754, 1980.

25. McGuire WL: Hormone receptors: Their role in predicting prognosis and response to endocrine therapy. Semin Oncol 5:428–433, 1978.
26. Durie BGM, Young LA, Salmon SE: Kinetics of myeloma stem cell culture: Correlations with in vitro and in vivo drug sensitivity. Br J Cancer (submitted, 1980).
27. Schabel IM, Skipper HE, Trader MW, Laster WR, Corbett TH, Griswold DP: Concepts for controlling drug resistant tumor cells. Eur J Cancer (in press).
28. Salmon SE, Alexanian R, Dixon D: Non-cross resistant combination chemotherapy improves survival in multiple myeloma. Blood 54(Suppl):207a, #552, 1979.
29. Drewinko B, Loo TL, Brown B, Gottlieb JA, Freireich EJ: Combination chemotherapy in vitro with Adriamycin. Observations of additive, antagonistic, and synergistic effects when used in two-drug combinations on cultured human lymphoma cells. Cancer Biochem Biophys 1:187–195, 1976.

19
Human Ovarian Cancer Colony Formation: Growth From Malignant Washings and Pharmacologic Applications

Robert F. Ozols, James K. V. Willson, and Robert C. Young

The ability of human ovarian cancer cells obtained from malignant effusions or solid tumors to form colonies in a double-layer agar system has been described by Hamburger and Salmon [1–3; and Chapter 6]. The effect of antitumor agents on colony formation in vitro has been used to individualize chemotherapy for patients with a variety of tumors, including ovarian cancer [4, 5; Chapters 10 and 18]. The initial results suggest that if a chemotherapeutic agent has little or no effect on tumor colony formation, regardless of the histologic type of cancer, it is extremely unlikely to produce a response in vivo. Conversely, if the drug results in significant reduction in survival of tumor colony-forming units (TCFUs) in vitro, the likelihood of a clinical response to that drug is approximately 60–70% [5; and Chapters 10 and 18].

Ovarian cancer patients who relapse after initial therapy may be potentially benefited by an in vitro chemotherapy sensitivity assay, since the response rate of most chemotherapeutic agents in this group of patients is less than 25% and their survival is characteristically short [6]. The maximum potential benefit of an in vitro chemotherapy sensitivity assay would likely be in patients with minimal residual disease, prior to the development of malignant effusions and large tumor masses, as the response to treatment is inversely related to the volume of residual cancer [7, 8]. For these reasons, the ability to grow colonies from malignant peritoneal washings in sufficient numbers for chemotherapy sensitivity testing would both increase the number of patients potentially benefited and allow for institution of individualized chemotherapy at a time when the tumor burden is small and the likelihood of a clinical response is greatest.

In addition to its use in the selection of active agents for individual patients, the stem cell assay can be used to investigate mechanisms of drug resistance in human tumors and to model experimental forms of treatment. Toward this end, we have used the assay to determine the effects of amphotericin B on human ovarian cancer colony formation in the presence of the cytotoxic agents Adriamycin, melphalan, and BCNU. Amphotericin B, a polyene antibiotic with no established antitumor activity, has been shown to ameliorate drug resistance and to potentiate the cytotoxicity of antitumor agents in certain cell culture lines [9, 10]. The effects of amphotericin B in the human tumor stem cell assay may have more clinical relevance than the effects observed in long-term tissue culture lines where clonogenicity is not the end point.

Finally, the human tumor stem cell assay offers the opportunity to model unique forms of treatment such as intraperitoneal chemotherapy for ovarian cancer patients. It has recently been demonstrated that cytotoxic agents can be administered to ovarian cancer patients intraperitoneally in large volumes at high concentrations by use of an in-dwelling semipermanent Tenckhoff dialysis catheter [11, 12]. The differential between the peritoneal and plasma clearances results in markedly higher intraperitoneal levels of drug than observed systemically [13]. We have used the tumor stem cell assay to determine if greater cytotoxicity results from the exposure of malignant clonogenic cells to the very high concentrations of Adriamycin and 5-fluorouracil achievable by intraperitoneal administration. The ability to increase local concentrations of anticancer drugs is of limited value unless there is an associated increase in cytotoxicity.

MATERIALS AND METHODS

Drugs

All chemotherapeutic agents were obtained from the Investigational Drug Branch, Division of Cancer Treatment, National Cancer Institute. Amphotericin B in the form of Fungizone was obtained from Grand Island Biological Co., Grand Island, NY.

Patient Specimens

Malignant effusions were collected under sterile conditions with 10 units of preservative-free heparin per milliliter of effusion. During initial staging or restaging with laparotomy or peritoneoscopy, peritoneal lavage was performed in those patients without ascites who had a histologic diagnosis of epithelial ovarian cancer. In patients who had a Tenckhoff catheter placed in the abdominal wall prior to chemotherapy, peritoneal lavage was performed with 1.5% Inpersol (Abbott Laboratories, Chicago, IL).

Soft Agar Culture System

The procedure of Hamburger and Salmon was used for the direct cloning of human ovarian cancer cells in a double-layer agar system [1, 2; and Chapter 6].

Conditioned medium was not used in the underlayers. The concentration of nucleated cells plated for malignant effusions varied from 0.5 to 1.0×10^6 cells per plate depending upon the percentage of viable cells (as determined by trypan blue exclusion) in the specimen. Peritoneal washings were processed in the same manner as malignant effusions. Occasionally peritoneoscopic washings contained less than a total of 1×10^6 cells, and consequently only 100–300,000 cells were plated in triplicate. Immediately prior to plating, the cell suspension containing enriched CMRL and agar was passed through 25-gauge needles to ensure a single cell suspension of tumor cells (Fig. 1). Colonies comprising 30 or more cells were counted 14–21 days after plating. Individual colonies were plucked for cytologic analysis, or the entire upper layer was fixed and stained as described by Salmon and Buick [14; and Chapter 12].

Effects of Chemotherapeutic Agents

Single cell suspensions were incubated with chemotherapeutic agents and Hank's balanced salt solution for one hour at 37°C. After incubation the cells were washed and plated in triplicate as described above. The results are expressed as percentage of control colonies ($\pm$ SE) at each drug concentration.

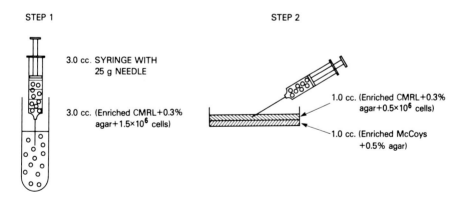

Fig. 1. Schematic representation of the plating of tumor cells in a double-layer agar system. A syringe is used to mix the cells, media, and agar immediately prior to plating. Expressing the cells through a 25-gauge needle helps break up clumps that may have formed during the drug incubation. One-milliliter aliquots are dispensed through the needle into 35-mm Petri dishes.

RESULTS

Characterization of Colonies

The morphology of human ovarian cancer colonies in soft agar was similar to that previously reported by Hamburger et al [2; and Chapter 6]. The cytologic characteristics of the individual cells in the colonies after Wright–Giemsa and Papanicolaou staining were similar to ovarian cancer cells as identified in the original specimen. The characteristics of ovarian cancer colonies grown from peritoneal washings in the absence of ascites were similar to those of colonies grown from malignant effusions.

Growth of Ovarian Cancer Colonies From Malignant Effusions and Peritoneal Washings

Malignant effusions were obtained from 20 patients with ovarian cancer, and colony growth was observed in 15 patients (75% success rate) (Table I). The average plating efficiency (number of colonies/number of nucleated cells plated × 100%) was 0.01% (range 0.001–0.04%).

Peritoneal washings, in the absence of ascites, were obtained in 20 patients (Table II). Ten of these washings on routine cytologic analysis (Cytopathology Section, Laboratory of Pathology, DCT, NCI) were found to contain malignant cells consistent with an ovarian cancer primary. Colony growth, defined as more than one colony per plate, was observed in nine of the cytologically positive specimens (90% success rate). There were no instances in which ovarian cancer colony growth was observed in cytologically negative peritoneal washings. The mean cloning efficiency of cells obtained at peritoneoscopy and from a Tenckhoff catheter was 0.004% and 0.005%, respectively, which reflects the greater number of macrophages and inflammatory cells in peritoneal washings compared to malignant effusions.

The mean volume of peritoneoscopic washings received for stem cell analysis was 125 ml (range 50–175 ml), although in most cases greater than 750 ml was recovered from the peritoneal lavage. In contrast, a mean of 1,700 ml was readily recovered from a Tenckhoff dialysis catheter after a 2-liter wash with Inpersol. The larger volume allowed sufficient clonogenic tumor cells to be harvested to enable chemotherapy sensitivity determinations (Fig. 2) in two of four patients.

Effects of Adriamycin and 5-Fluorouracil on Human Ovarian Cancer Colony Formation

The effects of Adriamycin and 5-fluorouracil on colony formation were examined at clinically achievable plasma levels following intravenous administration as well as at levels that can be reached only by intraperitoneal administration. Three patterns of in vitro sensitivity to Adriamycin were observed (Fig. 3).

TABLE I. Human Ovarian Cancer Colony Formation From Malignant Effusions

Number of specimens	Number with colony formation	Colonies/plate mean (range)	Plating efficiency mean (range)
20	15 (75%)	65 (20–150)	0.01% (0.001–0.04%)

TABLE II. Human Ovarian Cancer Colony Formation From Malignant Peritoneal Washings

Source	Number	Number with colony formation	Colonies/plate mean (range)
Peritoneoscopy	6	5 (83%)	20 (4–42)
Tenckhoff catheter	4	4 (100%)	41 (2–117)

1) Previously untreated patients (or those who were responding to a chemotherapy regimen that did not include Adriamycin [19]) demonstrated the greatest degree of in vitro colony inhibition.

2) Patients who had progressive disease on non-Adriamycin chemotherapy regimens had an average of more than 50% survival of TCFU after exposure to Adriamycin at 1.0 µg/ml, the peak plasma level attainable by intravenous therapy. These patients would be considered to be resistant in vitro to Adriamycin according to the criteria of Salmon ("sensitivity index" based on the survival curve) [4; and Chapters 16–18] and those of Von Hoff (a greater than 70% reduction in colony formation at a concentration that is 20% of the peak plasma level; Chapter 10). At an Adriamycin concentration of 10 µg/ml, however, there was less than 10% survival in vitro, suggesting that there may be therapeutic benefit if this concentration of Adriamycin or higher could be reached by intraperitoneal therapy.

3) The third pattern of sensitivity is seen in patients who had progressive disease while on intravenous Adriamycin. In these patients, even exposure to 10 µg/ml of Adriamycin in vitro resulted in an average survival of 40% of tumor colony-forming units (TCFUs), suggesting that intraperitoneal Adriamycin would not be as useful in this group of patients as in others.

The results of these studies have provided, in part, a rationale for an ongoing phase I trial of intraperitoneal Adriamycin in patients with refractory ovarian cancer [15]. Figure 4 depicts the in vitro sensitivity to 5-fluorouracil and Adriamycin in a patient who was treated with intraperitoneal chemotherapy. This patient had a greater than 70% reduction in tumor colony formation at 10 µg/ml of Adriamycin. Adriamycin was therefore selected, and she had a marked reduction in ascites formation while on therapy with 20 µg/ml of intraperitoneal Adriamycin.

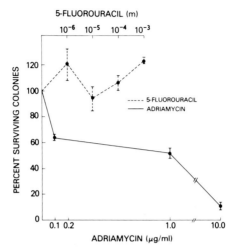

Fig. 2. The in vitro sensitivity of ovarian cancer cells to Adriamycin and 5-fluorouracil in a patient with a Tenckhoff catheter. The cells were obtained by peritoneal lavage in the absence of ascites. Untreated controls averaged 117 colonies/plate. Each point represents the mean ± SE of triplicate determinations.

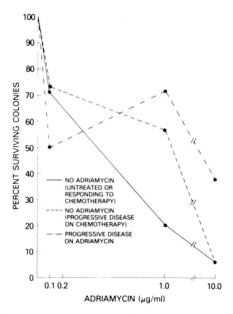

Fig. 3. In vitro sensitivity of ovarian cancer cells obtained from previously untreated patients (three patients), those refractory to a non-Adriamycin-containing regimen (five patients), and those refractory to intravenous Adriamycin (two patients). Each point is the mean of the determinations performed on individual specimens in each patient category.

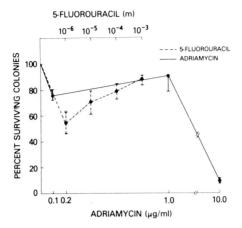

Fig. 4. In vitro sensitivity to 5-fluorouracil and Adriamycin in a patient who was treated with intraperitoneal Adriamycin with a partial resolution of malignant ascites.

Figure 5 shows the in vitro response curves to 5-fluorouracil in five patients with ovarian cancer, three who had progressive disease while on therapy with Hexa-CAF (*hex*amethylmelamine, *c*yclophosphamide, 5-fl*u*orouracil, and-methotrexate) [9] or Chex-UP (similar to Hexa-CAF but with cis-*p*latinum replacing methotrexate) [19], one who had no previous chemotherapy, and one who was responding to treatment with Chex-UP. Significant in vitro suppression (greater than a 70% reduction in colony formation) was observed in two patients: one patient who was responding to Chex-UP and another patient who had progressive disease after Chex-UP therapy. In addition, a second patient who had failed Chex-UP had a 60% reduction in colony formation at 10^{-3} to 10^{-4} M 5-fluorouracil, a concentration that cannot be maintained by intravenous therapy but that can be reached by intraperitoneal administration [14]. Although the numbers are small, the results suggest that some patients who have progressive disease while on a chemotherapy regimen that includes intravenous 5-fluorouracil might potentially benefit from the intraperitoneal administration of high concentrations of 5-fluorouracil.

Effect of Amphotericin B on Ovarian Cancer Colony Formation

Table III summarizes the effects of a one-hour exposure of chemotherapeutic agents plus amphotericin B on human ovarian cancer colony formation. Adriamycin, melphalan, and BCNU were tested at the peak plasma concentrations attainable by conventional intravenous administration. Amphotericin B was tested at 10 µg/ml, even though the peak plasma concentration after intravenous therapy is only 3–4 µg/ml [18]. If no increase in cytotoxicity could be observed at this concentration of amphotericin B, it would be unlikely that any beneficial effects could be detected at lower concentrations. However, if amphotericin B were to increase

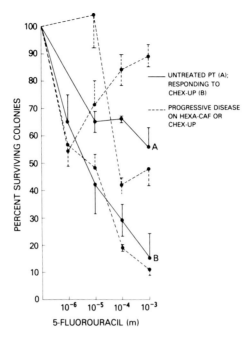

Fig. 5. In vitro sensitivity of ovarian cancer cells to 5-fluorouracil. Specimens were obtained from a previously untreated patient (A), a patient who was responding to therapy (B), and from patients who had progressive disease while on therapy with Hexa-CAF [7] (hexamethylmelamine, cyclophosphamide, methotrexate, and 5-fluorouracil) or Chex-UP [16] (similar to Hexa-CAF but with the substitution of cis-platinum for methotrexate).

TABLE III. Effect of BCNU, Melphalan, and Adriamycin Plus Amphotericin B on Ovarian Cancer Colony Formation

	Percent survival of TCFU	
	Chemotherapy alone	Chemotherapy plus amphotericin B 10 µg/ml
Adriamycin		
1 µg/ml	57 ± 6	38 ± 8
	21 ± 3	17 ± 1
	34 ± 6	47 ± 1
BCNU		
3 µg/ml	133 ± 10	126 ± 5
	94 ± 5	86 ± 5
	68 ± 5	27 ± 2
Melphalan		
0.5 µg/ml	82 ± 4	111 ± 7
	95 ± 6	89 ± 7
	69 ± 5	53 ± 8

drug cytotoxicity at 10 μg/ml, then a search for alternate mechanisms of amphotericin B administration (such as via a Tenckhoff catheter) would be indicated. Amphotericin B did not enhance the cytotoxicity of Adriamycin and melphalan under these conditions. However, amphotericin B plus BCNU did result in more in vitro cytotoxicity than BCNU alone in one of three patients. Unfortunately, when repeat studies were performed at 1.0 μg/ml of amphotericin B (corresponding to observed plasma levels after sustained intravenous administration), there was no enhancement of BCNU cytotoxicity.

DISCUSSION

The results presented in this study demonstrate that human ovarian cancer colonies can be grown from both malignant effusions and from peritoneal washings collected in the absence of ascites. Furthermore, the effects of Adriamycin, 5-fluorouracil, and amphotericin B (together with cytotoxic agents) on ovarian cancer colony formation provides experimental evidence that 1) intraperitoneal chemotherapy with high concentrations of drug, particularly Adriamycin, may be of therapeutic benefit in certain patients with ovarian cancer and 2) the addition of amphotericin B to the conventional chemotherapy of patients with Adriamycin, BCNU, or melphalan is unlikely to produce any increased therapeutic benefit.

The ability to grow human ovarian cancer colonies from cytologically malignant peritoneal washings can potentially save some patients from invasive procedures that might otherwise be necessary to obtain sufficient material for chemotherapy sensitivity testing. This technique of obtaining malignant cells for cloning studies at peritoneoscopy or with an in-dwelling catheter can likely be applied to the majority of patients with refractory ovarian cancer inasmuch as it has previously been demonstrated that up to 32% of patients undergoing a "second-look" peritoneoscopy at a time when they have no clinical evidence of disease have cytologically malignant washings [17]. In the present study we have demonstrated that sufficient numbers of malignant cells can be harvested using a Tenckhoff catheter to perform chemotherapy sensitivity studies in 50% of the cases. There is no technical reasons why 750–1,000 ml of washings cannot be collected through a peritoneoscope, which would allow for chemotherapy sensitivity studies to be performed without the necessity of an in-dwelling catheter. In particular, if drugs are tested at only one concentration, then five to ten chemotherapy sensitivity studies should be feasible on 1,000 ml of peritoneoscopic washings in the majority of patients. Such studies are in progress.

There is a technical advantage to the use of peritoneal washings instead of solid tumor masses in the cloning assay. After peritoneal lavage, ovarian cancer cells are already in a single cell suspension, and this obviates the need to make a single cell suspension from a solid tumor, which frequently is a difficult task.

A rationale for intraperitoneal chemotherapy in certain patients with intraabdominal malignancies has been developed on the basis of mathematical models of drug clearance from the peritoneal cavity and from systemic compartments

[13]. The finding that ovarian cancer cells obtained from patients who have relapsed on non-Adriamycin chemotherapy regimens are resistant in vitro to concentrations of Adriamycin that are comparable to peak plasma levels obtained after intravenous therapy but are sensitive at higher concentrations of Adriamycin (which can only be achieved by intraperitoneal administration) provides primary experimental support for the use of intraperitoneal chemotherapy.

Likewise, the observation that amphotericin B does not significantly enhance the cytotoxic effects of Adriamycin, melphalan, and BCNU suggests that clinical benefit is not likely to result from the use of amphotericin B, at least at concentrations comparable to those achievable by intravenous therapy. It is possible, however, that higher concentrations of cytotoxic agents and of amphotericin B (potentially reachable by intraperitoneal administration) could demonstrate a synergistic effect. However, there are no available toxicity data in humans regarding the administration of high doses of intraperitoneal amphotericin B. In addition, the effects of biological modifiers in the stem cell assay must be assessed with caution, since it is possible that agents have a biological effect that is mediated through a mechanism that does not lend itself to study in a clonogenic cytotoxicity assay. For example, amphotericin B is a potent immune stimulant in mice [19], and if similar effects exist in humans it is not likely that they would be detected in the stem cell assay where perturbations of the immune system may not be detected. However, in the case of amphotericin B, a recent clinical trial in refractory breast cancer and sarcoma patients could not demonstrate a beneficial effect from the use of amphotericin B in conjunction with chemotherapeutic agents [20].

In summary, the ability to grow ovarian cancer colonies from peritoneal washings increases the numbers of patients in whom individualized chemotherapy can be applied. Furthermore, the human stem cell assay is a useful model system in which to study new modalities of therapy such as intraperitoneal chemotherapy and the modification of drug resistance [21].

REFERENCES

1. Hamburger AW, Salmon SE: Primary bioassay of human tumor stem cells. Science 197: 461–463, 1977.
2. Hamburger AW, Salmon SE, Kim MB, Trent JM, Soehnlen BJ, Alberts DS, Smith HJ: Direct cloning of human carcinoma cells in agar. Cancer Res 38:3438–3444, 1978.
3. Hamburger, AW, Salmon SE, Alberts DS: Development of a bioassay for ovarian carcinoma colony forming cells. In Salmon SE (ed): "Cloning of Human Tumor Stem Cells." New York: Alan R. Liss, Inc, 1980 (Chapter 6). This volume.
4. Salmon SE, Hamburger AW, Soehnlen BJ, Durie BGM, Alberts DS, Moon TE: Quantitation of differential sensitivity of human tumor stem cells to anticancer drugs. N Engl J Med 298:1321–1327, 1978.
5. Alberts DS, Salmon SE, Chen H-SG, Moon TE, Surwit EA, Soehnlen BJ, Young L: Correlative and predictive accuracy of the human tumor stem cell assay (HTSCA) for anticancer drug activity in ovarian (Ov) cancer patients (Pts). Proc Am Assoc Clin Oncol 21: 431 (Abst C-445), 1980.

6. Stanhope RC, Smith JP, Rutledge F: Second trial drugs in ovarian cancer. Gynecol Oncol 5:52–58, 1977.
7. Young RC, Chabner BA, Hubbard SP, Fisher RI, Bender RA, Anderson T, Simon RM, Canellos GP, DeVita VT Jr: Advanced ovarian adenocarcinoma: A prospective clinical trial of melphalan (L-PAM) vs combination chemotherapy (Hexa-CAF). N Engl J Med 299:1361–1366, 1978.
8. Griffiths CT, Fuller AF: Intensive surgical and chemotherapeutic management of advanced ovarian cancer. Surg Clin North Am 58:131–142, 1978.
9. Medoff G, Schlessinger D, Kobayashi GS: Polyene potentiation of antitumor agents. J Natl Cancer Inst 50:1047–1050, 1973.
10. Medoff J, Medoff G, Goldstein MN, et al: Amphotericin B-induced sensitivity to actinomycin-D in drug resistant HeLa cells. Cancer Res 35:2548–2552, 1975.
11. Jones RB, Myers CE, Guarino AM, Dedrick RL, Hubbard SM, DeVita VT Jr: High volume intraperitoneal chemotherapy ("belly bath") for ovarian cancer. Pharmacologic basis and early results. Cancer Chemother Pharmacol 1:161–166, 1978.
12. Speyer JL, Collins JM, Dedrick RL, Brennan MF, Londer H, DeVita VT Jr, Myers CE: Phase I and pharmacologic studies of intraperitoneal (IP) 5-fluorouracil (5-FU). Proc Am Soc Clin Oncol 20:352, 1979.
13. Dedrick RL, Myers CE, Bungay PM, DeVita VT Jr: Pharmacokinetic rationale for peritoneal drug administration in the treatment of ovarian cancer. Cancer Treat Rep 62:1–12, 1978.
14. Salmon SE, Buick RN: Preparation of permanent slides of intact soft agar colony cultures of hematopoietic and tumor stem cells. Cancer Res 39:1133–1136, 1979.
15. Ozols RF, Young RC, Speyer JL, Weltz M, Collins JM, Dedrick RL, Myers CE: Intraperitoneal (IP) Adriamycin (Adr) in ovarian carcinoma (OC). Proc Am Soc Clin Oncol 21:425 (Abst C-423), 1980.
16. Young RC, Von Hoff DD, Gormley P, Makuch R; Cassidy J, Howser D, Bull JM: Cis-dichlorodiammineplatinum (II) for the treatment of advanced ovarian cancer. Cancer Treat Rep 63:1539–1544, 1979.
17. Ozols RF, Fisher RI, Makuch R, Anderson T, Young RC: Peritoneoscopy in the management of ovarian cancer. In Young RC (ed): "Peritoneoscopy in Medical Oncology." New York: Masson (in press).
18. Atkinson AJ Jr, Bennett JE: Amphotericin B pharmacokinetics in humans. Antimicrob Agents Chemother 13:271–276, 1978.
19. Stein SH, Plut EJ, Shine TE, Little JR: The importance of different murine cell types in the immunopotentiation produced by amphotericin methyl ester. Cell Immunol 40:211–221, 1978.
20. Krutchik AN, Buzdar AU, Blumenschein GR, Sinkovics G: Amphotericin B and combination chemotherapy in the treatment of refractory breast carcinoma and sarcoma. Cancer Treat Rep 62:1565–1567, 1978.
21. Ozols RF, William JKV, Grotzinger KR, Young RC: Cloning of human ovarian cancer cells in soft agar from malignant effusions and peritoneal washings. Cancer Res 40:2743–2747, 1980.

20
Chemosensitivity Testing for Human Brain Tumors

Mark L. Rosenblum

Analysis of the chemotherapeutic sensitivity of clonogenic cells derived from tumor biopsies may help to determine whether a tumor is likely to be resistant to chemotherapy. Testing of cell sensitivity entails a number of potential difficulties: selecting a representative cell population, understanding differences between treatment in vitro and in situ, interpreting studies to determine tumor cell sensitivity in vitro and tumor response in situ, and possibly encountering changes in tumor cell sensitivity both in vitro and in situ subsequent to the analysis. First, however, it is necessary to determine the feasibility of the methods of analysis; then it is necessary to demonstrate that the testing is applicable to the treatment of human tumors by establishing a correlation between results in vitro and actual patient response. This is a report of initial endeavors at developing an assay that is predictive of the in situ efficacy of chemotherapeutic agents against malignant gliomas, based upon a recently reported method of analyzing the colony-forming capacity of growing human brain tumors. The approach is analogous to, but differs in some technical and analytic details from, that discussed in other chapters of this text.

PRELIMINARY RESULTS OF TUMOR SENSITIVITY TESTING WITH BCNU

Methods

Tumor cell analysis. The procedure for obtaining monolayer clonal outgrowths of cells derived from human brain tumor specimens has been described by Rosenblum et al [1]. Briefly, it entails obtaining solid tumor biopsies from the operating room and mincing them into pieces less than 1 mm^3, which are treated with 0.25% trypsin (GIBCO, Santa Clara, CA) with 2 mg% EDTA at 37°C for ten minutes. After filtration to remove nondigested particles, the single cell suspension is aliquoted into 60 mm plastic Petri dishes at various dilutions. Semi-solid medium (agar or methyl cellulose) was not employed. Plates are incubated for four weeks at 37°C in a humidifed atmosphere with 5% CO_2, in an enriched medium containing 30% fetal calf serum and heavily irradiated feeder cells. The plates are examined, and a collection of more than 25 cells of similar morphology is considered to be a colony that has originated from a single "clonogenic" tumor cell from the initial biopsy specimen.

Biopsies were received fresh from the operating room and disaggregated within two hours. Single cells were treated in suspension with various concentrations of BCNU for two hours at 37°C while stirred continuously, then they were washed and plated. After incubation for four weeks under conditions optimized for in vitro growth [1], colonies developed.

Patient Population and Analysis

Brain tumor specimens obtained from ten patients harboring malignant gliomas treated at the University of California, San Francisco (UCSF), were analyzed. Table I shows the patient population, tumor location, treatment, and clinical response. Five patients were irradiated first, with concurrent administration of hydroxyurea, and subsequently received either BCNU or CCNU (which in the case of patient 6 was part of a combination drug regimen). Both nitrosoureas were administered in single doses [BCNU (carmustine), 180–210 mg/m^2 IV; CCNU (lomustine), 100–130 mg/m^2 po].

For the present study, the cell sensitivity and tumor response are considered to be equivalent for BCNU and CCNU inasmuch as 1) the clinical activity of the two nitrosoureas is similar in the treatment of primary malignant brain tumors [2], 2) they demonstrate cross-resistance [2], and 3) their mechanism of action is probably the same (personal communication from R. J. Weinkam and D. F. Deen, 1979).

All patients were treated with optimal standard care, and antitumor activity was assessed – without knowledge of the results of cell analysis – according to UCSF criteria based on changes in computerized tomographic (CT) and radionuclide (RN) scans and on the patient's neurological examination [3]. Figures 1 and 2 illustrate, respectively, clinical response and clinical progression following treatment of patients with BCNU.

Results

The BCNU dose-response curves for all ten patients are presented in Figures 3 and 4. Tumor cells were judged sensitive or resistant to BCNU according to the two criteria of cell response described below.

Method 1. Tumor cells were considered sensitive if a cell kill greater than 1 log was observed for doses up to 25 µg/ml (Fig. 3). Tumor cell kill and patient response were found to correlate directly in the majority of patients (Tables II and III), and there was a correlation between in vitro and clinical responses in four of

TABLE I. Response of Ten Patients With Malignant Gliomas to Nitrosourea and Radiation Therapy

Patient number	Age/sex	Location[a]	First treatment Therapy[b]	First treatment Response[c]	Second treatment Therapy[b]	Second treatment Response[c]	Survival Status	Survival Time (w.k)	Response to nitrosourea[c]
1	63/M	Lt-fr	BCNU	P	Rad	P	died	17	P
2	50/F	Lt-fr	BCNU	R	Rad	U	alive	58+	R
3	69/M	Rt-fr	Rad + Hu	P	BCNU	P	died	38	P
4	73/F	Lt-fr	Rad + Hu	P	CCNU	P	died	11	P
5	61/F	Rt-fr	Rad + Hu	R	BCNU	U	alive	49+	U
6	56/M	Lt-par	Rad + Hu	R	CCNU + Proc + VCR	U/P	died	36	U
7	55/F	Rt-par	Rad + Hu	R	BCNU	U	alive	33+	U
8	54/M	Rt-par	BCNU + FU	U	—	—	alive	13+	U
9	30/F	Medul	BCNU	R	—	—	alive	13+	R
10	22/F	Lt-fr	BCNU	R	—	—	alive	10+	R

[a]Lt = left; Rt = right; fr = frontal; par = parietal; Medul = medullary.
[b]Rad = radiation therapy; Hu = hydroxyurea; Proc = procarbazine; VCR = vincristine; FU = 5-fluorouracil.
[c]P = tumor progressed; R = tumor responded to therapy; U = tumor unchanged.

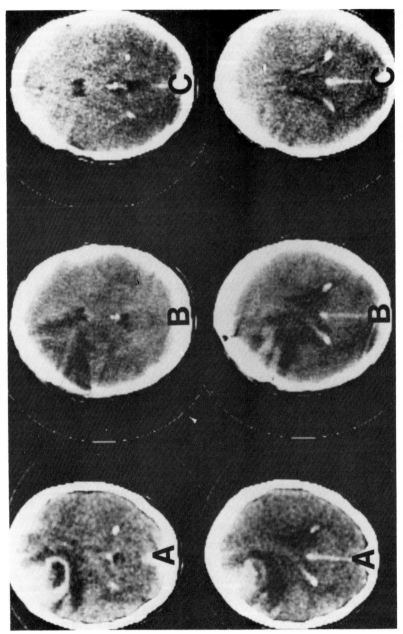

Fig. 1. Serial CT brain scans for patient 2 before (A) and after (B) subtotal resection of a left frontal glioblastoma multiforme, and six weeks after (C) a single dose of BCNU at 200 mg/m^2. The improved appearance in the scans demonstrates a dramatic tumor response to BCNU therapy.

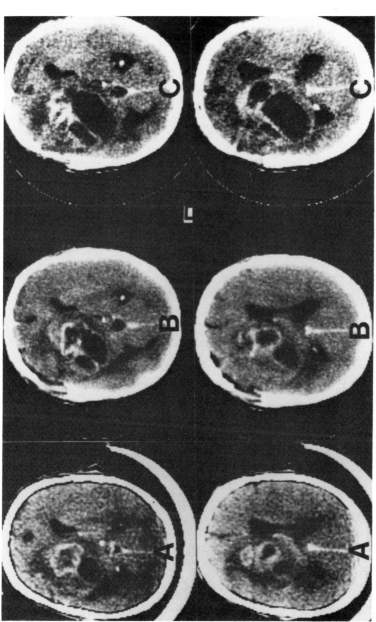

Fig. 2. Serial CT brain scans for patient 1 before (A) and after (B) subtotal resection of a left frontal glioblastoma multiforme, and six weeks after (C) a single dose of BCNU at 210 mg/m². The worsening appearance in the CT scans after BCNU corresponds to clinical deterioration and to an increase in the size of the defect in the blood-brain barrier shown on radionuclide scan. These studies illustrate tumor progression and insensitivity to BCNU therapy.

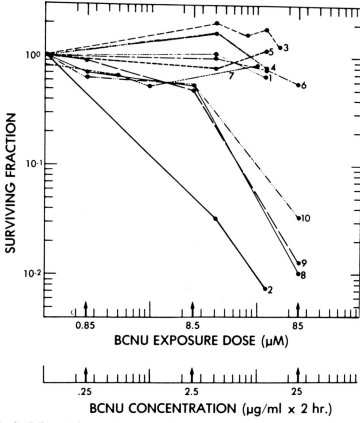

Fig. 3. Cell survival curves for tumor cells disaggregated from biopsies of ten human brain tumors and treated in vitro with various concentrations of BCNU for two hours at 37°C, then cultured under optimized conditions for four weeks. Cells derived from patients 2, 8, 9, and 10 appeared to be sensitive to BCNU in vitro; all others appeared to be resistant.

the five who received nitrosourea in the immediately postoperative period. None of the six patients whose cells were considered resistant to BCNU at these high dose levels in vitro showed a clinical response to nitrosourea in situ. The colony-forming efficiency (CFE) did not correlate with either cell sensitivity or patient response to BCNU (Table II).

Method 2. Alternately, sensitivity was evalutated on the basis of dose-response curves for cell exposure only up to the maximum anticipated dose of BCNU in vivo (Fig. 4), which was calculated according to standard formulas* [4] using a

*Exposure Dose = $Ae_{(o)}[1 - e^{-k_2 t}]$
$$k_2 = \frac{0.693}{T_{1/2}}$$

where $Ae_{(o)}$ = initial drug dose
$t = R_x$ time
$T_{1/2}$ = drug half-life

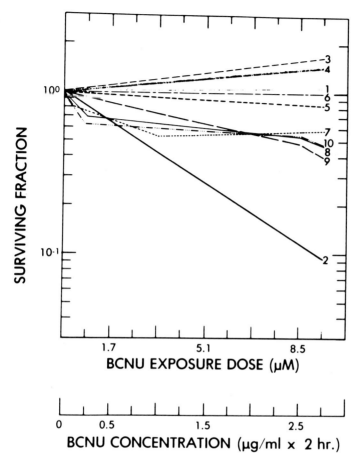

Fig. 4. Cell survival curves for tumor cells disaggregated from biopsies of ten human brain tumors and treated in vitro with clinically achievable exposure doses of BCNU.

BCNU half-life in phosphate-buffered saline of 64 minutes and a treatment time of two hours. The mean cell exposure dose to brain tumor cells in situ after administration of 180–220 mg/m^2 IV was calculated to be 8.5 μM (personal communication from V. A. Levin, 1979). The areas under the survival curves were calculated from linear plots of cell survival re BCNU dose (Chapter 17), for cell exposures of 0.85, 3.4, and 8.5 μM. Areas were related to the response of the respective patients using arbitrary cut-off points as described by Salmon et al [5; and Chapters 16–18]. Figure 5 provides an illustrative example of this method of analysis based on an exposure dose of 3.4 μM in vitro. On the basis of this method, cellular and patient responses correlated identically over the entire range of exposure doses analyzed (Table IV). It is important to note that, while this survival-concentration area is larger than the cut-off point used by Salmon et al [5] and

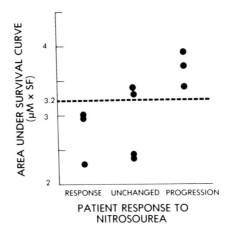

Fig. 5. Correlation of cellular and patient response to BCNU as determined by the area under the survival curves (linear plots of μM drug dose and cellular fraction surviving). An area defined by an arbitrary cut-off point of 3.2 appears to differentiate between patients demonstrating clinical response and progression after in situ administration of nitrosourea.

TABLE II. Cell Survival and Patient Response to Nitrosourea Therapy

Patient number	Culture				Patient response[a]
	Untreated CFE (%)	Maximum cell kill		Response	
		(%)	log		
1	0.174	33.0	0.17	−	P
2	0.149	99.3	2.15	+	R
3	0.008	0	0	−	P
4	0.513	10.0	0.05	−	P[b]
5	0.015	22.0	0.11	−	U
6	0.006	42.0	0.24	−	U[c]
7	0.015	46.0	0.27	−	U
8	0.101	99.0	2.00	+	U[d]
9	0.008	98.7	1.89	+	R
10	0.047	96.7	1.48	+	R

[a]P = tumor progressed; R = tumor responded; U = tumor unchanged.
[b]Received CCNU.
[c]Received CCNU, procarbazine, and vincristine.
[d]Received BCNU and 5-fluorouracil.

TABLE III. Correlation of Tumor Response to Nitrosourea Therapy in Culture (All Doses) and In Situ

Culture	Patient	
	First course	Survival (wks)
4 Responses	3 Responses	10+, 13+, 58+[a]
	1 Unchanged	13+
6 Nonresponses	3 Progression	11, 17, 38
	3 Unchanged[b]	33, 36, 49+[a]

[a]Second operation performed.
[b]Nitrosourea administered after a clinical response from radiation therapy.

TABLE IV. Correlation of Tumor Response to BCNU* in Culture (Achievable Doses) and In Situ

Culture	Patient
5 Responses	3 Responses
	2 Unchanged
5 Nonresponses	3 Progressions
	2 Unchanged

*BCNU exposure dose: $0.85-8.5$ μM ($0.25-2.5$ μg/ml $\times$ 2 hours).

discussed in Chapters 16–18, the in vivo doses of BCNU employed for brain tumor patients were also substantially higher than those used for myeloma and a different in vitro cloning technique was employed.

IMPROVED METHODS OF TUMOR DISAGGREGATION

Disaggregation of solid tumor specimens by mechanical means alone usually results in suspensions containing cell aggregates, the presence of which can markedly influence cell survival studies by affecting the CFE in monolayer and soft agar culture systems, especially when CFE is low. Various enzymatic methods of dissociating single cells from biopsies of animal and human brain tumors have been evaluated at our institution.

Specimens of rat 9L tumors were treated either with 0.25 trypsin for ten minutes at 37°C or with an enzyme cocktail for 30 minutes at 37°C. The cocktail (developed by M. Brown, personal communication, 1979) consists of pronase (0.05% of 45 PUK/mg, B grade), collagenase (0.02% of 125 units/mg), and DNAase

(0.02% of 7×10^4 dornase units/mg, B grade). The yield of clonogenic cells per mg is calculated as the product of the CFE and the total number of cells obtained per milligram. An increase in cell yield (total cells/mg) was responsible primarily for a fivefold greater yield of clonogenic cells with the use of the enzyme cocktail (11.1×10^4/mg) than with trypsin (2.3×10^4/mg) (Table V).

When biopsies from five human brain tumors were disaggregated by four different methods, the yield of clonogenic cells obtained using the cocktail was 4.5 times greater than that obtained with trypsin, 1.6 times greater than with trypsin plus DNAse, and 2.2 times greater than with mechanical methods alone. The increase obtained when the enzyme cocktail was used was due both to larger cell yields and to greater CFEs (Table VI). Furthermore, single-cell suspensions obtained with the cocktail for both 9L and human tumors showed the least cell aggregation and debris.

Disaggregation of malignant gliomas of various sizes using trypsin resulted in an apparent decrease in cell yield with large tumor specimens (>450 mg); but analysis of specimens of various sizes (50 to 1,350 mg) from six gliomas dissociated with the cocktail showed no significant change in cell yield (Table VII). As a result of these experiments, the enzyme cocktail is used in our laboratory to disaggregate solid tumor specimens for all studies of CFE.

CONSIDERATIONS REGARDING METHOD AND INTERPRETATION

Cell Selection

Heterogeneity of brain tumors has been demonstrated by several methods of analysis. The histology of several specimens from a single malignant glioma often demonstrates both comparatively benign areas and areas of cells with more malignant-appearing features [6]. Marked variation among different areas of a tumor has also been found by kinetic analysis of cells using autoradiography [7], as well as by flow cytofluorometric (FCM) analysis of the DNA content of nuclei derived from human brain tumors [8].

Furthermore, clonogenic analysis of specimens from various areas of individual tumors has shown CFEs of different specimens to differ by an average coefficient of variation of 33% (Table VIII). These observations serve to emphasize that any evaluation of a solid tumor must be made from multiple specimens from the tumor mass, endeavoring to represent all areas and to minimize cell selection.

Other factors probably contributing to tumor heterogeneity are differences among tumor cell clones and the presence of host cells within the tumor. Distinctly different tumor cell clones have been demonstrated within a murine mammary carcinoma and a human colon carcinoma by Calabresi et al [9] and within B16 melanoma by Fidler and Kripke [10]. Additionally, macrophages have been noted within 9L gliosarcoma tumors growing in the brain or flank (personal communication from K. Wheeler, 1979), in Lewis lung carcinoma [11], and in human brain tumors [12].

TABLE V. Disaggregation of 9L Tumors (Number of Clonogenic Cells/mg ($\times 10^3$))

Tumor number	Trypsin	Enzyme cocktail
1	7.2	104.7
2	14.8	123.1
3	20.2	114.2
4	49.0	101.5
Mean ± SD	22.8* ± 18.3	110.9* ± 9.8

*Significantly different ($P < 0.001$, Student's t-test).

TABLE VI. Clonogenic Cell Yield for Five Gliomas Using Various Disaggregation Methods

	Relative number clonogenic cells	
Method	Mean	± SD
Enzyme cocktail[a]	1.00[b]	0.92
Trypsin	0.22	0.14
Trypsin ± DNAse	0.64[b]	0.36
Mechanical alone	0.45	0.27

[a]Pronase, collagenase, and DNAse.
[b]Significantly different from trypsin ($P < 0.05$, Student's t-test).

TABLE VII. Effect of Specimen Size on Cell Yield for Human Gliomas

Cell yield (number cells $\times 10^4$/mg)	Specimen size (mg)		
	50–150	150–450	>450
Enzyme cocktail[a]			
Mean ± SE	4.4 ± 0.9	5.3 ± 1.2	3.6 ± 0.6
Trypsin			
Mean ± SE	2.0 ± 0.5	2.4 ± 0.6	0.8 ± 0.2
EC/T[b]	2.2	2.2	4.5

[a]Pronase, collagenase, DNAse.
[b]Mean cell yield with cocktail/mean cell yield with trypsin.

TABLE VIII. Variation in CFE Between Specimens of Nine Malignant Gliomas

Tumor number	CFE (% × 10^2)				Coefficient of variation (%)[a]	Fold differences[b]
	Specimen number					
	1	2	3	Mean ± SD		
87	1.50	2.37	1.30	1.72 ± 0.57	33	1.8
92	4.35	3.10	2.64	3.36 ± 0.88	26	1.6
98	0.58	0.24	1.23	0.68 ± 0.50	74	5.1
106	6.08	9.50	14.90	10.16 ± 4.45	44	2.5
114	3.19	3.52	2.03	2.91 ± 0.78	27	1.7
117	1.09	1.59	1.48	1.39 ± 0.26	19	1.5
118	0.37	0.22	0.31	0.30 ± 0.08	27	1.7
119	0.58	0.75	1.01	0.78 ± 0.22	28	1.7
121	0.49	0.61	0.99	0.60 ± 0.10	17	2.0
				Average	33	2.2

[a](SD/mean) × 100.
[b]Largest CFE/smallest CFE.

All investigators using monolayer or soft agar techniques for the clonal growth of human cells from tumor biopsies have noted low CFEs (or "plating efficiencies") ranging from 0.01% to 2% for specimens of bone marrow, ascitic tumor cell suspensions, and specimens of solid tumor [1, 5, 13–15; and Chapters 3–10]. In order to define more accurately the biology and chemosensitivity of human tissue and tumor specimens, it will be necessary to increase the CFE. This will inevitably aggravate the problem of tumor cell selection. The CFE should be increased by selecting the optimal duration of incubation in vitro and analyzing each constituent in the medium separately for its requirement and concentration dependence in supporting clonal growth of each different tumor line. The importance of concentration dependence is illustrated by our observation that a twofold increase in CFE is obtained by increasing the concentration of fetal calf serum from 10% to 30% [1]. Whenever possible, the optimal growth conditions should be determined on the basis of a plateau on dose-response curves in order to minimize the effects of small alterations in concentration. Cells may be sorted according to physical and chemical parameters such as cell size, density, and DNA content in order to enrich the clonogenic cell population.

Possible Problems With In Vitro Treatment

Exposure to a proteolytic enzyme may alter the cytotoxic activity of subsequently administered chemotherapeutic agents because changes in drug penetration or activity may occur or cellular repair processes may come into play. The possibility of this artifact must be taken into consideration in analyzing treatment results.

The differences in drug delivery to cells in vitro and to cells in situ must be taken into consideration if the chemosensitivity of cultured cells is to be related to tumor response in situ. Whereas drug testing in vitro involves the introduction of agents directly into clonogenic cells, drugs administered to a solid tumor must penetrate into the tumor, pass through the extracellular space, and enter the cells before it can produce cytotoxicity [16]. In situ, both clonogenic and nonclonogenic cells are available for drug incorporation; and nonclonogenic cells may absorb quantities of drug that would, in an in vitro assay system, be available for the clonogenic pool. If, however, drug resistance is noted in vitro, then tumor resistance in situ may be anticipated. Drug delivery may vary within an individual tumor as a result of variations in vascularity, cellular incorporation of drug, and metabolism. Moreover, the difficulty of drug access may be increased for subsequent drug courses, when tumor proliferation may be taking place primarily at sites not previously accessible to cytotoxic concentrations of drug.

An understanding of the pharmacokinetics of the chemotherapeutic agents tested is mandatory in order both to allow interpretation of cell survival analysis and to ensure that the drug concentrations tested in vitro are also attainable at the cellular level in situ. The exposure dose in vivo can be estimated from the parameters of drug breakdown (the half-life of the active agent), plasma-drug kinetics, and the agent's physical characteristics [4, 16, 17] and Chapter 16.

Evaluation of cell sensitivity in vitro is limited to drugs that may be administered to cells in their active forms, unless drugs such as cyclophosphamide and procarbazine are activated by exposure to microsome fractions or by other means before they are used in vitro.

Problems in Interpretation

Analysis in vitro. Nontumor cells may be disaggregated from solid tumors and form colonies in soft agar and monolayer cultures. Stevens et al [11] determined that macrophages accounted for 15% of the cells disaggregated from Lewis lung carcinoma and resulted in the formation of 8% of all the colonies attained with that tumor line in a soft agar culture system. To interpret dose-response studies accurately, therefore, it is important to differentiate the nontumor cells and the colonies they form from tumor cells and their colonies.

As a result of perturbation caused by the disaggregation process and the marked differences in the environment in vitro, disaggregated tumor cells probably display cell kinetics different from those of cells growing in vivo. For this reason, in vitro analysis of agents that demonstrate some dependency on cell cycle progression may give false indications of in situ tumor sensitivity.

A decrease in the number of colonies in a monolayer clonal assay after treatment may result from either cell kill or an alteration in plate adherence factors [18].

If, however, no decrease in colonies is noted after treatment in monolayer systems, cellular resistance to the chemotherapeutic agent could reliably be anticipated. Soft agar techniques for clonal growth and analysis would eliminate this dependence of colony growth on cellular plate adherence.

Analysis in vivo. The size of a solid tumor in situ is affected by many factors. Tumors will increase in size as a result of increasing numbers of cells, increasing cell size, and/or an increasing amount of tumor "matrix." The number of cells in a tumor will increase by division of both clonogenic and doomed tumor cells. In addition, host cells may infiltrate and multiply within the tumor; these may include fibroblasts (derived from vessels and normal tissue stroma), endothelial cells (from the incorporated vasculature), and white blood cells (such as monocytes, polymorphonuclear cells, and plasma cells). Cell size may increase as a result of amitotic division or increasing amounts of cytoplasmic constituents. Also, the tumor matrix may increase in quantity as a result of secretions from tumor cells or from the host cells (eg, collagen deposition from fibroblasts). Finally, after the administration of an effective chemotherapeutic agent, cells may swell prior to death and dissolution.

Decrease in tumor size depends not only on cell death but also on the removal of dead cells. Cell death results from inherent cell destruction (a phenomenon known as the "cell loss factor") as well as from therapy with cytotoxic agents. The cell loss factor is a product of several factors: abortive cell division resulting from aberrant tumor cell kinetics, tumor cell destruction by host cells, influences such as humoral factors from the immune system, and inadequate oxygenation and the build-up of toxic metabolites due to inadequate blood supply.

Measurement of the tumor burden by direct determination of solid tumor volume will be insensitive to small degrees of cell kill. The just mentioned factors influencing tumor size may limit the decrease in tumor size. For example, it was demonstrated that treatment of a 9L gliosarcoma in vivo with an LD_{10} dose of BCNU resulted in a 3.5 log cell kill within 30 minutes, yet after 14 days there was only a 60% reduction in tumor weight [19]. The comparative unresponsiveness of tumor size to change is a consequence of delayed removal of dead cells and the simultaneous proliferation of surviving clonogenic cells.

Indirect means of measuring tumor size include x-rays, computerized tomographic scans, radionuclide scans, and sonograms. These measurements have their own inherent limitations, including the difficulty of determining precisely the borders between the tumor and surrounding host tissues.

Tumor burden may also be monitored by measuring tumor secretions. A specific and sensitive evaluation of tumor burden might be obtained by measuring tumor-specific markers such as human chorionic gonadotropin for choriocarcinoma [20] and the level of polyamines in the cerebrospinal fluid in cases of medulloblastoma [21]. Indirect evidence of tumor activity might also be obtained from nonspecific factors such as pleural effusions and ascites from tumors adjacent to these visceral membranes, and increased production of cerebrospinal fluid with choroid plexus papillomas. Tumors will affect host organs and may produce quantifiable physiological dysfunction, such as chemical disorders from tumors growing in the liver, and neurological deficits from tumors in the brain. Finally, clinical changes, such as patients' subjective pain states, may relate to tumor size and growth rate as well as location. General metabolic changes in the host and distant effects of tumors (such as neuropathies) may also provide indirect evidence of tumor presence and progression.

In determining patient response, as many methods as possible should be applied by independent observers. The methods used in a particular patient will depend upon the tumor location and the sensitivity and accuracy of the assay method. A retrospective evaluation by Levin et al [3] of the reliability of CT and radionuclide brain scans, EEGs, and neurological examinations in the evaluation of brain tumor changes after chemotherapy determined that no single method of analysis was more than 80% accurate in predicting actual tumor progression. As a consequence, CT scan, radionuclide scan, and neurological examination are all used to evaluate patient response at UCSF [3].

The most accurate evaluation will be that of a drug administered alone, but tumors are usually treated with multiple chemotherapeutic agents, making it difficult to determine which drug actually produced a tumor response. Furthermore, delayed effects of therapy may confuse the evaluation of subsequent treatments. Hoffman et al [22] determined that brain irradiation resulted in occasional delayed, transient evidence of tumor progression. Approximately 25% of all patients with brain tumors who show indications of deterioration within 18 weeks after radiotherapy will improve spontaneously without the administration of further therapy. This transient progression may be documented on CT scan, radionuclide scan, or neurological examination and is a recognized phenomenon — the postradiotherapy slump. Finally, it is important to differentiate, as much as possible, the changes in scans that are attributable to the surgery itself in order that a subsequent improvement of these surgically induced abnormalities is not interpreted as a response to chemotherapy.

Changes in Tumor Cell Growth In Vitro and In Situ Subsequent to Sensitivity Analysis

Selection or adaptation of cell populations is observed after serial passage in vitro of cells derived from primary tumor cell suspensions. We have observed increases in CFE from tenfold to 10,000-fold for eight of nine tumors after one to 15 passages in culture (Table IX). Furthermore, CFEs obtained from a meningioma and a glioblastoma that had been serially transplanted in immune-deficient mice were found in their fifth passage to have increased markedly — to 41% and 11%, respectively. Chemotherapeutic sensitivity of cells after serial passage in vitro or in vivo may differ from that of the primary explant.

TABLE IX. Serial Culture of Human Tumors

	CFE (%)							
Tumor number	66	85	86	87	89	106	126	159
Diagnosis[a]	GM	Astro	Mening	AA	GM	GM	GM	GM
Original	0.0006	0.014	2.17	0.013	0.211	0.102	0.149	0.10
Passage number								
3–4	–	0.003	–	0.12	1.02	32.0	1.98	12.7
5–7	74.2	–	29.6	–	–	–	5.02	–
9–10	68.7	–	56.7	–	–	–	17.1	–

[a]GM = glioblastoma multiforme; Astro = astrocytoma; Mening = meningioma; AA = anaplastic astrocytoma.

At the time of the second and subsequent drug cycles, tumor resistance may develop in situ that would not be expected on the basis of the cell sensitivity analysis. If a biopsy is feasible at that later stage — which frequently it is not — then it might be possible to select alternative agents by a second cell sensitivity analysis. To understand and combat this phenomenon, basic investigations into the biology of tumor resistance will be necessary.

FUTURE PROSPECTS

Despite all the potential limitations, the clinical application of the cell assay system may be possible if a correlation can be demonstrated between in vitro drug sensitivity testing and patient response in vivo. The direct correlation between malignant glioma cell sensitivity to BCNU in vitro using a monolayer colony-formation assay, and tumor response in situ, as demonstrated in the present study, is a preliminary result but is encouraging.

In the future, methods of cloning human tumor cells should also be applied in the investigation of the biology of human tumors. The methods discussed here should permit an evaluation of the clonogenic tumor population, its heterogeneity, kinetics, and drug sensitivity. An evaluation of tumor resistance may be possible. Investigations in this regard should improve the results of therapeutic tactics against human tumors.

SUMMARY

A method for analyzing clonogenic cells is described. Single cells obtained from biopsies of ten malignant gliomas were treated with 1,3-bis(2-chloroethyl)-1-nitrosourea (BCNU) in vitro, and tumor cell survival was compared to patient response to nitrosourea therapy. There was a direct correlation between cell sensitivity to nitrosourea and patient response to nitrosourea therapy. The limitations of in vitro and in situ methods and their interpretations are discussed.

In addition, an improved method is described for disaggregating single cells from specimens of solid tumor with the use of an enzyme cocktail consisting of pronase, collagenase, and DNAse.

ACKNOWLEDGMENTS

The author wishes to acknowledge the excellent technical assistance of D. A. Dougherty, the helpful editorial assistance of Don Shevlin, and the invaluable general support of Drs. Charles B. Wilson and Marvin Barker.

REFERENCES

1. Rosenblum ML, Vasquez DA, Hoshino T, Wilson CB: Development of a clonogenic cell assay for human brain tumors. Cancer 41:2305–2314, 1978.
2. Fewer D, Wilson CB, Boldrey EB, Enot KJ: Phase II study of 1-(2-chloroethyl)-3-cyclohexyl-1-nitrosourea (CCNU; NSC-79037) in the treatment of brain tumors. Cancer Chemother Rep 56:421, 1972.
3. Levin VA, Crafts DC, Norman DM, Hoffer PB, Spire JP, Wilson CB: Criteria for evaluating patients undergoing chemoterahpy for malignant brain tumors. J Neurosurg 47:329–335, 1977.
4. Weinkam RJ, Deen DF: Chemically activated alkylating agents and cytotoxic response relations in cell culture. Proc Natl Acad Sci USA (in press).
5. Salmon SE, Hamburger AW, Soehnlen BJ, Durie BGM, Alberts DS, Moon TE: Quantitation of differential sensitivity of human tumor stem cells to anticancer drugs. N Engl J Med 298:1321–1327, 1978.
6. Russel DS, Rubinstein LJ: "Pathology of Tumors of the Nervous System," Ed 3. Baltimore: Williams and Wilkins, 1971.
7. Hoshino T, Wilson CB, Rosenblum ML, Barker M: Chemotherapeutic implications of growth fraction and cell cycle time in glioblastomas. J Neurosurg 43:127–135, 1975.
8. Hoshino T, Nomura K, Wilson CB, Knebel KD, Gray JW: The distribution of molecular DNA from human tumor cells – Flow cytometric studies. J Neurosurg 49:13–21, 1978.
9. Calabresi P, Dexter DL, Heppner G: Clinical and pharmacological implications of cancer cell differentiation and heterogeneity. Biochem Pharmacol 28:1933–1941, 1979.
10. Fidler JJ, Kripke ML: Metastasis results from preexisting variant cells within a malignant tumor. Science 197:893–895, 1977.
11. Stephens TC, Currie GA, Peacock JH: Repopulation of x-irradiated Lewis lung carcinoma by malignant cells and host macrophage progenitors. Br J Cancer 38:573–582, 1978.
12. Morantz RA, Wood GW, Foster M, Clark M, Gollahon K: Macrophages in experimental and human brain tumors, part 2: Studies of the macrophage content of human brain tumors. J Neurosurg 50:305–311, 1979.
13. Wells J, Berry RJ, Laing AH: The chemosensitivity of freshly explanted tumor cells of various origins as determined by clonal assay and six-day growth in vitro and variation in chemosensitivity with subsequent subculturing. In Dendy PP (ed): "Human Tumors in Short Term Culture – Techniques and Clinical Applications." London: Academic Press, 1976, pp 158–164.

14. Hamburger AW, Salmon SE: Primary bioassay of human tumor stem cells. Science 197: 461–463, 1977.
15. Courtenay VD, Selby PJ, Smith IF, Mills J, Peckham MJ: Growth of human tumor cell colonies from biopsies using two soft-agar techniques. Br J Cancer 38:77–81, 1978.
16. Levin VA, Patlak CS, Landahl HD: Heuristic modeling of drug delivery to malignant brain tumors. J Pharmacokinet Biopharm (in press).
17. Wheeler KT, Levin VA, Deen DF: The concept of drug dose for in vitro studies with chemotherapeutic agents. Radiat Res 76:441–458, 1978.
18. Good M, Lavin M, Chen P, Kidson C: Dependence on cloning method of survival of human melanoma cells after ultraviolet and ionizing radiation. Cancer Res 38:4671–4675, 1978.
19. Rosenblum ML, Knebel KD, Vasquez DA, Wilson CB: In vivo clonogenic tumor cell kinetics following 1,3-bis(2-chloroethyl)-1-nitrosourea brain tumor therapy. Cancer Res 36:3718–3725, 1976.
20. Bagshawe KD, Harland S: Immunodiagnosis and monitoring of gonadotropin-producing metastases in the central nervous system. Cancer 38:112–118, 1976.
21. Marton L: The potential of cerebrospinal fluid polyamine determination in the diagnosis and therapeutic monitoring of brain tumors. In Campbell RA et al (eds): "Advances in Polyamine Research," vol 2. New York: Raven Press, 1978, pp 257–264.
22. Hoffman J, Levin VA, Wilson CB: Evaluation of malignant glioma patients during the post-irradiation period. J Neurosurg 50:624–628, 1979.

21

Further Experience in Testing the Sensitivity of Human Ovarian Carcinoma Cells to Interferon in an In Vitro Semisolid Agar Culture System: Comparison of Solid and Ascitic Forms of the Tumor

Lois B. Epstein, Jen-Ta Shen, John S. Abele, and Constance C. Reese

INTRODUCTION

Several years ago, the main source of the world's supply of human leukocyte interferon came from the laboratory of Dr. Kari Cantell, in Finland [1]. Dr. Cantell, working with the Finnish National Red Cross, which provided him with the large amounts of blood necessary for the isolation and partial purification of the interferon, was able to produce approximately 10^{11} international units of the interferon per year. As interferon is one of the most potent naturally occurring substances known to man, with a specific activity of $> 1 \times 10^9$ units/mg for the material purified to homogeneity, 10^{11} units corresponds to only approximately 100 mg, and at the doses then employed, was enough to treat only 150 patients a year. There was considerable interest throughout the world in using interferon not only as an antiviral agent (for review see [2, 3]), but also as an antitumor agent (for review see [4]). At that time some success had been observed using interferon to treat patients with osteosarcoma [5], and clinical trials using interferon to treat patients with lymphoma [6] and multiple myeloma [7] were in progress. Only scant information was available about the response of other types of human malignancies to interferon [8], and much of that information was derived from the use of long-term tumor cell lines [9].

At that time a new technique developed by Dr. Sydney Salmon and his colleagues for growing human tumor cells from individual patients in vitro was described [10–12]. The ability of these investigators to test the sensitivity of the resulting tumor colonies to numerous conventional cancer chemotherapeutic agents was well documented [13], and their success in using their in vitro results as predictive of in vivo response to individual agents was eventually substantiated [14]. Additional evidence for the predictive power of this assay is provided in Chapters 10 and 18 of this text. Recognizing that the same system could be used to evaluate the response of individual tumors to interferon as well as to more conventional agents, and perhaps to be reflective of in vivo response, we set about adapting the system to make this possible.

We chose ovarian carcinoma for study because it was known to grow well in the semisolid agar culture system [11], because it was a likely candidate for future clinical trials with interferon, and because there was absolutely no information concerning its sensitivity to interferon in vitro or in vivo. In our first studies we examined the growth characteristics and response to interferon of 18 samples of ascitic fluid obtained from 15 patients with ovarian carcinoma [15]. Cultures were examined at weekly intervals after initiation and the number, size, and degree of degeneration of the colonies noted. Growth was defined in the initial studies as an increase in total tumor colony number and/or by an increase in colonies with $>$ 30 cells/colony with increasing duration of culture, and this was observed in 67% of the ascitic fluid samples with tumor cells demonstrated by Pap smear. No growth occurred in samples from patients treated with chemotherapeutic agents within four weeks of the paracentesis or in samples devoid of tumor cells as assessed by Pap smear [15].

Response to interferon was measured in cultures in which the interferon was directly incorporated into the agar, or preincubated with cells prior to their inclusion in culture. Response was defined as reduction in total tumor colony number by $\geqslant$ 50% and partial response by $\geqslant$ 25% reduction. The response rate for samples, the non-treated controls which showed evidence of growth throughout the duration of the culture, was 71%. The response rate for samples which had distinct tumor colonies, but whose non-treated controls showed no increase in tumor colony number after the first week in culture was 75%. The time course of the response of tumor colonies derived from ascitic fluid samples to interferon incorporated into the agar for the duration of culture is depicted in Figure 1.

Sensitivity to interferon was not related to the histology or grade of the tumor or to the stage of disease. In general, the responsiveness of the tumor cells to interferon ran parallel to their overall responsiveness to a variety of other chemotherapeutic agents [15], but mg for mg, the amount of interferon employed was far less than that employed for the other antitumor agents.

The purpose of the present report is to update our experience with ovarian carcinoma and interferon. Specifically we will present new data on the growth patterns and sensitivity to interferon of solid tumor samples from patients with ovarian carcinoma and compare them with data previously obtained from ascitic fluid samples. In addition we will present our thoughts on questions concerning the semisolid agar technique which are yet to be resolved.

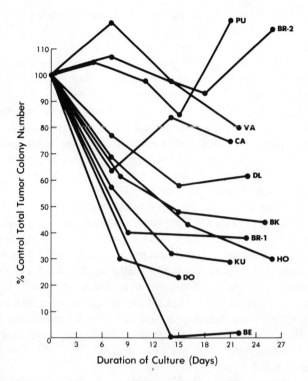

Fig. 1. Ovarian carcinoma colonies derived from ascitic fluid: Time course response to interferon incorporated into the agar of cultures. Concentration of interferon 300 units/ml. Values depicted are the mean values for at least 3 cultures at each time point. Additional clinical information concerning these patients may be found in [15].

METHODS

Solid tumor samples were obtained at the time of surgery and placed in sterile containers in Hanks balanced salt solution, which contained penicillin and streptomycin. Cell suspensions were prepared by teasing and mincing the tissue with sterile needles and scalpel blades. The cell suspensions were then passed through sterile #22 and #25 gauge needles to exclude tissue chunks and partially freed of nonviable cells by passage through a Hypaque-Ficoll gradient during centrifugation at 1,450 rpm for 20 min [16]. After the cells were washed, triturated, and counted, aliquots were prepared for determination of viability by trypan blue dye exclusion and for Pap smears as an independent analysis of the proportion of tumor cells.

The cells were then cultured according to the technique of Hamburger and Salmon [10] as described in our previous report [15]. The washed cells were incorporated into 0.3% agar with CMRL 1066 medium enriched with 15% horse serum to yield a final cell concentration of 2×10^5/ml. One-ml aliquots of the agar-cell mixture were then pipetted into 35mm plastic Petri dishes previously prepared with an underlayer which contained 0.5% agar enriched with conditioned medium. The latter was obtained from the supernatant culture fluids of adherent murine spleen cells previously primed by the intraperitoneal injection of mineral oil.

Interferon and the other agents to be evaluated for their antiproliferative capacity were incorporated into the upper layer of the culture system for the duration of the culture. Human leukocyte interferon (specific activity 2.3×10^5 units/mg protein) was obtained through the courtesy of the Viral Resources Branch, NIAID, NIH, Bethesda, Maryland, and its antiviral titer checked in a virus plaque reduction assay [17] prior to its use in the culture system. The other antitumor agents were obtained in pure form through the courtesy of the Drug Synthesis and Chemistry Branch and the Natural Products Branch of the Division of Cancer Treatment of NCI, NIH, Bethesda, Maryland. The drugs were dissolved in either sterile buffered saline or water and stored at $-70°C$ in small aliquots. The final concentrations of each agent employed were: Interferon, 30, 100, and 300 units/ml; Adriamycin, 0.05, 0.1, and 0.5 µg/ml; and cis platinum, 0.01, 0.1, and 1.0 µg/ml. Three culture dishes were prepared for each concentration of agent employed, and six were prepared as controls, with no antitumor agent added.

The cultures were maintained at $37°C$ in a humidified CO_2 incubator, and the plates examined with a Tiyoda inverted microscope at weekly intervals after initiation of the cultures. Tumor-cell colonies were scored in three categories: 1) those which contained 10–30 cells; 2) those which contained > 30 cells; and 3) those which contained 10 or more cells of which more than half showed evidence of degeneration by virtue of their darkened appearance.

RESULTS

Patient Characteristics

A summary of the important clinical and histologic data concerning the patient source of the solid tumor tissue employed in this study is given in Table I. The 10 patients ranged in age from 23–88 years and presented in either stage I, III, or IV of the disease. With one exception, all samples represented either tumor taken from the primary site or from omental nodules at the time of initial laparotomy or omental nodules taken at later exploratory operations. The one exception, sample FE, was taken from a biopsy of lesions metastatic to the skin. One of the tumors was undifferentiated, one was classified as mucinous, two as endometrioid, and six were classified as serous with either good, moderate, or poor degree of differentiation.

TABLE I. Solid Ovarian Tumor Samples Used in Semisolid Agar Cultures

Patient	Age	Stage of disease[a]	Histology	Degree of cellular differentiation	Therapy prior to present biopsy[b]
AR	23	III	serous	good	Act D, FU, C, Meth, HU, V, Chlor > 1 yr rad > 1 yr
DA	88	I	endometrioid	good	none
BA	77	I	endometrioid and serous	good	none
DL	57	III	serous	moderate	FU, C, < 1 mo
FE	62	IV	–	poor	A, CP, C, Meth > 1 mo
CA	72	IV	mucinous	moderate	none
VA	65	IV	serous	moderate	none
UI	66	IV	serous	moderate	CP < 3 wk
BK	58	IV	–	undifferentiated	none
CO	45	IV	serous	moderate	none

[a]Stage I, growth limited to the ovaries. Stage II, growth involving one or both ovaries with pelvic extension. Stage III, growth involving one or both ovaries with widespread intraperitoneal metastasis to the abdomen. Stage IV, growth involving one or both ovaries with distant metastases outside the peritoneal cavity.
[b]Act D, Actinomycin D; A, Adriamycin; CP, Cis Platinum, C, Cytotoxan; Chlor, Chlorambucil; FU, 5-Fluorouracil; HU, Hydroxyurea; Meth, Methotexate; V, Vincristine.

As these studies were run concurrently with those described in our previous report [15], opportunity was provided us to study simultaneously both solid and ascitic fluid samples from five patients, DL, CA, VA, UI, and BK.

Growth Patterns of Cultures

The patterns of growth of the solid tumor samples from individual patients are depicted in the upper half of Figure 2. Tumor cell colonies were observed in 8 of 10 samples, but sufficient numbers of tumor colonies to allow meaningful statistical evaluation of effects of interferon and other agents were obtained in only four samples.

For comparison, we have also included in the bottom half of Figure 2, the growth patterns of ascitic fluid samples obtained and cultured at the same time as solid tumor samples from the same patients. However, before comparing the results obtained from the solid and ascitic samples, important differences between these two forms of tumor samples must be considered (Table II). When the data from our total experience with ascitic fluid and solid tumor samples are considered (Table II), it is apparent that the mean value for the percent viable cells in the ascitic samples, 89%, is considerably higher than that observed in the solid-tumor samples, 39%. By contrast, the mean percent tumor cells in the ascitic fluid samples, 28%, is considerably lower than that observed for the solid tumor samples, 81%.

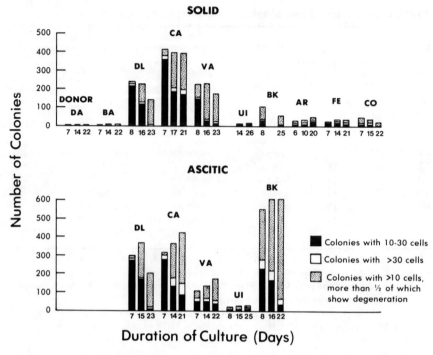

Fig. 2. Comparison of growth patterns of solid and ascitic forms of ovarian carcinoma. Five of the patients studied (DL, CA, VA, UI, and BK) had both solid tumor and ascites available for the serial growth measurements.

Thus, although an equal number (2×10^5) of viable cells from ascitic and solid-tumor samples were used to seed the semisolid agar cultures, the cell composition of the inocula from the solid tumor samples differed considerably from those of the ascitic fluid. In the solid-tumor samples in which viable cell number is low, large numbers of nonviable tumor cells are brought along in the inoculum, and in the ascitic samples in which the tumor cell number is low, many of the other cell types found in ascitic fluid, ie, lymphocytes, macrophages, polymorphonuclear leukocytes, and mesothelial cells, are included in the inoculum. Thus one would not expect the colony numbers of solid and ascitic fluid samples from the same donors to be comparable, and in fact that is what we observed (Fig. 2, bottom). Theoretically, the only definitive way to achieve equal number of colonies from ascitic and solid tumor samples would be if one could adjust the inoculum to contain comparable numbers of viable tumor cells and/or tumor stem cells. This problem has also been discussed by Buick in Chapter 11.

TABLE II. Comparison of Cell Suspensions Prepared From Solid Ovarian Carcinoma Samples and From Ascitic Fluid

Donor	% Viable cells		% Tumor cells	
	Solid	Ascitic	Solid	Ascitic
AR	20	–	94	–
DA	71	–	–	–
DA	90	–	–	–
DL	16	94	–	65
FE	32	–	99	–
CA	72	93	90	84
VA	21	97	35	38
UI	6	97	–	–
BK	19	97	89	27
CO	39	–	–	–
PU	–	–	–	4
BE	–	97	–	20
KU	–	93	–	15
BR-1	–	–	–	79
DO	–	98	–	39
SL-1	–	86	–	1
SL-2	–	91	–	1
BR-2	–	91	–	50
KL-1	–	23	–	0
KL-2	–	89	–	0
RI	–	94	–	0
Mean values	39	89	81	28

In our previous studies [15], we had established the criteria for growth in culture as an increase in the total colony number during the entire period of culture and/or an increase in the number of colonies containing > 30 cells. We now realize, from an examination of the data depicted in Figure 2 and data included in our previous report [15] that these criteria were too stringent, and that, in fact, the greatest increment in tumor-colony number most often occurs within the first week of culture. Our scoring system, in which we take into account those colonies with 10 or more cells, of which more than half show evidence of degeneration, also gave supportive evidence that, from the first week of culture on, increasing amounts of degeneration were seen. We now feel that any culture which has at least 50 colonies by the first week of culture has shown evidence of growth and can be used successfully for the evaluation of antitumor agents.

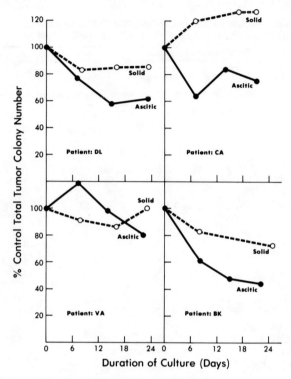

Fig. 3. Comparison of response of solid and ascitic forms of ovarian carcinoma to interferon. Values depicted are the mean values of at least 3 cultures at each time point, using a concentration of interferon of 300 units/ml incorporated into the agar for the duration of culture.

Comparison of the Response of Solid and Ascitic Forms to Interferon, Cis Platinum, and Adriamycin

The responses of tumor colonies derived both from solid and ascitic fluid samples to a final concentration of 300 units/ml of interferon incorporated into the agar for the duration of culture are depicted in Figure 3. Cultures from solid and ascitic fluid samples were grown simultaneously and similar experiments were performed with final concentrations of interferon of 100 and 30 units/ml, but the data are not shown. It is apparent that in 3 of 4 experiments with 300 units/ml cultures derived from ascitic fluid were more sensitive to the antiproliferative effects of interferon than were the cultures derived from solid-tumor samples.

To determine whether cultures derived from ascitic fluid samples were also more sensitive to other antitumor agents, similar experiments were performed using cis platinum and Adriamycin incorporated for the duration of culture, and

TABLE III. Comparison of Response to Cis Platinum and Adriamycin by Solid and Ascitic Forms of Ovarian Carcinoma

Agent	Final concentration (μg/ml)	Solid tumor			Ascitic form		
		Donor	Day of culture	Response[a]	Donor	Day of culture	Response
Cis Platinum	1.0	CA	7	+	CA	7	−
			17	+		14	−
		VA	8	−	VA	7	−
			16	+		14	−
			23	+		22	+
		BK	8	+	BK	8	−
			25	++		25	+
Adriamycin	0.5	CA	7	−	CA	7	−
			17	+		14	+
		VA	8	++	VA	7	+
			16	++		14	+
		BK	8	+	BK	8	−
			25	++		25	+

[a]+, ≥ 25% reduction in total tumor colony number as compared with the non-treated control examined on the same day; ++ ≥ 50% reduction in total tumor colony number; −, no response to agent. Values reflect the mean of at least 3 cultures for each time point studied.

the results are depicted in Table III. It is apparent that the reverse situation occurred with both cis platinum and Adriamycin. The cultures derived from solid-tumor samples were more sensitive to the antiproliferative effects of cis platinum in 3/3 experiments and to Adriamycin in 2/3 experiments, than were the cultures derived from ascitic fluid samples.

DISCUSSION

Early in the present studies it became apparent that the viability of the cell suspensions derived from solid-tumor samples was far less than was observed with those derived from ascitic fluid samples, even when the same patient was used on the same occasion to donate both samples. More rapid processing of the solid-tumor samples within minutes of obtaining them, the use of Ficoll Hypaque gradients to remove nonviable cells, and attempts at enzymatic disruption of the samples with trypsin-versene did not help in improving the viability of the cell suspensions prepared from solid tumor samples.

Also, although in our previous studies with ascitic fluid samples increase in total tumor colony growth throughout the entire culture period was seen in 67% of samples studied, this never occurred in our cultures derived from solid-tumor cells. In both systems maximum increase in total tumor-colony number occurred within the first week of culture, but beyond that, greater increases in total-colony number were observed in the cultures derived from ascitic fluid samples. Whether this phenomenon is related to our observation that fewer macrophages were present in the solid-tumor samples remains to be proven. Salmon and his colleagues have demonstrated that depletion of the macrophage population from the ascitic fluid of patients with ovarian carcinoma has been shown to greatly reduce the growth of tumor colonies [11], and they have suggested that such macrophages might play an important role in affecting tumor growth [18] and Chapters 11 and 23.

Precisely because the growth pattern of each individual tumor sample is unique, and because growth appears to be more sustained in samples obtained from ascitic fluid as compared with solid tumors, we feel that enumeration of colony number at only one time point, as is the case in several other laboratories, is inadequate. We advocate the use of serial evaluations of each culture that take into account the number, approximate size, and degree of degeneration of each colony. The recent development of an automated electronic image analysis apparatus, the Omnicon Feature Analysis System II (FASII), as described in Chapter 15, will make evaluation of cultures in this manner a routine procedure.

As was our experience with cell suspensions prepared from ascitic fluid samples [15], there is still a question in our minds as to whether single-cell preparations can be obtained uniformly at the initiation of cultures derived from solid-tumor samples. Perhaps the remedy for this would be to do colony counts early in the culture period, ie, at day 1 or 2, so that large clumps included at the time of seeding can be recognized and excluded from future counts.

Also, as we had observed with ascitic fluid samples, the resultant total tumor colony number was not proportional to the number of viable tumor cells unidentified in Pap smears of the cell suspension prepared from the solid tumor samples. This perhaps was a reflection of the unique clonigenic potential of individual tumors, and we recognize that colony number and size reflect the interplay of intrinsic cloning efficiency, growth rate, and effects of prior therapy on viability during culture. We also recognize that the histology of ovarian carcinoma stem cells is yet known, and that when tumor cells are identified on Pap smear, it is not possible to know if any or all are stem cells.

Another issue of concern in the design of experiments to evaluate the antitumor effects of interferon or other agents is the adequacy of exposure of the agent to the tumor cells. Knowledge concerning the uptake by individual tumor cell types of conventional antitumor agents is scanty, and for interferon, nonexistant. We really do not know that a one-hour in vitro exposure to a drug, as was the initial design of some of our own experiments and the experiments of other investigators, is sufficient or that it approximates the situation in vivo. Therefore, for the present studies and some previous ones [15], we chose to incorporate these agents into the agar for the duration of culture. This is especially relevant for interferon, for it, unlike the other antitumor agents, is usually administered either on a daily or thrice-weekly schedule for a month when used as an antitumor agent. The potential value of continuous contact of drugs in the agar for more general new drug screening efforts is also discussed in Chapter 22.

It is not yet known whether serum levels of interferon accurately reflect the amount of interferon that might reach a tumor either at its primary or metastatic sites. Similarly, the dose of interferon that is optimum for the response of a given tumor also is not yet known. In our present and previous studies [15] we employed interferon concentrations equivalent to the serum levels obtainable in patients with other diseases who are currently on therapeutic trials with interferon [19]. Yet it is conceivable that higher doses than those employed in our studies would result in much greater than 50% reduction in colony number. Certainly more than 50% reduction in tumor cell number would be desirable if interferon were to be considered as a therapeutic tool in the management of patients with ovarian carcinoma.

Our current data indicate that ovarian carcinoma colonies derived from tumor cells adapted to growth in the ascitic fluid of the patient are more sensitive to interferon in vitro than are tumor cell colonies derived from solid tumor samples. Our previous studies in which antiserum to interferon neutralized the antiproliferative effects of interferon help to prove that the effects observed were due to interferon and not a contaminant in the partially purified interferon preparation. However, our studies do not indicate whether interferon is acting as a tumoricidal or tumoristatic agent. The nature of the culture system, especially when interferon is incorporated into the agar and could not be completely removed by elution, precludes obtaining such information. However it would be possible to use this culture system to sort out whether interfreron has a direct effect on the tumor cells or acts via its effects on the lymphocytes or macrophages, which are associated with the tumor cells in ascitic fluid samples. There are numerous precedents in the literature demonstrating the multifaceted effects of interferon on various aspects of the immune response and its cellular components, the lymphocyte and the macrophage [20–23].

There are several possible explanations for our observation that colonies derived from ascitic fluid are more sensitive to the antiproliferative effects of interferon than are colonies derived from solid tumor cells. The first is the issue mentioned above that lymphocytes and macrophages are far more abundant in the cultures derived from ascitic fluid than from solid tumors and that the antitumor functions of these cells may be potentiated by interferon. A second explanation might be that the presence of exogenous interferon may prime for the production of additional interferon by the lymphocytes, as a part of their reaction to the tumor cells. The priming by interferon for additional interferon production is a well known phenomenon [24], and models for lymphocyte-tumor cell interactions which involve interferon production have also been reported [25]. Finally, a third explanation might be that cells adpated for growth in the ascitic fluid of the host might in fact be more susceptible in their own right than tumor cells found in large solid masses.

Confirmation by other laboratories of our observation of enhanced sensitivity of ascitic-fluid-derived tumor colonies would be very important for physicians involved with clinical trials of interferon. Such an observation would provide a rationale for using interferon intraperitoneally for the treatment of patients with widespread intraperitoneal metastases in addition to using it systemically.

Although the production of human leukocyte interferon is now receiving the support and attention of numerous major pharmaceutical companies, governmental agencies, and private philanthropic sources throughout the world, it still will be several years before adequate supplies of interferon will be available for all. Similarly, the use of recombinant DNA technology [26] coupled with knowledge of the partial amino acid sequence of several forms of interferon [27–29] will ultimately permit production of interferon by bacteria on a preparative scale, but not until several years in the future. Therefore, information gained from our own and others' studies concerning the sensitivity of individual patients' tumors to interferon, the tumor types that are responsive to interferon, and the definition of the mechanism of the antitumor effects of interferon for each tumor will help in the judicious use of currently existing small supplies of interferon and the wise use of future, more abundant supplies.

SUMMARY

We studied the in vitro growth characteristics of 10 solid-tumor samples of patients with ovarian carcinoma using a semisolid agar culture technique. Tumor cell colonies were observed in 8 of the 10 samples, but sufficient number of tumor colonies to evaluate the effects of interferon and other antitumor agents were obtained in only four samples.

As compared with cell suspensions prepared from ascitic fluid samples, solid-tumor samples had markedly lower viability, 39% vs 89%, and had more tumor cells, 81% vs 28%. Also, whereas the maximum increase in tumor-colony number occurred during the first week of growth in both solid- and ascitic-fluid-derived samples grown concurrently from the same donors, increase in tumor colony number was sustained for longer periods in ascitic-fluid-derived cultures. The ascitic-fluid-derived tumor colonies were more sensitive to the antiproliferative effects of interferon than colonies derived from solid-tumors. At a concentration of 300 units/ml incorporated into the agar for the duration of the culure, three of four ascitic fluid samples showed a reduction in tumor colony number by $\geqslant 25\%$, whereas none of the solid-tumor samples were affected by the interferon to that degree. In contrast, solid-tumor samples showed greater response to cis platinum and Adriamycin than did ascitic-fluid-derived cultures. Such studies and observations are critical in designing clinical trials for the use of interferon in the treatment of malignancy and the judicious selection of patients and route of administration most likely to provide optimal results, especially in view of present critical shortages in availability of interferon.

ACKNOWLEDGMENTS

We thank Drs. Howard Jones, Louis Bartolucci, Gilbert Webb, Hunter Cutting, Paul Fraser, Michael Friedman, George Winch, and Edward Hill for helping us to obtain the solid-tumor samples.

We thank Dr. Charles Epstein for patience and wisdom and Mary Evelyn Rose for typing the manuscript.

This work was supported by a grant from the Cancer Research Coordinating Committee of the University of California and by NIH grant CA 27903 (formerly AI 12481.

REFERENCES

1. Cantell K, Hirvonen S: Preparation of human leukocyte interferon for clinical use. Tex Rep Biol Med 35:138–144, 1977.
2. Galasso GJ, Dunnick JK: Interferon, an antiviral drug for use in man. Tex Rep Biol Med 35:478–485, 1977.
3. Greenberg HB, Pollard RB, Lutwick LI, Gregory PB, Robinson WS, Merigan TC: Effect of human leukocyte interferon on hepatitis B virus infection in patients with chronic active hepatitis. N Engl J Med 295:517–522, 1976.
4. Strander H: Anti-tumor effects of interferon and its possible use as an antineoplastic agent in man. Tex Rep Biol Med 35:429–435, 1977.
5. Strander H, Cantell K, Ingimarsson S, Jakobsson PA, Nilsonne U, Soderberg G: Interferon treatment of osteogenic sarcoma – a clinical trial. Fogarty Int Center Proc 28:377–381, 1977.
6. Merigan TC, Sikora K, Breeden JH, Levy R, Rosenberg SA: A tumor reducing effect of human leukocyte interferon in non Hodgkin lymphoma. N Engl J Med 299:1449–1453, 1978.
7. Mellstedt H, Ahre A, Bjorkholm M, Holm G, Johannson B, Strander H: Interferon therapy in mycelomatosis. Lancet 1:245–247, 1979.
8. Gresser I: Antitumor effects of interferon. In Becker F (ed): "Cancer – A Comprehensive Treatise." New York: Plenum, 1977, pp 521–571.
9. Hilfenhaus J, Damm H, Johannsen R: Sensitivities of various human lymphoblastoid cells to the antiviral and anticellular activity of human leukocyte interferon. Arch Virol 54:271–277, 1977.
10. Hamburger AW, Salmon SE: Primary bioassay of human tumor stem cells. Science 197:461–463, 1977.
11. Hamburger AW, Salmon SE, Kum MB, Trent JM, Soehnlen BJ, Alberts DS, Schmidt HG: Direct cloning of human ovarian carcinoma cells in agar. Cancer Res 38:3438–3444, 1978.
12. Hamburger AW, Salmon SE: Primary bioassay of human myeloma stem cells. J Clin Invest 60:846–854, 1977.

13. Salmon SE, Hamburger AW, Soehnlen B, Durie BGM, Alberts DS, Moon TE: Quantitation of differential sensitivity of human tumor stem cells to anticancer drugs. N Engl J Med 298:1321–1327, 1978.
14. Von Hoff DD: Clinical correlations of drug sensitivity in tumor stem cell assay. Proc Am Assoc Cancer Res Am Soc Clin Oncol 21:134 (Abstr 535), 1980.
15. Epstein LB, Shen JT, Abele JS, Reese CC: Sensitivity of human ovarian carcinoma cells to interferon and other antitumor agents as assessed by an in vitro semisolid agar technique. Ann NY Acad Sci, vol 350, September, 1980.
16. Boyum A: Isolation of mononuclear cells and granulocytes from human blood. Scand J Clin Lab Invest 21(Suppl 97):77–83, 1968.
17. Epstein LB, McManus NH: Macro and microassays for the antiviral effects of human and mouse interferons. In Rose NR, Friedman H (eds): "Manual of Clinical Immunology," 2nd ed. Washington, DC: Am Soc Microbiol, in press, 1980.
18. Salmon SE, Hamburger AW: Immunoproliferation and cancer: A common macrophage-derived substance, Lancet 1:1289–1290, 1978.
19. Jordan GW, Fried RP, Merigan TC: Administration of human leukocyte interferon in herpes zoster. I. Safety, circulating antiviral activity, and host responses to infection. J Infect Dis 130:56–62, 1974.
20. Epstein LB: Effects of interferon on the immune response in vitro and in vivo. In Stewart WE, II (ed): "Interferons and Their Actions." Cleveland: CRC Press, 1977, pp 91–132.
21. Herberman RB, Djeu JY, Ortaldo JR, Holden HT, West WH, Bonnard GD: Role of interferon in augmentation of natural and antibody-dependent cell-mediate cytotoxicity. Cancer Treat Rep 62:1893–1896, 1978.
22. Herberman RB, Ortaldo JR, Bonnard GD: Augmeniation by interferon of human natural and antibody-dependent cell-mediated cytotoxicity. Nature 277:221–223, 1979.
23. Lee SHS, Epstein LB: Reversible inhibition by interferon of the maturation of human peripheral blood monocytes to macrophages. Cell Immunol 50:177–190, 1980.
24. Stewart WE, II, Gosser LB, Lockhart RZ Jr: Priming: A nonantiviral function of interferon. J Virol 7:792–801, 1971.
25. Trinchieri B, Santoli D, Knowles BB: Tumor cell lines induce interferon in human lymphocytes. Nature 270:611–612, 1977.
26. Nagata S, Taira H, Hall A, Johnsrud L, Streuli M, Ecsodi J, Boll W, Cantell K, Weissman C: Synthesis in E coli of a polypeptide with human leukocyte interferon activity. Nature 284:316–320, 1980.
27. Knight E Jr, Hunkapiller MW, Korant BD, Hardy RWF, Hood LE: Human fibroblast interferon: Amino acid analysis and amino terminal amino acid sequence. Science 207:525–526, 1980.
28. Taira H, Broeze RJ, Jayaram BM, Lengyel P, Hunkapiller MW, Hood LE: Mouse interferons: amino terminal amino acid sequences of various species. Science 207:528–530, 1980.
29. Zoon KC, Smith ME, Bridgen PJ, Anfinsen CB, Hunkapiller MW, Hood LE: Amino terminal sequence of the major component of human lymphoblast interferon. Science 207:527–528, 1980.

22
Applications of the Human Tumor Stem Cell Assay to New Drug Evaluation and Screening

Sydney E. Salmon

INTRODUCTION

Development of new anticancer drugs has thus far been a very difficult and time-consuming procedure [1]. Although there were some early successes in cancer chemotherapy, including alkylating agents [2] and folate antagonists [3], this area was not extensively exploited at that time by the pharmaceutical industry, largely because of the combination of high cost and relatively low return from the development of anticancer drugs. This problem led to the creation a quarter of a century ago of the Drug Development Program in the Division of Cancer Treatment (DCT) of the National Cancer Institute [4]. The DCT program has, to a large extent, coordinated new drug development in the United States. The major steps required to assure the flow of drugs from experimental systems into the clinic are summarized in Figure 1, and often described as the "Linear Array" [5]. New drugs can enter this process from the pharmaceutical industry, universities and research institutes, and various governmental agencies, including the National Cancer Institute itself. The NCI uses a special committee structure (Decision Network) to reach scientific judgements with regard to considering advancement of a new drug towards general medical use [5]. The development of new drugs can begin from rational design or concept (eg, methotrexate, cytosine arabinoside, 5-fluorouracil), serendipitious discovery (eg, cis-platinum, vinblastine), or random screening (eg, BCNU, hexamethylmelamine) in mouse tumor systems.

Mouse tumor systems have played a central role in anticancer drug development in the United States, which has had an intensive search underway for new anticancer drugs as one of the objectives of The National Cancer Program. The major screening strategies used by the National Cancer Institute prior to and since 1975 are summarized in Figure 2. At present, approximately 15,000 synthetic com-

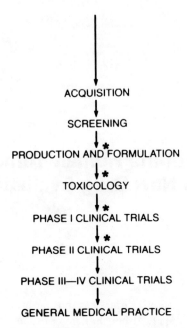

Fig. 1. Stages in new drug development (*DCT Decision Network points). Reproduced from [5] with permission of the publishers.

pounds and 400 purified natural products enter the standard screen of the Division of Cancer Treatment (DCT) of the National Cancer Institute each year [5]. In this screening system, unknown compounds are subjected first to a pre-screen in murine P388 leukemia. If they are inactive in this system, they are not pursued any further unless sufficient rationale or ancillary evidence of activity in another biologic or biochemical system exists to advance the testing of the compound to a tumor panel including 5 transplantable mouse tumors plus some human tumor xenografts in nude mice. Human tumor xenograft testing has involved three specific human tumor cell lines (lung, breast, and colon). Until recently, subcutaneous transplants had been utilized; however, a high infectious death rate, expense, and other logistic problems have limited the feasibility and amount of data obtained with that model, and the use of implants in the subrenal capsule of mice (which can be assessed more quickly) is now being tested by DCT as an alternative in vivo human tumor model in the mouse. All of the above in vivo models have certain practical and conceptual limitations for standard screening. For example, it is likely that the use of mouse leukemia (eg, L1210 or P388) as a pre-screen or screen influences the type of drugs selected as "actives." The overwhelming majority of drugs identified with the mouse leukemia screens manifest bone marrow sup-

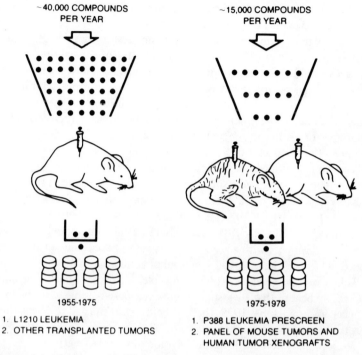

Fig. 2. Historical development of DCT mouse-tumor-screening systems. Reproduced from [5] with permission of the publishers.

pression as one of their major toxic manifestations. This conceivably may reflect the use of a bone marrow neoplasm as the screen rather than any a priori reason that myelosuppression would be a requisite toxicity for an anticancer drug. In fact, the use of the pre-screen may, by design, reject many drugs which are not myelosuppressive. Of equal concern is the limitation to one cell line from each of several tumor types. Conceptually, each cell line is the equivalent of a single subclone derived from a tumor from one individual patient whose cells may not be typical or might even already be resistant if the biopsy came from a heavily pretreated patient. Available data from clinical trials of active agents, studies of a series of murine myelomas [6], and studies with tumors of various histopathology in the human tumor stem cell assay provide strong evidence that there is marked heterogeneity of response of tumors of the same histopathology to currently available agents. Survival curves obtained with the stem cell assay suggest that even within a single biopsy sample, there often are drug sensitive and drug resistant TCFUs present. The tumor type used for screening selective agents is clearly important. For example, the highly active anti-estrogen tamoxifen, which is quite useful in breast cancer, would have been missed if it were screened only against mouse leukemia and would require testing in an estrogen-receptor-positive breast cancer or related hormone sensitive neoplasm. Testing of an estrogen-receptor-negative breast cancer (which occurs in about 40% of clinical specimens) [7] would also yield a negative result. We are fortunate to know something about potential mechanisms of resistance to hormones in breast cancer, but in most instances we are unaware of the nature of potential resistance mechanisms or the degree of clonal heterogeneity among tumors of the same type. Therefore, the importance of testing an adequate number of tumor specimens of a given histopathology (as discussed by Dr. Moon in Chapter 17) cannot be overemphasized.

Following that line of reasoning, we might give the example of a mouse tumor screen which incorporated a single transplantable colon cancer. It would only probably miss activity of a drug such as 5-fluorouracil which is active in 20% of patients. Of interest, a "20% response rate" has also been observed in vitro with the tumor stem cell assay [8]! Relatively few single drugs induce remission in over 50% of patients with a given tumor type, and even some of those might be missed unless they had a broad spectrum of action covering many tumor types.

While extensive screens involving multiple isolates are impractical and expensive in mice, we believe that they should prove far more feasible with the in vitro tumor stem cell assay. In June, 1979, the NCI held a workshop to review various alternatives to the current screening system. After considering the potential of this in vitro screening system, in 1980 the Division of Cancer Treatment of the National Cancer Institute initiated testing of the in vitro system to determine whether this assay is feasible for large-scale screening and attempt to determine whether it might prove superior to the current pre-screen and tumor panel in mice (Figure 3). If successful, the culture approach could lead to a radical revision of testing methodology in relation to new drug development by the National Cancer Institute, the pharmaceutical industry, and various cancer centers.

In this chapter, I will discuss several potential new applications of the in vitro clonogenic assay for human tumor cells for several steps in the Linear Array: 1) practical and design aspects to be considered for exploratory studies testing the human tumor stem cell assay preclinically, to see whether it might provide a more effective primary screen for identification of entirely new compounds; and 2) use of the tumor stem cell assay for "in vitro Phase II clinical trial" [9, 10] as secondary screening in which new agents already developed and approved for clinical trial are tested in the assay system to predict which human cancers are likely to be susceptible to these agents.

METHODS

Tumor biopsies were obtained, processed, and cultured as described elsewhere in this text. Additionally, cryopreserved tumor cells which had been frozen in 10% DSMO were also utilized in developmental studies to define usable techniques

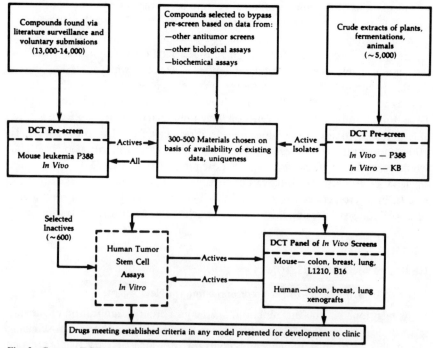

Fig. 3. Current DCT standard screen. In addition to the in vivo murine tumor screen, the in vitro human tumor stem cell assay is undergoing comparative evaluation.

for primary screening of new agents. Drug incubations with cryopreserved cells have used continuous contact with the drug in the agar rather than one-hour exposure prior to plating. This minimizes problems with clumping which occur quite commonly with repeated washings of frozen cells. Viability of cryopreserved cells is improved by first incubating them in medium for two hours at 37°C prior to plating [11]. In Chapter 16, Alberts et al have discussed preliminary results on the comparison of one-hour and continuous contact experiments with cycle-active agents. Drug concentrations of Phase I–II agents for new drug screening were based on in vivo pharmacokinetic data when this was available, and the dose range studied was designed to include the maximum pharmacologically achievable concentration-time product in vivo. Two representative Phase I–II agents studied included the cytotoxic drugs (methansulfonamide, N(4-(9-acridinylamino)-3-methoxyphenyl) AMSA, dihydroxyanthracenedione (DHA), phosphonacetyl 1-aspartate (PALA), as well as the biological response modifier, interferon. Studies with other Phase I–II agents have been reported elsewhere [9, 12]. Culture procedures were carried out as described in Chapter 18. Thus far, we have focused on those agents which are soluble in standard solvents for tissue culture (eg, water, saline, ethanol, dimethylsulfoxide). Appropriate solvent controls are always used in these experiments.

Calculations of drug-induced lethality were performed as described in Chapters 16–18. Data on new Phase I–II agents were entered in a Wang 2300 laboratory computer, which was used for data analysis and graphic output (Chapter 18). Criteria for in vitro sensitivity for new Phase I–II agents for which in vivo correlations were not available were operationally defined as sensitive when at least a 70% reduction in survival of tumor colony-forming cells was observed at a relatively low-dose exposure (eg, 10%–20% of the achievable concentration-time product in vivo). Dose-finding studies for new agents on which pharmacokinetic data were not available were carried out over a 3-log concentration range from 1.0 to 100 μg/ml. The 100 μg upper limit seems reasonable inasmuch as on a pharmacokinetic basis, it would generally correspond to a single dose ranging up to 7 gm/M^2 in vivo as a single IV dose. Very few drugs can, in fact, be given in vivo in dosages of that magnitude. Interferon concentrations and schedules were selected with a combination of dose-finding studies and information on achievable serum concentrations in vivo.

RESULTS AND DISCUSSION

I. Use of In Vitro Assay for Primary Screening for New Agents

As mentioned in the Introduction, one major potential application of the in-vitro clonogenic assay procedure for human tumor cells is in the area of screening and evaluation of entirely new compounds for anticancer activity. There are a number of factors which have led us to consider that this assay might serve as a favorable alternative to the standard mouse screening systems. Many of these are summarized in Table I. It is quite conceivable that the assay system could detect entirely new classes of compounds that have been missed by the conventional murine screening system. Such compounds might have fewer or different toxic side effects than most of the standard active agents. The proposal that the in vitro assay might detect useful compounds appears to be reasonable because the TCFUs which clone in the assay appear to reflect many features of human tumor biology, including the heterogeneity of response within individual specimens as well as patient-to-patient differences with standard agents. Furthermore, there is now good evidence that the assay is predictive of clinical response in vivo. Nonetheless, it must be recognized that it remains to be proven that the in vitro assay will be useful for successfully identifying entirely unknown agents (not first selected by in vivo mouse tumor systems) which prove to be clinically useful. In Chapter 17, Dr. Moon has addressed statistical issues relating to the number of tumors of each histopathologic type that would be needed to make successful identification of useful agents a predictable event. Matters of assay design and logistics are of equal importance, and it would not be surprising if the ultimate in vitro system which evolves for primary screening differs substantially in design from the standard system described in Chapter 18 for standard drug testing. However, it may be of developmental value to review some of our current concepts on approaches to primary screening in relation to the tumor stem cell assay.

TABLE I. Potential Advantages of Human Tumor Colony Assay for New Drug Screening

1) Simple, relatively rapid assay with defined reagents, quantitative results in useful biological terms.
2) Assay directly applicable to fresh or frozen biopsies, containing clonogenic human tumor cells.
3) Dose-response curves for established anticancer drugs predictive of clinical response.
4) Standard panels of cryopreserved cells can be established and banked for assay use and quality control.
5) Assay has sensitivity to detect activity in submicrogram amounts in fermentation broths or with scarce compounds.
6) Automated counting permits standardization of assay results and reporting from multiple laboratories.

Logistic considerations. Logistical problems unique to the tumor stem cell assay warrant discussion as do techniques which might prove useful to increase the likelihood of detecting active compounds. At the current time, the standard cell source upon which the tumor stem cell assay is based is composed of fresh human tumor tissue. This presents a logistic problem not encountered with current murine tumor systems, which are maintained by means of the passage of cloned cell lines in mice by transplantation or in spinner culture.

It is evident that there will be an ongoing need for fresh tumor tissue, preferably from patients who have not previously received chemotherapy or radiotherapy. Inasmuch as there are currently 700,000 newly diagnosed cancer patients in the United States alone, there theoretically should not be a shortage of tumor tissue. However, it is likely that significant coordination of tumor procurement for screening will involve hospital pathologists who would be needed to keep sample referral adequate to the screening laboratories. Similar logistic problems have been successfully met in areas such as pituitary collection for extraction of human growth hormone, and corneal and kidney banking for transplantation which still requires ongoing cooperation of surgeons and hospital pathologists. With such logistical support, even if only 10% of specimens were available and proved useful, a city of 500,000 could house a regional screening laboratory for major tumor types, as it could have 600 tumors available annually for screening purposes. Location of screening laboratories in cities with large cancer referral hospitals, as well as facilitated shipping, would further increase tissue availability for screening purposes. We have already established that heparinized malignant effusions and bone marrows can be shipped rapidly by air and delivered to our laboratories in viable form from numerous cities in the United States. Such samples grow well despite up to a 24-hour delay (at room temperature or $2-4°C$) between sample collection and plating. When heparinized vacuum bottles are used for collection of effusions, we believe it important to vent the residual vacuum prior to shipment to prevent anoxia. Unfortunately, many solid tumor specimens do not maintain viability adequately for much more than 2–3 hours after surgery. It is probable that this time limit for solid tumors could be extended if simple techniques for rapid tissue processing could be developed so that they could be carried out by the surgical pathologist. Such initial processing could then be used in conjunction with "transport media" (as are used in bacteriology) or with cooling or cryopreservation prior to shipment.

Cryopreservation. Development and application of an effective cryopreservation protocol for maintaining viability of clonogenic tumor cells could enormously simplify new drug screening efforts. Implementation of cryopreservation would not only facilitate specimen shipment with less stringent time limits, but it could facilitate development of large-scale drug screening laboratories and the coordination of a national screening program. Some clear advantages to be accrued from cryopreservation include the following: 1) Tumors of the required histopathology could be "banked" until they are needed. Thus, for example, 5–10 breast cancers could routinely be set up simultaneously for a given new agent or agents. Additionally, rarer tumor types could also be banked and used selectively in testing agents for which there might be theoretical rationale for efficacy in a rare tumor type (eg, an adrenal steroid synthesis antagonist for adrenal carcinoma). 2) The drug sensitivity phenotype of any given specimen for standard drugs could be established in advance of testing new agents. Potentially, clonogenic tumor cells might be classified as inherently resistant or sensitive to specific classes of agents (eg, anthracyclines, vinca alkaloids, etc). Such information might prove also valuable in testing analogs designed to overcome resistance mechanisms. 3) Quality control could be improved by permitting repeated testing on the same specimens when a positive result is obtained. Such repetition could thus be used to validate that a detected "sensitivity" is inherent for any given tumor and not due to day-to-day fluctuation in laboratory conditions. A new compound which is in crude form could, of course, also be tested repeatedly at various steps in purification against the same cell frozen cell sources. Additionally, cryopreserved reference tumor cells could be shipped to various screening laboratories to test their comparability for given standard compounds. 4) Laboratory schedules for initiation of assays could be standardized, rather than being dependent on the vicissitudes of time of arrival of the specimens.

While all of the above theoretical reasons for preferring cryopreservation over fresh specimens appear to be clear, it is also apparent that cryopreservation cannot be substituted for fresh sample testing until it is established that identical or similar drug sensitivity profiles are obtained with cryopreserved clonogenic cells as are obtained when the same specimens are tested in identical fashion prior to freezing. It is also difficult to exclude, a priori, that cryopreservation procedures themselves might influence apparent sensitivity or resistance to a given agent. Cell recovery from cryopreservation must also be adequate to permit sufficient growth of control colonies for quantitative assays to be achievable. While the percentage recovery of TCFUs may vary from specimen to specimen, the fractional survival profiles for any given specimen tested repeatedly should be similar for any given drug nonetheless. Such considerations led Buick (Chapter 2) to comment that fractional survival curves are more important than the absolute number of TCFUs counted in any given series of experiments.

In order to evaluate such questions, and to test the logistics involved in using cryopreserved materials, my laboratory has carried out a preliminary series of experiments with cell freezing, recovery, and drug testing. The cryopreservation protocol which we currently use is detailed in Appendix 1. Table II provides examples of viable cell recovery from various tumors tested, while Figure 4 provides comparisons of several reference cytotoxic drugs on fresh and frozen cells from the same tumor specimens. Thus far, our cryopreservation studies have been limited to continuous drug exposure (rather than 1 hour), because cryopreserved cells often clump irreversibly with repeated centrifugation and washing. However, the thawed cells are first exposed to the drug in suspension culture at 37°C for 1 hour before the molten agar is added and the cells plated. Fresh and frozen controls are, of course, prepared in the same fashion.

As can be assessed from Figure 4, drug sensitivity profiles for melphalan, vincristine, and methotrexate on fresh and frozen cells do compare favorably. However, it is to be emphasized, such studies are only preliminary and it is by no means a certainty that the current cryopreservation protocol will prove satisfactory for new drug screening. Nonetheless, effective cryopreservation and recovery should be achievable and in my opinion has a good chance of being a mainstay for drug screening in the future.

Optimizing sensitivity of screening assays. In all screening efforts, it is important to keep testing as simple as possible while still maintaining specificity and sensitivity for detecting active new compounds. Table III summarizes some potential strategies to optimize detection of active compounds. The number of tumors which are re-

TABLE II. Recovery of TCFUs From Cryopreservation

Patient	Tumor type	No. clusters–colonies/5×10^5 cells		Percent[a] recovery of TCFUs
		Fresh	Cryopreserved	
1	Ovarian	30	70	233
2	Ovarian	34	36	106
3	Ovarian	37	57	154
4	Ovarian	37	70	190
5	Melanoma	47	28	60
6	Melanoma	46	45	97
7	Lung	105	62	59
8	Lung	105	94	90

[a]Percent recovery = $\dfrac{\text{No. colonies from frozen TCFUs}}{\text{No. colonies from fresh TCFUs}} \times 100$.

Note: While colonies were generally observed with fresh cells, growth often halted at the cluster stage with cryopreserved cells.

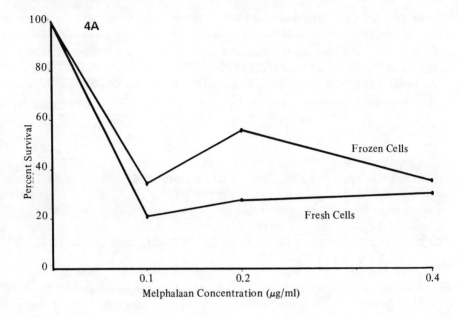

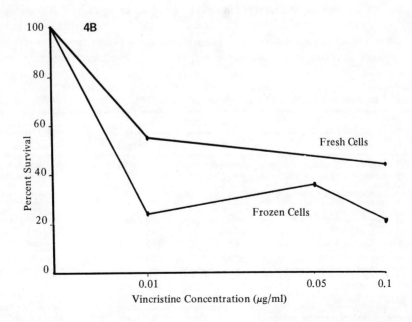

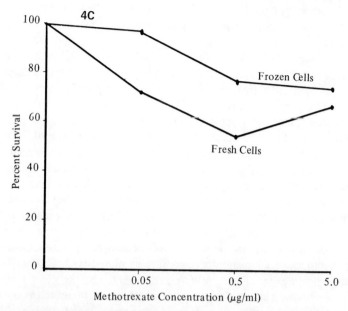

Fig. 4. In vitro response of fresh and cryopreserved TCFUs from the same patients to 3 standard anticancer drugs A) melphalan, B) vincristine, and C) methotrexate.

TABLE III. Strategies to Optimize Detection of Active Compounds With the Human Tumor Stem Cell Assay

1) Use multiple human tumor stem cell samples from untreated patients including the "major killers" (eg, lung, breast, colon, melanoma, ovary).
2) Test drug or broth by continuous contact in culture plate:
   a) Maximizes concentration-time product achievable.
   b) Cell-cycle-dependent cytotoxicity will not be missed.
   c) Permits detection of drug activity against G_0 cells which proliferate in culture.
3) Test drug over 3-log dose range (1.0–100 μg/ml).
   a) Discard if inactive.
   b) Test lower doses if active in this range.
   c) Retest purified or modified compounds during drug development.
4) Incorporate "bioactivation" testing of new compounds.

quired of a given type to maintain adequate specificity as well as probability of detecting an active compound has already been addressed by Dr. Moon in Chapter 17. However, this assumes that the in vitro dose range and exposure time have been optimized. Conceptually, the dose-finding problem appears to be fairly straightforward. The in vitro tumor stem cell assay already shows substantial evidence that antitumor activity for active compounds is clearly identified at very low dosage (eg, 0.01–0.1 μg·hours of exposure). This degree of sensitivity makes it likely that trace quantities of very active compound (eg, an antibiotic present in a crude fermentation brew) could be detected with the assay system even if the dosage were unknown but in the nanogram range. On the other hand, how high a concentration exposure is needed to reject an inactive agent, and can the dosage range be standardized? In my opinion, a potentially optimal solution to this latter problem would be to bracket a 3-log dose range of 1.0–100 μg/ml, using continuous exposure of the new agent. The experimental protocol could also include a 1-hour exposure of the new agent to tumor cells (in a panel) at 37°C in suspension culture followed by plating the cells in agar in the presence of the drug being tested. Continuous contact should assure adequate drug exposure by more than bracketing the entire pharmacologic range for all currently available agents and likely also allow detection of drugs having a cell-cycle-specific mode of action. Inasmuch as the clonogenic cells must traverse the cell cycle to form clusters and colonies, cycle-active drugs (even with a narrow specificity for G_1, S phase, G_2, or mitosis) should be successfully identified. This, of course, assumes that the action of the new agent is not inhibited by the presence of agar and that it is stable in vitro.

If continuous contact is used, is a 3-log dose range required? That remains an uncertainty, but would approach the ideal. Use of a single concentration (eg, 10 μg/ml) might potentially miss a few of the agents which are clinically administered in dosages of >5 gm/M^2; however, these are unusually high doses for the currently useful anticancer drugs. The dose range, therefore, represents an area wherein some compromise (on the number of doses tested) could be considered. Drugs found to be active at a relatively high dose against any given tumor type would ideally be retested against a broadened panel of tumors of the same, and other, histologies with lower doses (and with 1 hour exposure) on the positives to establish the dose-response range.

Bioactivation. Several very active anticancer drugs (eg, cyclophosphamide) [13] require bioactivation in vivo by the liver microsomal enzymes to express cytotoxicity. In vitro screening for new anticancer drugs should include the capability for detecting agents which require metabolic activation. An analogous problem has been dealt with relatively effectively in Ames testing for putative carcinogens. The technique uses liver microsomal cocktail (S9 mix) as devised to active putative procarcinogens (or mutagens) in vitro [14]. Several groups have applied similar metabolic activation techniques for in vitro screening of antineoplastic agents [15–17]. At the Second Workshop on Human Tumor Cloning Methods, Lieber et al [18] presented evidence that the same type of "S9 mix" could be used to activate compounds with anticancer activity. Using this system, they not only activated and showed in vitro activity of standard drugs such as cyclophosphamide in the tumor stem cell assay, but also showed that a pyrolizidine alkaloid (heliotrine) could be activated to express cytotoxicity in vitro. We recently tested the microsomal system and confirmed that it could be used successfully in conjunction with the clonogenic assay. Such testing would ideally be included in the ultimate in vitro screening program. However, Meltz and Winters [16] argue that bioactivation adds additional cost and operations which they believe make it unwarranted for large-scale screening. They were using only a single cell line (KB) and apparently assumed that negatives would be tested in the mouse system in any case. In my opinion, this equation would likely shift should in vitro systems prove to be the mainstay in future screening programs. Bioactivation could be applied routinely and in detail with drugs rationally designed to be bioactivated, but also likely should be used with selected natural products thought otherwise to be inactive in vitro.

In vitro activity levels required for positive compound identification. While sensitivity guidelines are somewhat difficult to select for totally unknown agents, it appears reasonable to propose that putative active new agents induce a similar magnitude of reduction in survival of TCFUs as is observed with known active agents. Given that proviso, such agents should cause at least a 70% and preferably at least a 90% reduction in survival of clonogenic human tumor cells in multiple samples from a given histology or several histologic types of cancer. The exact value for percent reduction in survival and relative concentration required to designate a compound as active will likely be adjusted with increasing knowledge and tempered with practical considerations relating to the number of drugs that can be effectively processed through the toxicology testing phase.

What step follows in vitro screening? From the perspective of the author, the in vitro system as described would not serve as a "pretest" for murine systems, as inactivity in the mouse would not mean that a drug would be inactive in man. Thus, the best test for new agents classified as active in the human tumor stem cell assay would be to test them for anticancer activity in cancer patients. It would, therefore, appear reasonable to bring those agents considered by the Decision Network to be "promising positive compounds" to the next step in the Linear Array as delineated in Figure 1, and advance the compound to toxicology testing. Those agents which pass toxicology would next be entered into Phase I–II clinical trial. As discussed below, clinical trials would ideally be done in conjunction with selection of potentially sensitive patients with the tumor stem cell assay. The reader should recognize that a primary purpose of current Phase II clinical trials is to estimate the response rate for a given agent. Treating only in vitro sensitive patients would not permit an accurate estimate of the in vivo response rate in the general cancer patient population. It would, however, permit an estimate of the predictive value for sensitivity for the in vitro assay. A purpose of classical Phase II trials may thus be modified. This approach, while markedly different from the current one, would provide the clearest test of this system and should permit a clear decision to be made concerning the applicability of the tumor stem cell assay to effective new drug screening. "The proof of the pudding will be in the testing" and, in my opinion, such a developmental screening program will prove to be a challenging (and hopefully successful) effort.

II. Secondary Screening of New Agents

Phase I—II drugs. In concept, a Phase II in vitro trial could potentially obviate the need for a significant component of large-scale clinical trials wherein new drugs are given to patients to find out whether they have antitumor effect. A major component of expense and toxicity of administration of new agents in clinical trials unfortunately involves patients who achieve no benefit from the agent administered. If patients received the agent only if their TCFUs were sensitive, then the agent would not have to be given to patients whose TCFUs were resistant. A number of new Phase I—II agents have been investigated in a preliminary fashion with the in vitro clonogenic assay. Suitable examples of our initial studies of several current Phase I—II drugs have been selected to illustrate applications of the tumor stem cell assay for in vitro Phase II clinical trial. This concept has also been detailed elsewhere [9, 10]. Strategically, it would be of value to test each patient's tumor cells with as many Phase II drugs as possible (eg, 10—12) as this would maximize the chance of identifying a useful drug for the patient's treatment.

AMSA. The new Phase II agent acridinylamino-methansulfon-m-anside (AMSA) [19] has shown activity against epithelial tumors such as breast carcinoma and some other neoplasms. While animal studies suggested that this agent in cross-resistant with adriamycin, some clinical observations in acute leukemia suggest that the drug may prove useful in patients who relapse from adriamycin. Thus far, in our in vitro Phase II study of AMSA, 17 patients with ovarian carcinoma (Fig. 5A), 16 patients with melanoma (Fig. 5B), and 15 patients with other neoplasms (breast, cervix, colon, lymphoma, myeloma, and sarcoma) that grew in the culture system were tested with AMSA at doses ranging from 0.05 to 10 μg/ml for 1 hour. We consider that 1 μg · hr of exposure should be pharmacologically achievable in vivo. Additionally, four patients with melanoma were studied in vitro with continuous exposure to AMSA in the agar. Substantial heterogeneity in response to this agent was observed, and only two patients with ovarian carcinoma had reduction in cell survival to 30% of control. However, five patients with melanoma had less than 30% survival of TCFUs with the 1 μg · hr exposure to AMSA (one of whom had less than 10% survival) (Fig. 5B). That patient was treated with this agent clinically and achieved a mixed response clinically with dramatic reduction in mas-

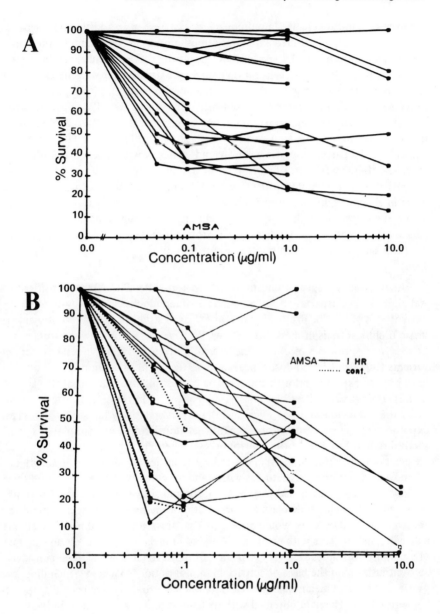

Fig. 5. Survival of TCFU after in vitro exposure to AMSA. •——• 1 hour contact, ... continuous exposure. A) Ovarian carcinoma, B) melanoma. Reproduced from [9] with permission of the publishers.

sive liver size, but no change in pulmonary metastases. In an additional four patients with melanoma studied with continuous contact to AMSA, no marked increase in lethality to TCFUs was evident, suggesting that resistance may be more on a cellular uptake or biochemical than on a cytokinetic basis. Inasmuch as AMSA binds to melanin granules, even a one-hour exposure may provide "continuous contact" in melanoma. Of the other tumor types tested, one patient with diffuse histiocytic lymphoma had reduction in survival of TCFUs to 20% of control at the 0.1 µg/ml dose (but without increasing lethality at the higher doses) and achieved a partial response when treated clinically with this agent. Both of these patients were part of an in vitro/in vivo Phase II clinical trial with AMSA recently reported from this institution by Ahmann et al [20]. As with many of our other in vitro drug studies, AMSA was evaluated mainly on cells from patients who had received prior chemotherapy, including alkylating agents and anthracyclines. Additional in vitro studies on cells from patients who have not had previous drug exposure would be important to evaluate fully the potential role of this new agent.

PALA. The new agent n-phosphonacetyl-1-aspartate (PALA) is a transition state inhibitor of aspartate transcarbamylase which blocks de novo pyrimidine biosynthesis [21]. The drug has undergone Phase I study [22] and recently entered Phase II clinical trials in the United States. We have carried out in vitro studies on TCFUs from various (8 ovarian, 6 melanoma, 2 myeloma, and 1 breast) cancer patients (Fig. 6). A broad dose range was investigated (1–100 µg/ml) with one-hour in vitro exposure prior to culture. The highest in vitro concentration (100 µg/ml) approximates the achievable plasma concentration attainable after a maximally tolerated clinical dose of 7.5 gm/M^2 [22]. At relatively high in vitro dosage exposures, the TCFUs from 1 myeloma, 1 melanoma, and 2 ovarian patients had survival reduced to 30% of the control or less. Whether such high concentrations are achievable intratumorally remains to be established. As discussed in Chapters 16 and 18, in our predictive studies with standard cytotoxic drugs, we found that for in vivo activity, drugs had to exhibit substantial activity in vitro at a fraction of the clinically achievable plasma concentration-time product. If such criteria also are relevant for PALA, we would anticipate that this agent should be relatively inactive in most patients with ovarian cancer and in melanoma. PALA is an example of a non-myelosuppressive agent that was advanced to clinical trial with considerable ehthusiasm on the basis of activity in mouse tumor models [5], including ovarian cancer and melanoma. Clinical data available to date have been disappointing with PALA. The data obtained with melanoma and ovarian cancer in the human tumor stem cell assay are, thus, more consistent with the in vivo data with humans than with the animal tumors. The explanation for this species difference is obscure.

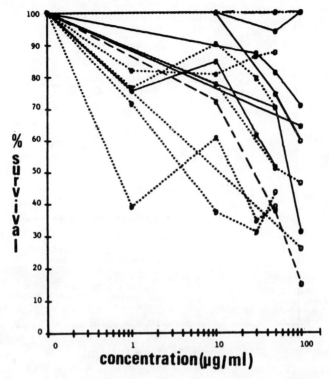

Fig. 6. Effect of a one-hour exposure to PALA on survival of TCFUs from 15 patients with myeloma (○— —○), melanoma (●——●), and ovarian carcinoma (○·····○). Reduction in survival to 30% of the control was observed only at the 100 μg/ml dose level.

Dihydroxyanthracenedione. Certain anthraquinones have been studied as potential model compounds analogous to anthracyclines such as adriamycin and daunomycin [23]. These new compounds have been designed with the thought that cardiac toxicity might be averted and therapeutic efficacy enhanced. Recently, a bis (hydroxy-ethylamino-ethylamino) anthraquinone was found to possess antitumor activity in several mouse tumors [24]. The new agent, dihydroxyanthracenedione, has recently entered Phase I clinical trial in San Antonio [25] and Tucson. In conjunction with our Phase I trial, we have carried out preliminary observations of the in vitro lethality of dihydroxyanthracenedione in the tumor stem cell assay on cells from various neoplasms. Results of these in vitro survival curves are summarized in Figure 7. Survival of TCFUs after a one-hour exposure was reduced to 35% of the control as observed in one breast and one endometrial

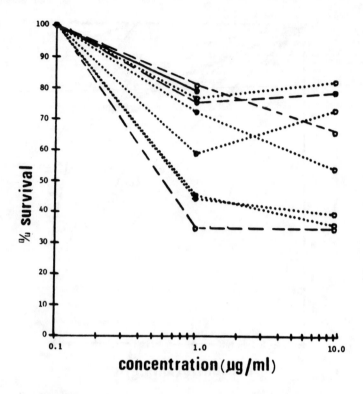

Fig. 7. Effect of a one-hour exposure to dihydroxyanthracenedione on survival of TCFUs from 10 patients with breast (o------o), lung (o———o), or gynecologic cancers (o······o).

cancer tested. Other tumor types and a larger panel of tumor types are needed, as are kinetic data and results of Phase I studies, in order to determine the achievable concentration-time products in vivo. However, in this pilot in vitro study, a log dose response covering doses of dihydroxyanthracenedione from 0.1 to 10 μg hours of exposure should more than bracket the achievable dosage exposure in vivo based on available preliminary clinical pharmacology data.

Interferon. Interferons are glycoproteins capable of inducing an antiviral state in cells undergoing RNA and protein synthesis. Human leukocyte interferon is a biological response-modifying agent with known antiproliferative activity [26]. It is currently very expensive to produce, and clinical dose schedules are limited to $3 \times 10^6 - 1 \times 10^7$ units per day with serum levels generally reaching a peak of 100–300 units/ml [27]. Early observations have suggested activity in myeloma [28], breast cancer, and lymphoma [29]. The American Cancer Society, the National Cancer Institute, and pharmaceutical manufacturers have recently started major programs to expand interferon production and test it clinically. In vitro

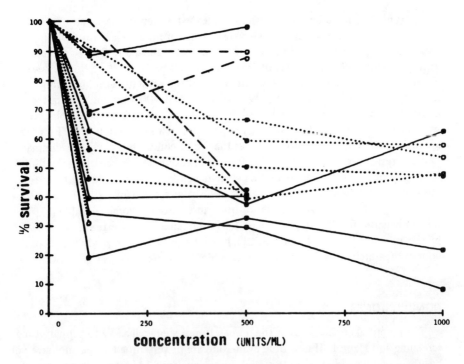

Fig. 8. Effect of continuous contact with interferon on survival of TCFUs from 14 patients with breast (o - - - - - - o), gynecologic (o · · · · · · o), and miscellaneous (melanoma, thyroid, lung) cancers (o ——— o). Reduction in TCFU survival to 20% of the control at the 100-unit dose was observed from one adenocarcinoma of the lung, while the 1,000-unit dose reduced TCFU survival from 1 prostate cancer patient to 8% of control.

study would appear to be ideal for this agent because of its current expense and uncertainty concerning required serum concentrations or dosage exposures. In Chapter 21, Dr. Epstein and her associates have provided an updating of their study on the effects of relatively low doses of human leukocyte interferon on ovarian TCFUs in the tumor stem cell assay. Although their criteria of sensitivity vary slightly from ours, the observations provide clear evidence for biological activity of interferons in this assay system. We, therefore, followed Dr. Epstein's lead. In our initial in vitro dose-finding studies with human leukocyte interferon, we observed that with one-hour exposure, dosages of up to 50,000 units per ml often had minimal effect. More significant effects were seen at lower dosage with continuous exposure in the agar, suggesting either slow uptake or a cell-cycle-dependent mode of action. These observations thus confirmed those of Dr. Epstein. We recently began a broad Phase II in vitro trial with human leukocyte interferon at dosages of 100, 500, and 1,000 units per ml (by continuous contact in the agar) to simulate daily injections. Figure 8 depicts the initial dose-response curves we have observed and indicates evidence of in vitro activity of interferon in various tumor types. Such studies could give useful leads for identifying specific tumor types and patients for selective in vivo treatment. Of note, TCFU survival of cells from one lung cancer patient (adenocarcinoma) was reduced to 20% of the control at the currently achievable level of 100 units/ml. Of perhaps greater interest was the response observed in one adenocarcinoma of the prostate with increasing lethality at increasing dose with only 8% survival at 1,000 units/ml. This latter dosage level has not yet been achieved in vivo. The plateaus in response observed with many specimens tested suggests the presence of a subfraction of TCFUs that are inherently resistant to interferon, as the use of continuous exposure eliminates "kinetic resistance" from accounting for this phenomenon. Based on the preliminary one-hour experiments suggesting cycle-dependent activity and these continuous contact experiments, further studies of resistance mechanisms to interferon will be of great interest. As supplies of other interferons become available (eg, lymphoblastoid B-cell [30], fibroblast [31], and leukocyte interferon synthesized by recombinant DNA techniques [32]), it will be of interest to study these compounds separately as well as in various combinations.

CONCLUSIONS

The potential applications of the human tumor stem cell assay to primary drug screening and Phase I–II new drug evaluation are of very broad scope and have not been completely addressed in this exploratory analysis. As in most other areas of science, careful and stepwise testing in many laboratories will be required to confirm or refute the utility of these approaches. Of the two areas discussed, the applications to new drug screening are still most uncertain and subject to change. From my perspective, in vitro growth conditions, for clonogenic human tumor cells, are perhaps at a similar stage of development as culture techniques for bacterial cells were in the 1930s. Availability of those techniques, and a combination of good ideas, serendipidy, and hard work led to the subsequent development of penicillin and other antibiotics which now can be used to cure most forms of infectious disease. The value of a simple and reproducible assay for screening therefore cannot be overemphasized. Applications of the tumor stem cell assay to Phase I–II drug evaluation already show preliminary signs of utility. Given this background and the more limited preclinical data on other agents now under study, broadened testing of this approach to new drug screening appears warranted. If use of an in vitro testing approach such as this proves successful for both primary and secondary screening, it would not be surprising if significant qualitative and quantitative changes occurred in the areas of new drug development and cancer clinical trials.

REFERENCES

1. Zubrod CG: Historic milestones in curative chemotherapy. Semin Oncol 6:490–505, 1979.
2. Goodman LS, Wintrobe MM, Dameshek W et al: Nitrogen mustard therapy. Use of methyl-bis (beta-chloroethyl) amine hydrochloride and tris (betachloroethyl) amine hydrochloride for Hodgkin's disease, lymphosarcoma, leukemia and certain allied and miscellaneous disorders. JAMA 132:126–132, 1946.
3. Farber S, Diamond LK, Mercer RD et al: Temporary remissions in acute leukemia in children produced by folic acid antagonis, 4-aminoptenyl-glutamic acid (aminopterin). N Engl J Med 238:787–793, 1948.
4. Zubrod CG, Schepartz S, Leiter J et al: The chemotherapy program of the National Cancer Institute: History, analysis and plans. Cancer Chemother Rep 50:349–540, 1966.
5. DeVita VT, Oliverio VT, Muggia FM et al: The drug development and clinical trials programs of the Division of Cancer Treatment, National Cancer Institute. Cancer Clin Trials 2:195–216, 1979.
6. Ogawa M, Bergsagel DE, McCulloch EA: Chemotherapy of mouse myeloma: Quantitative cell culture predictive of response in vivo. Blood 41:7–15, 1973.
7. McGuire WL: Hormone receptors: Their role in predicting prognosis and response to endocrine therapy. Semin Oncol 5:428–433, 1978.
8. Salmon SE, Von Hoff DD: In vitro evaluation of anticancer drugs with the human tumor stem cell assay. Semin Oncol 7, in press, 1980.
9. Salmon SE, Meyskens FL, Alberts DS, Soehnlen BJ, Young L: New drugs in ovarian cancer and malignant melanoma: In vitro Phase II screening with the humor tumor stem cell assay. Cancer Treat Rep, in press, 1980.
10. Salmon SE, Alberts DS, Meyskens FL et al: A new concept: In vitro Phase II trial with the human tumor stem cell assay (HTSCA). Proc Am Assoc Cancer Res and Am Soc Clin Oncol 21:329, 1980.
11. Law P, Lepock JR, Kruuv J: Effect of protective agents on amount and repair of sublethal freeze-thaw damage in mammalian cells. Cryobiology 16:430–435, 1979.
12. Meyskens FL, Salmon SE: Inhibition of human melanoma colony formation by retinoids. Cancer Res 39:4055–4057, 1979.
13. Cohen J, Jao JY: Enzymatic basis of cyclophosphamide activation by hepatic microsomes of the rat. J Pharmacol Exp Ther 174:206–210, 1970.
14. Ames BN, McCann J, Yamasaki E: Methods for detecting carcinogens and mutagens with the Salmonella/mammalian-microsome mutagenicty tests. Mutat Res 31:347–364, 1975.
15. Dolfini E, Martini A, Dorelli MG et al: Method for tissue culture evaluation of the cytotoxic activity of drugs active through the formation of metabolites. Eur J Cancer 9:375–378, 1973.
16. Meltz M, Winters D: Utilization of metabolic activation for in vitro screening of potential antineoplastic agents. Cancer Chemo Rep 63:1795–1801, 1979.
17. Arnold H, Bourseaux F, Brock N: Neuartige Krebs-Chemotherapeutica aus der Gruppe der Zyklischen N-Losi-Phosphamidester. Naturwissenschaften 45:64–66, 1958.
18. Lieber MM, Ames MM, Powis GP, Kovach: Use of a liver microsome activating system in determining the sensitivity of human tumor cells to chemotherapeutic agents in vitro (abstr). Proceedings at 2nd Workshop on Human Tumor Cloning Methods, University of Arizona Cancer Center, Tucson. Also see: Kovach JS, Ames MM, Powis G, Lieber MM (1980): Use of a liver microsome system in testing drug sensitivity of tumor cells in soft

agar. Proc Am Assoc Cancer Res and Am Soc Clin Oncol 21:257, 1980.
19. Von Hoff DD, Howser P, Gormley P, Bender RA, Claubiger P, Levine AS, Young RC: Phase I study methansulfonamide, N-(4-(9-acridinylamino)-3-methoxyphenyl)-(m-AMSA) using a single dose schedule. Cancer Treat Rep 62:1421–1426, 1978.
20. Ahmann F, Meyskens FL, Jones SE et al: A broad Phase II trial of AMSA with in vitro stem cell culture drug sensitivity correlation. Proc Am Assoc Cancer Res and Am Soc Clin Oncol 21:369, 1980.
21. Johnson RK, Inouye T, Goldin A, Stark GR: Antitumor activity of N-(phosphonacetyl)-L-aspartic acid, a transition-state inhibitor of asparate transcarbamylase. Cancer Res 36:2720–2752, 1976.
22. Erlichman C, Strong J, Wiernik P, Edwards L, Cohen M, Levine A, Hubbard S, Chabner B: Phase I trial of PALA (N-phosphonacetyl-L-aspartate). Proc Am Assoc Cancer Res and Am Soc Clin Oncol 20:C-98, 1978.
23. Zee-Cheng RKY, Cheng CC: Antineoplastic agents. Structure activity relationship study of bis (substituted aminoalkylamine) anthraquinones. J Med Chem 21:290–294, 1978.
24. Johnson RK, Zee-Cheng RKY, Lee WW, Acton EM, Henry DW, Cheng CC: Experimental antitumor activity of aminoanthraquinones. Cancer Treat Rep 63:425–439, 1979.
25. Von Hoff DD, Pollard E, Kuhn J et al: Phase I clinical investigation of dihydroxyanthracenedione (NSC 301739). Proc Am Assoc Cancer Res and Am Soc Clin Oncol 21:349, 1979.
26. Cantell K, Hirvonen S, Mogensen KE, Pyhala L: Human leukocyte interferon: Production purification, stability, and animal experiments. The production and use of interferon for the treatment of human virus infections. In Vitro Monogr 3:35–38, 1974.
27. Cantell K, Pyhala L, Strander H: Circulating human interferon after intramuscular injection into animals and man. J Gen Virol 25:453–458, 1974.
28. Mellstedt H, Bjorkholm M, Johansson B, Ahre A, Holm G, Strander H: Interferon therapy in myelomatosis. Lancet 1:245–247, 1979.
29. Gutterman J, Yap Y, Buzdar A et al: Leukocyte interferon (IF) induced tumor regression in patients (PTS) with breast cancer and B-cell neoplasms. Abstr no. 674, Proc Am Assoc Cancer 20:167, 1979.
30. Havell EA, Yip YK, Vilcek J: Characteristics of human lymphoblastoid (Namalva) interferon. J Gen Virol 38:57–60, 1978.
31. Knight E Jr: Interferon purification and initial characterization from human diploid cells. Proc Natl Acad Sci USA 73:520–523, 1976.
32. Negata S, Taira H, Hall A, Johnsrud L, Strenli M, Ecsodi J, Boll W, Cantell K, Weissmann C: Synthesis in E coli of a polypeptide with human leukocyte interferon activity. Nature 284 (5754):316–320, 1980.

V. Afterword

23
Perspectives on Future Directions

Sydney E. Salmon

New developments in science and technology can often lead to more new questions than answers. In that context, the current studies of cloning of human tumors already suggest a number of areas for future exploration. Recognizing that I must be somewhat selective, there are several clinical and biological areas on which I would like to comment.

DIAGNOSTIC APPLICATIONS OF THE IN VITRO ASSAY

While a major emphasis of in vitro tumor cloning studies has been directed towards evaluation of drug sensitivity testing, it is clear that the assay system may have value for other diagnostic purposes. The ability to cultivate various types of cancer from micrometastatic clones from bone marrow, peritoneal, bronchial, or bladder washings is clearly documented in this text. In some instances, routine cytology or pathology on direct specimens has been reported as negative for neoplasm while tumor clones subsequently grew out on the agar plates prepared simultaneously on these samples. Such disparate results could, of course, be due to sampling error or a real difference in sensitivity of the two diagnostic approaches. The extent to which diagnostic applications of the assay will improve clinical diagnosis remains unclear and future studies of such applications will be of great interest. For example, should cloning from peritoneal washings prove more sensitive and definitive than cytology for evaluation of persistent peritoneal disease in ovarian cancer, then repeated assays after cessation of chemotherapy might eliminate some unnecessary second look laparotomies for patients in whom bioassayable neoplasm is still present. Studies such as those performed by Dr. Ozols (Chapter 19) might also provide more effective ways to erradicate persisting intraperitoneal disease as well. Dr. Stanisic's observations in bladder cancer (Chapter 7) also provide considerable encouragement for the view that both diagnosis and treatment of this difficult neoplasm might change radically in the future. In many ways, bladder cancer represents almost an ideal model for carrying out repeated clonogenic assays on bladder irrigations or perhaps even voided urine specimens. Should transitional cell carcinoma clone well from the urine (and be distinguishable from cloning of the normal urothelium), then a new approach to early diagnosis of bladder cancer may emerge. Similar applications to serial bronchial washings also warrant testing in the diagnosis and follow-up of patients with lung cancer.

APPROACHES TO THERAPEUTIC TRIALS WITH THE DRUG SENSITIVITY ASSAY

As clinical investigators begin to incorporate in vitro clonogenic assay into their research programs, it is of value to delineate several stages of clinical trial development in relation to in vitro assay (Table I). The earliest level of clinical application of the in vitro assay could be classified as a *retrospective correlative trial.* It is relatively unplanned at its outset. Clinicians arrange to have tumor specimens submitted to the laboratory where a series of in vitro sensitivity tests are performed without awareness of the clinician's therapeutic plans for the patient. Standard anticancer drugs are generally tested in vitro. Subsequently, for patients on whom an evaluable in vitro assay is obtained, the physician is contacted, and it is determined whether the patient was treated with the drug tested in vitro. If so, and if the results of clinical treatment can be evaluated, correlations between in vitro and in vivo results are made. This approach is often wasteful of tumor specimens and laboratory resources (the wrong drugs may be tested). The approach also suffers from a lack of standardization of the clinical treatment protocol and assurance that clinical dosing is adequate even when the drugs coincide with those tested in vitro. Nonetheless, some useful correlations will emerge. The retrospective correlation approach should be considered as no more than the pilot phase for checking on the validity of the in vitro assay. The *prospective correlative trial* represents a more valuable approach for evaluating the validity and specificity of results of the in vitro assay. In this trial design, which may incorporate either a single treatment plan (eg, use of a phase II agent) or involve a randomized clinical trial design for clinical treatment, the trial plan dictates which drug or drugs patients with a given type of cancer are to receive on a standardized protocol dosing schedule and also dictates which drugs are to be tested in vitro. Subsequently, the clinical results for each patient tested both in vitro and in vivo are compared and correlations made.

The *prospective correlative trial* should be considered as an essential step in trial development for any given tumor type and/or new drug until the validity

TABLE I. Stages of Development in Clinical Trials With the Tumor Stem Cell Assay

1. Retrospective Correlative Trials – Clinical therapy and the in vitro testing are carried out independently.
2. Prospective Correlative Trials – The clinical trial design dictates which specific agent or agents all patients are to receive clinically and to have tested in vitro.
3. Prospective Decision-Aiding Trials – The results of testing a large battery of drugs in vitro leads to the selection of a single agent or drug combination for clinical trial.

of the assay system can be assessed for the specific therapeutic modality being tested. Once clear evidence is available that the assay result correlates well with clinical results of sensitivity or resistance in such prospective trials, then the final level of testing can be entertained. The *prospective decision-aiding trial* is undoubtedly the most appealing approach for clinical application of the in vitro assay system as it more closely corresponds to that which clinicians would ultimately want to have available. In this trial design, each patient's tumor is tested in vitro against a standardized large battery of drugs (eg, 10 or more), and the patient's clinical treatment plan is specifically based on the results of the in vitro assay. Protocol dosing schedules for the clinical trial in this circumstance must be every bit as stringent as in the prospective correlative trial; however, the number of potential treatment plans which could be implemented will undoubtedly be greater. Initially, such decision-aiding trials should involve the use of the best single agent, although in some instances simple combinations could be considered. Unlike the other trial designs, decision-aiding trials may potentially involve a delay of 1–3 weeks before treatment can be initiated due to the time required for the in vitro cultures to grow. Additionally, in some instances, no growth will be obtained. The lag of 1–3 weeks often corresponds to the recovery time from major surgery, and may not constitute "lost time" in that setting. With rapidly progressive cancers (eg, acute leukemia, lung cancer with SVC syndrome, breast cancer with hypercalcemia) such delays are unacceptable, and the trial design would require initiation of a standard therapeutic program with a switch to the program considered optimal by in vitro assay once the results are available. In surgical adjuvant trials, this latter approach may also be preferable, as it minimizes delay in initiation of systemic therapy when curative treatment is intended, and wherein only micrometastases remain, which cannot be directly assessed for tumor regression during therapy.

These general perspectives on predictive drug testing must of course still be viewed with some caution with regard to their widespread applicability. Culture conditions are still clearly inadequate for some tumor types and it is still uncertain as to how many types of cancer and clinical settings will prove amenable to predictive chemosensitivity testing.

APPLICATIONS OF THE SOFT AGAR CLONOGENIC ASSAY TO STUDIES OF RADIOSENSITIVITY

Although the major therapeutic applications of the tumor colony assay discussed in this text have related drug sensitivity plus some study of interferon, retinoids, and other biological response modifiers, it is apparent that this assay system should prove quite suitable for the study of radiation sensitivity. In Chapter 8, Dr. Meyskens has provided examples of radiation survival curves he has obtained in vitro with clonogenic melanoma cells from human tumors. It is evident from these curves that there is apparent heterogeneity in response to ionizing radiation that is not dissimilar to that which has been observed with drugs. In preliminary studies, we have observed a survival fraction of greater than 50% in the agar culture system with ovarian carcinoma at single radiation dosages up to 5,000 rads. Such resistance has been observed whether thiols (2-mercaptoethanol) were included in the medium or not. Colonies formed by surviving clonogenic cells often become large (>60 cells), suggesting that true resistance rather than a last few doublings of lethally injured cells is being observed. Similar observations have been made by Dr. Von Hoff (personal communication). These preliminary observations are of considerable theoretical and practical interest. While conventional concepts in radiobiology would make such marked heterogeneity in response to high dose radiation seem unlikely, it is conceivable that hypoxia, repair of sublethal or potentially lethal damage, or radioprotection by undefined ingredients of the culture system might be playing a role. Previously Good et al [1], from Brisbane, have reported marked difference in radiosensitivity of certain tumor cell lines plated in agar as contrasted to the same cell types plated in the conventional fashion for radiobiology experiments, namely as adherent cells on Petri dishes supported with liquid medium. Inasmuch as there are some baterial species which can survive massive doses of irradiation (eg, 100,000 rads), it is not beyond possibility that some heteroploid clonogenic tumor cells from fresh biopsies that are cultivated in agar might also have higher degrees of radioresistance than established tumor cell lines. Close collaboration of radiophysicists and radiobiologists will be essential to better define this phenomenon. Should these initial observations be substantiated, clinical studies correlating in vitro radiosensitivity or radioresistance with clinical response to radiotherapy will be of obvious importance.

FURTHER DEVELOPMENT OF THE ASSAY SYSTEM: PRACTICAL CONSIDERATIONS

While progress in understanding of growth regulation and tumor biology is being made in many research laboratories at the present time, the tumor stem cell assay has only just begun to become clinically useful. In view of the many clinical correlations to disease status and treatment response that have already been made with this system, it is reasonable to project that this assay system or a simplification of it could become a standard diagnostic procedure in the future. Practical considerations usually weigh heavily on the transition of a research procedure into a routine clinical tool. While the semisolid culture systems support growth of a wide variety of tumors in vitro, it is apparent that the cluster/colony growth currently observed is still suboptimal for most tumor types. The studies of completely artificial media that have been conducted by Sato's investigative group [2] provide substantial encouragement for the view that in the future, groupings of specific growth factors, hormones, and transport proteins may be incorporated along with more common medium constituents so that chemically defined media can be standardized for cloning various tumor types in vitro. Ideally, such media would not require serum supplement, as serum clearly has many unknown components which may lead to variability in results as a function of the serum source. The relative serum content of various substrates may also influence clonogenicity in vitro. However, in the near future, culture conditions may become more complex rather than simpler, as various cultural additions are identified which enhance growth. Recognizing that completely defined media are still a long term goal, what steps might be worth taking in the near future? Clearly, there is much promise in studying tissue extracts and conditioned media as well as specific growth factors for various tumor types. For example, epidermal growth factor (EGF) has been reported to enhance the growth of a variety of neoplastic cell lines [3]. At the 1980 Worshop on Human Tumor Cloning Methods, Hamburger reported on her preliminary observations in which a 2–3-fold enhancement of in vitro clonogenicity of some breast and colorectal carcinomas could be obtained with EGF (Appendix 1). Sarcoma growth factor, nerve growth factor, T cell growth factor and somatomedin are other specific factors which also warrant detailed study in the clonogenic assay with various tumor types, as do specific prostaglandins and cyclic nucleotides. Detailed studies of the macrophage-elaborated tumor growth factor from BALB/c mouse spleen cells (Chapters 1 and 3) and the factor or factors elaborated by human macrophages in malignant effusions (Chapters 4 and 11) are clearly of high priority.

It should be evident to the reader that at the current time, the in vitro tumor stem cell assay is a labor-intensive procedure which could benefit enormously from the availability of standardized commercial reagents and automated instrumentation. Various practical areas for reagent development or simplification are summarized in Table II. In addition to needs for standard and nontraumatic techniques for tumor disaggregation, commercialized production of transport media, prepackaged growth media, feeder layers, and drug assay reagents would provide the basis for quality control between laboratories which would be quite beneficial. As discussed in Chapter 15, use of multichamber plates for drug assay would facilitate integration of the automated tumor colony counter into routine

TABLE II. Technological Developments Which Will Facilitate Clinical Use of In Vitro Clonogenic Assay

1. Areas for Reagent Development
   a. "Transport medium" to maintain viability between time of biopsy and culture.
   b. Enzymatic cocktail for disaggregation of solid tumors without compromising clonogenicity or drug assay. (Chapter 20 and Appendix 2).
   c. Prepacked complete growth media and feeder layers containing necessary growth factors.
   d. Drug assay kits with calibrated reagent standards and diluents for sensitivity testing. (Stabilized drugs in active form highly desirable).
2. Areas for Automated Instrument Development
   a. Dispersion of tumor specimens into single cell suspension with washing, counting, and preparation of standard cell dilutions.
   b. Preparation of drug dilutions, timed incubating of drugs with cells, cell washing and plating on feeder layers, and sample marking.
   c. Automated tumor colony counter with data analysis capability (Chapter 15).

clinical application as well as for new drug screening efforts. Should continuous-contact drug assay (as discussed in Chapters 16, 18, and 22) prove predictive of clinical response to treatment, then either incorporation of stabilized standard anticancer agents (in active form) into prepoured feeder layers or simple addition of the drugs in a small amount of liquid medium after plating the tumor cells could provide substantial simplification in assay preparation. With either of these approaches, preincubation and several centrifugation and washing steps could be eliminated, and the disaggregated cells plated with fewer handling steps. Alternatively (Table II), instrumentation to automate the initial plating procedures could include component steps required for sample aliquoting, drug incubation, washings, and plating. Due to the relatively low cloning efficiencies currently observed, simple disc sensitivity assays, as are carried out in bacteriology, are unfortunately not feasible with the human tumor clonogenic assay. Another problem with clonogenic assays is the time required for the TCFUs to form colonies. However, there is reason to anticipate that a better understanding of early events associated with proliferation of clonogenic cells might permit substantial shortening of total culture incubation times. For example, if sensitivity patterns at the cluster stage (3–4 days) prove predictive of clinical response, then the entire culture preparation, incubation, and quantitation might prove amenable to larger scale integration of automated instrumentation for clinical and new drug screening laboratories.

SPECULATIONS ON THE BIOLOGY OF TUMOR GROWTH

Some of the most important applications of the tumor stem cell assay are undoubtedly in studies of tumor biology. It is of interest to reflect on the major approaches that have been applied to studies of tumor cell biology over the past few decades. Studies of cell lines and transplantable tumor models have predominated. These approaches have been quite valuable for subcellular studies, extraction of antigens and definition of molecular mechanisms within cells, and studies of inherent mechanisms of proliferation of relatively autonomous and homogeneous tumors [4]. As evidenced by data presented in this text and elsewhere [5], the use of clonal assays of tumor growth of spontaneous tumors are anywhere near as autonoous or homogeneous as prior experimental models might imply. The clonal assay approach also provides an excellent means for studying effects of growth-modulating hormones (eg, steroids, insulin, thyroxin, and pituitary factors) and growth factors (EGF, NFG, etc). Additionally, the studies on cell cooperation in support of tumor growth conducted by Drs. Hamburger, Buick, and myself (as discussed in Chapters 3, 6, and 11), provide clear indication that non-neoplastic host mononuclear cells play major growth modulating roles. In most instances, the host cell component (which is often macrophage-rich) appears to support rather than inhibit clonal growth of the neoplastic cells in vitro; however, dose-response effects of these cells and conditioning factors derived from them clearly have complex effects. Although the modulators involved and their mechanisms of action may differ, interactions between host and neoplastic cells in the agar culture system are in many ways analogous to those observed in studies of in vitro hemopoiesis. In both hemopoiesis and lymphopoiesis, macrophages can exert either stimulatory or inhibitory effects on clonogenic proliferation of progenitor cells. With respect to cancer, the observation that adherent and phagocytic macrophages are often required for in vitro growth of not only ovarian carcinoma (Chapter 6), but a number of other tumors as well (Chapter 11), has provided the basis for developing a still-evolving hypothesis on the growth-promoting interaction between macrophages and clonogenic tumor cells in human cancer and on the relation of this interaction to the cellular immune response [6]. Based on other evidence, Prehn [7] had previously suggested that there is a "lymphodependent" phase of tumor growth (in which the immune system supports the growth of a cancer). On the basis of our macrophage depletion studies in ovarian cancer, and of macrophage-conditioned medium stimulation of clonal growth of myeloma and other evidence, we proposed an initial "macrophage hypothesis" in 1978 [6]. We suggested (Fig. 1) that one or more macrophage-derived promoter substances facilitated clonal growth of human cancers in vivo, as well as having a stimulatory effect on B-lymphocyte proliferation. We also proposed that such macrophage effects were critical in the promotional step in the two-step model of skin carcinogenesis [8] and in the induction of myeloma in the BALB/c mouse [9]. Of interest, tumor promoters are now known to exert potent effects on macrophages and to stimulate their production of prostaglandins [10, 11]. Additional data supportive for the concept that autologous macrophages can potentiate tumor growth include the observation in an in vivo tumor model by Evans [12].

The cell separation and reconstitution studies designed by Dr. Buick (Chapter 6 and [8]) to further define the host cell requirement for in vitro clonal growth of human neoplasms clearly established that growth promotion was conveyed by soluble factors from the adherent cells which diffused through the agar, and the finding that elaboration of prostaglandins by host macrophages was associated with this process. These findings, and the knowledge that macrophage-derived prostaglandins can potently inhibit the proliferation of T cells [13, 14] as well as modulate fetal development, now leads me to propose an even broader hypothesis: that the macrophage is a promoter of both fetal and cancer growth as well as immunological tolerance to both of these growth processes. Furthermore, it seemed reasonable to postulate a feedback loop, in which macrophage secretion of prostaglandins and other growth factors occurred in response to a fetal antigen or hormonal signal. This more generalized hypothesis is depicted in Figure 2.

It has long been recognized that there are similarities between maternal tolerance to the growing fetus and the immunological tolerance of a patient's immune system to a progressively growing cancer [15]. In both circumstances,

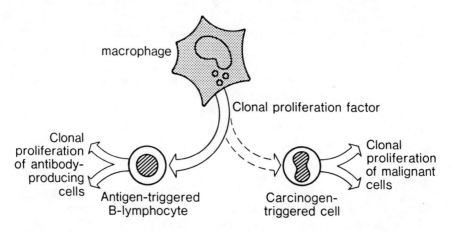

Fig. 1. The first macrophage hypothesis (1978): The macrophage as a two-edged sword in immunity and cancer. A proliferation factor elaborated to support clonal multiplication of antigen-triggered B-lymphocytes also incidentally supports proliferation of neoplastic cells. Membrane proteins from dying cancer cells further stimulate elaboration of this factor by macrophages. Reproduced from Lancet [6] with permission of the publisher.

progressive growth of the immature cells occurs while the cellular immune response to histocompatibility or tumor antigens is inhibited. With respect to cancer, a central question has been to identify the mechanism by which cancers can growth progressively despite the presence of demonstrable tumor immunity in the host lymphoid cells. Certain biological markers also appear to be expressed in both fetal and tumor tissues: these include embryonic antigens and prostaglandins.

The central postulate upon which the new hypothesis is based is that a normal function of macrophages is to respond to the presence of fetal antigens or hormones by secretion of prostaglandins and growth factors which normally stimulate the proliferation of immature cells (expressing fetal properties) while simultaneously inhibiting the immune response. Prostaglandins appear to be important in normal fetal development and parturition, and it is conceivable that embryonic primordia are stimulated to proliferate in response to both prostaglandins and

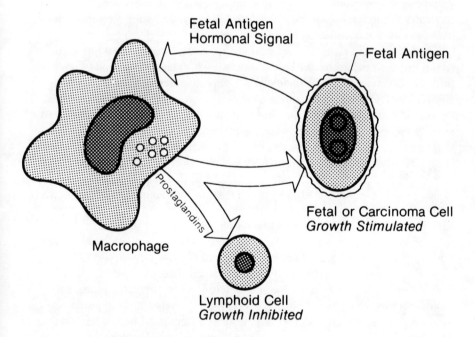

Fig. 2. The second macrophage hypothesis (1980): The macrophage as a promoter of fetal and cancer growth and immunological tolerance: relation to fetal antigens and prostaglandins. Factors secreted by the responding activated macrophage stimulate fetal (or carcinoma) growth. The secretion of prostaglandin E also initiates immunological tolerance by inhibiting the T-lymphocyte response to histocompatibility or tumor antigens. Thus, in cancer, the macrophage is "tricked into thinking" that a fetus is present.

growth factors elaborated by endogenous macrophages which can migrate to appropriate sites. Fetal cell proliferative responses to such growth factors would likely be dose dependent, and not all tissues would respond positively to macrophage signals. For example, macrophage-derived prostaglandins would be anticipated to inhibit allogeneic immunoproliferative responses in both mother and fetus, thus contributing directly to maternal-fetal tolerance. Presumably other macrophage-secreted factors (eg, lymphocyte-activating factor) are down regulated by similar fetal signals. While the specific fetal biological signals which we postulate remain to be directly identified, they might well be some of the already recognized "oncofetal antigens" (eg, α-fetoprotein, AFP; human chorionic gonadotropin, HCG, and carcinoembryonic antigen, CEA). HCG [16] and AFP [17] have been reported to suppress cellular immunity in vitro; these effects could be macrophage mediated. Activated macrophages secrete high levels of prostaglandin E which, in turn, potently suppresses the proliferation of normal T-lymphocytes [18], as well as modulating other aspects of immunity. Prostaglandin F2α, which can also be produced by the macrophage, is known to rise significantly in concentration in fetal blood prior to normal delivery, and both prostaglandin have been used as abortifacients.

This hypothesis on the role of macrophages in fetal growth would be quite relevant to the cancer problem as the expression of fetal antigens and other features of "immaturity" are characteristic abnormalities associated with neoplastic clones. One would have to assume that macrophages retain the potential to respond to fetal hormonal signals throughout life. Retention of the ability to recognize a fetal antigenic or hormonal signal and respond by secreting immunosuppressive as well as growth factors would provide the environment which would allow an evolving carcinoma arising from a transformed cell to have a growth advantage. Recent investigations support the concept that fetal antigens can stimulate tumor growth. Gershwin et al [19] found that the induction of myeloma in the BALB/c mouse can be facilitated by the administration of AFP. Furthermore, administration of irradiated normal fetal liver cells has been shown to enhance the growth of a rat fibrosarcoma while blocking the expression of immune cytotoxicity [20]. If the various fetal signals and macrophage responses which we hypothesize occur in the fetus also occur in patients harboring neoplastic cells, then both growth of the carcinoma and inhibition of the immune response to tumor antigens would be anticipated. Thus, this relatively simple hypothesis would provide simultaneous explanations for two long-standing biological puzzles: 1) maternal-fetal tolerance, and 2) the mechanism by which a carcinoma can "slip through" the immune response.

With respect to cancer growth, we would propose that the early growth of a carcinoma requires both a transformed clonogenic cell expressing one or more fetal antigens and an appropriate microinductive environment (eg, macrophage-containing soil). Plescia et al [21], have observed that small numbers of tumor cells can induce immunosuppression in association with PGE_2 production. Tumors are known to be rich in prostaglandins and the studies by Buick et al (Chapter 11 and [8]) would suggest that these are secreted primarily by functional macrophages which have migrated into spontaneous human tumors. The use of compounds which block either the elaboration of or the response to prostaglandins and other growth factors secreted by macrophages could have an anti-promotional effect, either before or after the neoplastic transforming event has occurred. In the two-step model of carcinogenesis in the mouse skin, prostaglandin synthesis inhibitors block the promotional step [22] which is known to require a mononuclear inflammatory response to the tumor promoter. Clinically, macrophages present in the lymph node, liver, or lungs of a patient might provide the local environment for colonization by circulating clonogenic tumor cells by suppressing local tumor immunity mediated by lymphoid cells. How might this vicious circle be reversed? A two-pronged approach to cancer treatment might be reasonable. Inhibitors of macrophage function and growth factor secretion or effect would be used to block the promotional step, while conventional cytotoxic or hormonal agents would be used to directly attack the neoplastic cells. Selection of the appropriate agents might eventually be accomplished routinely by in vitro assay. The clinician may have already discovered the two-pronged approach to cancer treatment empirically by combining biological response modifiers (eg, BCG, levamisole, corticosteroids) along with conventional cytotoxic drugs. Perhaps of relevance, BCG treatment of animals has recently been reported to significantly reduce the stimulated secretion of prostaglandins by macrophages [11]. Since macrophages in high concentration have also been reported to have suppressive effects on tumor growth [23], the possibility remains that macrophage subpopulations could be identified which could be induced to suppress tumor growth at clinically achievable macrophage-tumor cell ratios.

The proof of any given hypothesis is, of course, in the testing, and a variety of relevant biological experiments can be designed which may either support, refute, or lead to modification of this macrophage-oncofetal tolerance hypothesis.

CONCLUSION

In the last analysis, empirical testing of the in vitro clonogenic assays (rather than hypotheses) will provide the best measure of their value in cancer research. While a number of applications already appear clear-cut, and there definitely have been worthwhile achievements and many areas of promise, there are still many obstacles to overcome. It is, in fact, the unanswered questions and deficiencies in our current knowledge that will undoubtedly provide investigators and clinicians with the greatest challenge for the future.

REFERENCES

1. Good M, Lavin M, Chen P, et al: Dependence on cloning method of survival on human melanoma cells after ultraviolet and ionizing radiation. Cancer Res 38:4671–4675, 1978.
2. Bottenstein J, Izumi H, Hutchings S, Masui H, Mather J, McClure DB, Ohasa S, Rizzino A, Sato G, et al: The growth of EGF cells in serum-free, hormone supplemented media. In Jakoby WB, Pastan IH (eds): "Methods in Enzymology." Vol. LVIII (Cell Culture), New York: Academic Press, 1979, pp 94–109.
3. Fox CF, Vale R, Peterson SW, Das SM: The EGF receptor: identification and functional modulation. In Sato G, Ross R (eds): "Hormones and Cell Culture." Book 'A'. New York: Cold Spring Harbor Conference on Cell Proliferation, Cold Spring Harbor, 1979, pp 143–158.
4. Buick RN, Fry SE, Salmon SE: Application of in vitro soft agar technique for growth of tumor cells to the study of colon cancer. Cancer 45:1238–1242, 1980.
5. Buick RN, Fry SE, Salmon SE: Effect of host-cell interactions on clonogenic carcinoma cells in human malignant effusions. Br J Cancer 41:695–704, 1980.
6. Salmon SE, Hamburger AW: Immunoproliferation and cancer: a common macrophage-derived promoter substance. Lancet 1:1289–1290, 1978.
7. Prehn RT: Immunostimulation of the lymphodependent phase of neoplastic growth. J Natl Cancer Inst 59:1043, 1977.
8. Boutwell RK: The function and mechanism of promoters of carcinogenesis. CRC Crit Rev Toxicol 2:419–443, 1974
9. Potter M, Cancro M: In Saunders GF (ed): "Plasmacytogenesis and the differentiation of immunoglobulin-producing cells in cell differentiation and neoplasia." New York: Raven Press, 1978, pp 145–161.
10. Humes JL, Davies P, Bonney RJ, Kuehl FA Jr: Phorbol myristate acetate stimulates the release of arachidonic acid and cyclo-oxygenase products by macrophages. Fed Proc 37:1318, 1978.
11. Humes JL, Burger S, Galavage M, Kuel FA Jr, et al: The diminished production of arachidonic acid oxygenation products by elicited mouse peritoneal macrophages: possible mechanisms. J Immunol 124:2110–2116, 1980.
12. Evans R: Macrophage requirement for growth of a murine fibrosarcoma. Br J Cancer 37:1086–1089, 1978.
13. Kurland JI, Bockman RJ: Prostaglandin E production by human blood monocytes and mouse peritoneal macrophages. J Exp Med 147:952–957, 1978.
14. Bockman RJ, Rothschild M: Prostaglandin E inhibition of T-lymphocyte colony formation. J Clin Invest 64:812–819, 1979.
15. Uriel J: Fetal characteristics of cancer. In Becker FF (ed): "Cancer, a Comprehensive Treatise." Vol. 3, New York: Plenum Press, 1975, pp 21–49.
16. Lange PT, Hakala TR, Fraley EE: Suppression of antitumor lymphocytic mediated cytotoxicity by human gonadotropins. J Urol 115:95–98, 1976.
17. Gershwin ME, Castles JJ, Makishima AA: The influence of α-fetoprotein on Maloney sarcoma virus oncogenesis: evidence for generation of antigen nonspecific suppressor T cells. J Immunol 121:978–985, 1978.
18. Pelus L, Strausser H: Prostaglandins and the immune response. Life Sci 20:903–913, 1977.
19. Gershwin ME, Castles JJ, Makishima R: Accelerated plasmacytoma formation in mice treated with α-fetoprotein. J Natl Cancer Inst 64:145–150, 1980.
20. Keller R: Competition between foetal tissue and macrophage-dependent natural tumor resistance. Br J Cancer 40:417–423, 1979.

21. Plescia OJ, Smith AH, Grinwich K: Subversion of the immune system by tumor cells and the role of prostaglandins. Proc Natl Acad Sci USA 72:1848–1851, 1975.
22. Verma AK, Rice HM, Boutwell RK: Prostaglandins and skin tumor promotion: inhibition of tumor promoter-induced ornithine decarboxylase activity in epidermis by inhibitors of prostaglandin synthesis. Biochem Biophys Res Commun 79:1160–1166, 1977.
23. Hibbs JB: Discrimination between neoplastic and non-neoplastic cells in vitro by activated macrophages. J Natl Cancer Inst 53:1487–1492, 1974.

VI. Appendices

1
Standard Laboratory Procedures for In Vitro Assay of Human Tumor Stem Cells

Barbara Soehnlen, Laurie Young, and Rosa Liu

In this appendix we will summarize practical aspects relating to clonogenic assay methods that we use routinely in our laboratories at the University of Arizona Cancer Center. Techniques which will be summarized include: (A) Preparation of the spleen-conditioned medium (as discussed in Chapter 3), (B) Preparation of the underlayers and upper layers for agar culture of TCFU's, (C) standard drug sensitivity assay preparation and plating, (D) methods for cryopreservation and thawing human TCFUs for culture, and (E) supplies and suppliers we have used in these procedures.

A. PREPARATION OF OIL-PRIMED MOUSE SPLEEN-CONDITIONED MEDIUM FOR THE MYELOMA COLONY ASSAY

Materials

Sterile scissors and forceps; alcohol burner with 95% alcohol for flaming; sterile tissue culture dishes (#3002); and three to five Balb-c mice (female only) primed a minimum of four weeks to three months previously with 0.2 ml of mineral oil. The mice should be at least six weeks old.

Media

RPMI 1640 containing heat-inactivated fetal calf serum (HI-FCS); 1% L-glutamine; 1% penicillin/streptomycin; and *cold* RPMI 1640 are used.

Procedure

Remove spleens aseptically and tease apart with 18g needles in a sterile petri dish containing some RPMI medium. Pass the cell suspension through a series of needles (21g, 23g, and 25g). The cells are then counted and stained for viability with trypan blue. The final cell concentration should be adjusted to 1×10^6 cells per ml of RPMI medium. Pipette 5 ml of the suspension into #3002 tissue culture dishes.

Incubate the plates for a minimum of two hours at 37°C to allow for adherence. The plates should then be examined under an inverted microscope to assure that good cellular adherence has been achieved. Wash the plates three times with 3-ml volumes of RPMI 1640. Each 3-ml volume is added, the plate is swirled briefly, and the medium is decanted. After the final wash, add 5 ml of RPMI medium containing HI-FCS, L-glutamine, and penicillin/streptomycin to each plate and allow the cells to condition the medium for three days. After three days, pipette off the medium and centrifuge it at 200g for ten minutes, filter through a nalgene filter, and store at −20°C in 10-ml aliquots. Date and label the tubes. Each batch of conditioned medium requires testing (on myeloma cells) to establish that it is stimulatory for growth and to determine the optimal dilution required.

B. PREPARATION OF AGAR LAYERS FOR CULTURE OF TCFUs

Underlayer or Feeder Layer (Using Conditioned Medium)

Enriched McCoy's. Ingredients are: 500 ml McCoy's, 25 ml horse serum, 50 ml HI-FCS, 5 ml Na pyruvate (2.2%), 1 ml L-serine 21 mg/ml, 5 ml glutamine 200 mM, and 5 ml penicillin/streptomysin 10,000 units/ml.

Plating media. Ingredients are: 40 ml enriched McCoy's, 10 ml tryptic soy broth (3% in density H_2O), 0.6 ml asparagine (6.6 mg/ml), and 0.3 ml DEAE dextran (50 mg/ml).

Three percent agar. The conditioned medium from spleen-adherent cells is diluted with the plating medium so that 0.25 ml of conditioned medium is in each 1 ml of agar feeder layer. This should be made in small batches to prevent the agar from "lumping." All ingredients must be warm. The agar should be boiled until it is a clear liquid and placed in a 50°C water bath during use. The final concentration of agar in the underlayer is 0.5%.

Suggested amounts are 11.5 ml plating medium, 5.0 ml conditioned medium, and 3.5 ml agar (3%), totaling approximately 20 plates.

Plain Underlayer (Without Conditioned Medium)

All ingredients are prepared as above.

The underlayers are prepared as above except that the conditioned medium is omitted. The final concentration of agar in the underlayer is 0.5%. (Other growth factors are sometimes added to the underlayer to enhance clonogenicity.

In a recent modification introduced by Dr. A. Hamburger (personal communication, 1980), 30 ng of epidermal growth factor (EGF) per ml is added to the plating layer mixture. Dr. Hamburger found that this concentration of EGF increased clonogenicity twofold to threefold in various epithelial neoplasms. Thus a total of 600 ng (0.6 µg) of EGF is included in the 16.5 ml of plating medium listed above.)

Cell-Containing Upper Layer

Enriched CMRL 1066. Ingredients are: 100 ml CMRL 1066, 15 ml horse serum, 4 ml $CaCl_2$ (100 mM), 2 ml insulin (100 units/ml), 1 ml vitamin C (30 mM), 1 ml penicillin/streptomycin, and 2 ml glutamine. Just before the cells are plated in the upper layer, add: asparagine (6.6 mg/ml) (0.6 ml/40 ml medium), DEAE dextran (50 mg/ml) (0.3 ml/40 ml medium), and mercaptoethanol (1/100 dilution of a 0.5×10^{-3} M stock prepared fresh each week) added to the enriched CMRL 1066 medium. Three percent agar is added to yield a final agar concentration in the upper layer of 0.3%. The appropriate concentration of cells (usually 0.5×10^6 cells/plate) is then mixed quickly with the agar and medium (at 45–50°C) and plated. The plates are then checked to assure that they contain a single-cell suspension and monitored twice weekly for growth.

C. DRUG SENSITIVITY ASSAY PREPARATION

All tests are done in triplicate — 500,000 cells/plate; stock cell concentration 3×10^6 cells/ml; drug concentration: one-hour exposure = drug per incubated volume (in tissue culture tubes in suspension culture). Continuous exposure ÷ drug per volume of upper layer of each plate. (Note: After equilibration in plate, the final drug concentration in the upper layer in continuous contact experiments is only half that stated, as the total volume in the plate including the under layer is 2.0 ml.) Example: control tubes, 1.0 µg desired drug concentration, stock drug concentration 0.01 mg/ml of 10 µg/ml.

One-Hour Exposure

Control tube	Drug tube
0.5 ml stock cell concentration ($1.5 + 10^6$ cells)	0.5 ml stock cell concentration
	0.15 ml drug stock (equals 1/0.0 dilution of drug stock)
1.0 ml McCoy's 5A medium + 10% HI-FCS	0.85 ml McCoy's 5A medium + 10% HI-FCS
Total volume 1.5 ml	Total volume 1.5 ml

Incubate for one hour at 37°C. Wash the cells two times by centrifugation at 1,200g with replacement of the medium. Add 2.7 ml of enriched CMRL to the cell button, MIX. Add 0.3 ml of 3% agar to the tube (final agar concentration 0.3%). Mix and plate.

Three control plates, 500,000 cells/plate, 1 ml/plate.

Three drug plates, 500,000 cells/plate, 1 ml/plate.

Continuous Exposure

Cells are preincubated for one hour at 37°C prior to plating.

Control tube	Drug tube
0.5 ml stock cell concentration ($1 \times 5 \times 10^6$ cells) 2.2 ml enriched CMRL 0.3 ml 3% agar	0.5 ml stock cell concentration 0.30 ml drug stock (equals 1/10 dilution of stock) 1.9 ml enriched CMRL 0.3 ml 3% agar
Total volume 3.0 ml	Total volume 3.0 ml
Three plates each upper layer containing 1 ml volume with 500,000 cells.	Three plates each upper layer containing 1 ml volume with 500,000 cells and a drug concentration of 1 µg.

D. METHODS FOR PREPARING AND USING CRYOPRESERVED CLONOGENIC HUMAN TUMOR CELLS

Freezing Procedure

Materials and Media Include: Freezing tubes, dimethyl sulfoxide (DMSO), ice bucket and ice, McCoy's 5A media containing 10% HI-FCS, hemocytometer and microscope, and freezer (−80°C or liquid nitrogen).

Label and date freezing tubes. Determine the cell count and trypan blue viability. Suspend the cells in McCoy's 5A medium containing 10% HI-FCS in the concentration of $4-20 \times 10^6$ cells/ml. Cool in an ice bucket for 15 minutes or more. Add DMSO to a final concentration of 10% while gently shaking the cell suspension, invert the cell suspension tube two times for thorough mixing, and quickly pipette into the freezing tubes (1–3 ml/tube). Label the tubes with the final cell concentration and store in −80°C deep freezer. (Note: The freezing rate can be slowed by placing the freezing tubes in a pre-chilled styrofoam container immediately before transfer to the freezer. Twenty-four hours later the tubes can be removed from the styrofoam container and placed in standard racks or boxes in the freezer. Alternatively, a controlled-rate freezing apparatus can be used.)

Thawing Procedure

Materials and media include: 50-ml sterile plastic tubes, centrifuge (for washing cells), 37°C water bath, McCoy's 5A medium containing HI-FCS, and plating medium (CMRL or other selected medium).

Remove the tubes from the freezer and thaw the frozen cells quickly in a 37°C water bath without agitation to the point of transition to the liquid phase. Then, transfer the cells quickly into a 50-ml tube and dilute tenfold with McCoy's 5A medium containing 10% HI-FCS. Wash two times by centrifugation at 200g for ten minutes. Add plating medium to the cell bottom after second wash. Perform a cell count and determine viability (trypan blue) on each sample and calculate percentage recovery. Incubation of the cells at 37°C for two hours prior to plating appears to enhance clonogenicity. Subsequent steps for cell plating are performed as described in this appendix and the text. To minimize cell clumping, drug sensitivity assays are generally carried out by continuous contact as described in Chapter 21.

E. SUPPLIES AND SUPPLIERS

I. Tissue Culture Media and Sera

Hank's Balanced Salt Solution 10X
without Ca, Mg, NaHCO$_3$
with Phenol Red
Cat #310-4180

Grand Island Biological Company
(GIBCO)
3175 Stanley Rd, Grand Island, NY 14772
or 519 Aldo Ave, Santa Clara, CA 95050

Hank's Balanced Salt Solution 10X
without Ca, Mg, NaHCO
(GIBCO)
Cat #310-4185

RPMI Medium 1640
with L-glutamine
without NaHCO$_3$, antibiotics
(GIBCO)
Cat #430-1800 — powdered
Cat #320-1875 — liquid

McCoy's 5A medium (dry powder
prepared according to directions on
box, at this time NaHCO$_3$ added)
(liquid contains NaHCO$_3$)
with L-glutamine/without NaHCO$_3$
without antibiotics
(GIBCO)
Cat #430-1500 — powdered
Cat #320-6600 — liquid

CMRL medium with L-glutamine
(GIBCO)
Cat #320-1535 — liquid

L-glutamine 200 mM (110X)
(GIBCO)
Cat #320-5030

Penicillin/streptomycin solution
10,000 units
(GIBCO)
Cat #600-5140

Na pyruvate
(GIBCO)
Cat #130-840

L-serine
(GIBCO)
Cat #11010

Asparagine — for 6 mg/ml
(GIBCO)
Cat #10130

Fetal calf serum — heat inactivated
(FLOW)
Cat #29-102-49 — 100 ml
Cat #29-102-54 — 500 ml

Horse serum
(FLOW)
Cat #29-211-49 — 100 ml
Cat #29-211-54 — 500 ml

Flow Laboratories (FLOW)
936 W Hyde Park Blvd, Inglewood,
CA 90302
or 1710 Chapman Ave,
Rockville, MD 20852

Bacto agar
(DIFCO)
Cat #0140-01

Difco Laboratories (DIFCO)
Detroit, MI 48232

Tryptic soy broth 1 lb
(DIFCO)
Cat #0370-01 (for 1 lb)

II. Chemicals and Drugs

Mineral oil
(SIGMA)
Product #400-5

Sigma Chemical Company (SIGMA)
PO Box 14508
St. Louis, Mo 63178

Mercaptoethanol
(SIGMA)
Cat #M6250

L-ascorbic acid (vitamin C)
(GIBCO)
Cat #30080

DEAE dextran
(PHARMACIA)
Cat #17-0350-01

Pharmacia Fine Chemicals
(PHARMACIA)
800 Centennial Ave, Piscataway, NJ 08854

Epidermal growth factor (EGF)
Cat #40001

Collaborative Research, Inc
1365 Main St, Waltham, MA 02154

Ether — anesthesia
(Scientific Products)

Scientific Products, General Offices,
1430 Waukegaw Rd, McGaw Park, IL 60085

Heparin without preservative
(O'Neill, Jones & Feldman)
Cat #NDC 0456-0778-26

O'Neill, Jones & Feldman
2510 Metro Blvd.,
Maryland Heights, MD 63043

Insulin — 100 units/cc U-100 R
(LILLY)
Cat #NDC 0002-1135-01
M 210

Eli Lilly & Company
Indianapolis, IN 46206

Dimethyl sulfoxide (DMSO)
($(CH_3)_2SO$)
(Mallinckrodt)

Mallinckrodt Chemical Company
2nd & Mallinckrodt Sts,
PO Box 5439, St. Louis, MO 63147

III. Labware

Freezing tubes, sterile
12 × 75 mm
polypropylene with cap
(Scientific Products)
Falcon Cat #2005

Scientific Products, General Offices,
1430 Waukegaw Rd
McGaw Park, IL 60085

Conical graduated tubes, sterile,
50 ml, with positive-seal cap
(Scientific Products)
Falcon Cat #2070

Nalgene filter units
(Scientific Products)
Cat #3200-1

35- × 10-mm petri dish with 2-mm grid
(FLOW)
Cat #LUX 5217

35- × 10-mm petri dish
(Scientific Products)
Falcon Cat #3001

60- × 15-mm petri dish with lids
(Scientific Products)
Falcon Cat #3002

100- × 20-mm petri dish with lid
(Scientific Products)
Falcon Cat #3003

IV. Animals

Female — Balb-c mice
Jackson Laboratories

Jackson Laboratories
Bar Harbor, ME 04609

2
An Enzymatic Method for the Disaggregation of Human Solid Tumors for Studies of Clonogenicity and Biochemical Determinants of Drug Action

Harry K. Slocum, Z.P. Pavelic, and Y.M. Rustum

Two basic approaches towards the production of cell suspensions have been taken taken: Mechanical separation of cells, and enzymatic treatment of samples. Mechanical methods seem the most straightforward and simple, but often result in poor cell yields and suspensions that contain many dead cells [1–4]. Enzymatic treatments, however, may have deleterious effects on cells [5–8] and these may be subtle and difficult to control. Due to the fact that a large number of functionally viable, human solid tumor cells are required for the purpose of conducting simultaneously biochemical, biological, and pharmacological studies, methods have been developed utilizing mechanical and enzymatic approaches for disaggregation [9–11]. These methods were evaluated primarily in three human solid tumor types: Melanoma, sarcoma, and pulmonary carcinoma. In this brief communication, the results are compared in terms of yield of viable cells per gram of tissue, by their clonogenic potential in a soft agar system, and by their ribonucleotide pools.

After an initial trial with different enzymes and combination of enzymes under different conditions, the disaggregation procedure was standardized to the method outlined in Figure 1: Slicing of tissue with the Stadie-Riggs microtome and incubation of slices for two hours at 37°C in 5% CO_2/95% air atmosphere in RPMI 1640 medium with 10% fetal bovine serum, 0.8 gm/100 ml of collagenase II, and 0.002 gm/100 ml of deoxyribonuclease I. Repeated washing of tissue in a petri dish with fresh medium followed by pouting through a 100-mesh screen completed the disaggregation procedure. The entire procedure, from receipt of the surgical speci-

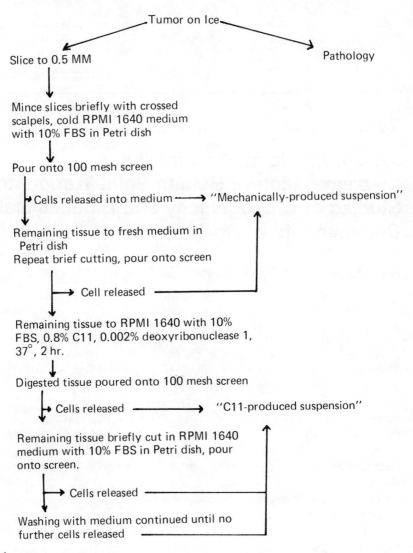

Fig. 1.

men in the laboratory to obtaining the final cell suspension, was generally completed in three hours.

Direct comparison of cell yield and dye exclusion capabilities of suspensions produced mechanically and by collagenase II from the same specimen for human melanoma, sarcoma, and lung tumor are shown in Table I. The data indicate that the median yields were from two- to seven-fold greater by enzymatic method, and suspensions substantially free of trypan blue staining cells were obtained only by the enzymatic method.

TABLE I. Mechanical and Enzymatic Disaggregation of Same Specimen*

Disease	Yield (millions of viable cells[a] per gm tissue)		Viability[a] (%)	
	Mechanical	Enzymatic	Mechanical	Enzymatic
Melanoma	9.2 (0.53–49.0)	21 (6.6–241.0)	29 (1.0–57.0)	85 (54.0–94.0)
Sarcoma	4:7 (0.11–18.0)	35 (7.6–189.0)	9 (0.8–31.0)	96 (84.0–98.0)
Lung tumo	8.1 (0.5–29.0)	34 (1.2–241.0)	29 (4.0–65.0)	90 (9.0–97.0)

*Median values with ranges in parentheses.
[a]By trypan blue exlcusion.

Colony growth in soft agar of cells disaggregated mechanically and enzymatically from 59 different solid human malignant tumors was evaluated. Tumor cells were cultured in a two-layer agar system [12] in enriched medium with or without adding conditioned medium from human sarcoma cell lines. In most trials, conditioned medium did not increase colony growth of either mechanically or enzymatically disaggregated tumor cells. A linear relationship was observed between the number of cells plated and the number of colonies obtained.

Enzymatic disaggregation had advantages over mechanical methods in most cases of sarcomas and malignant melanomas, and yielded more colonies and increased the probability of achieving growth in soft agar as shown in Table II. Enzymatically released pulmonary carcinoma cells had lower clonogenic potential especially in the case of anaplastic carcinomas. This type of tumor may require other conditions for cell disaggregation.

The results of the ribonucleoside triphosphate pools (RNTP) of cells disaggregated from human melanoma tissues are summarized in Table III. These indicate that the pools of CTP, UTP, ATP, and GTP were the highest for cells obtained from tissues disaggregated by collagenease II method. In 3/4 cases disaggregated mechanically, the pools of CTP were less than < 0.1 pmole/10^7 cells. In the cases investigated, the ratio of ATP/ADP ranged from four to eight indicating high viability in enzymatically disaggregated cells.

In summary, the enzymatic method described herein provides a means of obtaining sufficiently large amounts of functionally viable single-cell suspension from human solid tumors for biochemical, biological, and pharamcological characterization and for cell suspension studies which are underway. The clonogenic results indicated that the plating efficiency of enzymatically released cells are as good if not better than those released by mechanical means. In addition, the RNTP in melanoma

TABLE II. Effect of Mechanical and Enzymatic Disaggregation on Colony Formation of Tumor Cells

Sources of tissue specimen	Specimen producing colonies/total specimens tested (%)			
	Enzymatic[a]		Mechanical[b]	
Solid tumors				
Malignant melanoma	13/17	(76)	9/17	(53)
Pulmonary carcinoma	12/14	(86)	11/14	(79)
Soft tissue sarcoma	9/11	(82)	5/11	(45)
Malignant lymphoma	1/2		1/2	
Malignant schwannoma	0/1		0/1	
Cancer of unknown site	3/3	(100)	3/3	(100)
Giant cell tumor	1/1		0/1	
Epiglottis cancer	1/1		1/1	
Islet cell tumor	1/1		1/1	
Colon cancer	4/4	(100)	3/4	(75)
Breast cancer	1/4	(25)	1/4	(25)

Viability: [a]78% and [b]59% — overall success in cloning.

TABLE III. RNTP of Cells Disaggregated From Human Melanoma Tumor Tissues by Mechanical and Enzymatic Methods.*

Patient No.	Cell type	nmoles/10^7 Cells			
		CTP	UPT	ATP	GTP
1	Mechanical	<0.1	0.112	2.34	0.793
	Enzymatic	0.209	2.22	20.4	4.01
2	Mechanical	0.132	0.704	3.57	0.690
	Enzymatic	0.629	3.93	27.8	4.31
3.	Mechanical	<0.1	0.548	4.44	0.267
	Enzymatic	<0.1	0.652	12.0	1.84
4	Mechanical	<0.1	0.562	12.2	1.27
	Enzymatic	0.680	2.00	20.5	4.56

RNTP = ribonucleotide triphosphate pools.

*Separation and quantitation of these components were carried out as described by Rustum [12].

cells and their ratio are similar to those found in highly viable leukemia L1210 cells. The apparent advantage of the enzymatic method is that a large yield of functionally viable cells can be obtained with a relatively low yield of dead cells. It remains to be determined whether or not the enzymatic procedure alters the response of clonogenic tumor cells to cytotoxic drugs or other agents.

ACKNOWLEDGMENTS

This work was supported in part by program grant CA-21701 and core grant CA-24538 from the National Cancer Institute, U.S. Public Health Service.

REFERENCES

1. Grobstein C, Cohen J: Science 150:626–628, 1965.
2. McLimans WF, Mount DT, Bogitch S, Crouse EJ, Harris G, Moore GE: Ann NY Acad Sci 139:190–213, 1966.
3. Rinaldi LMJ: Int Rev Cytol 7:582–647, 1958.
4. Rosenberg IL, Russell CWS, Giles GR: Br J Surg 65:188–190, 1978.
5. Speicher DW, McCarl RL: In Vitro 10:30–41, 1974.
6. Waymouth C: In Vitro 10:97–111, 1974.
7. Wessells NK, Cohen JH: Dev Biol 18:294–309, 1968.
8. Pavelic ZP, Rustum YM, Creaven PJ, Karakousis C, Takita H: Proc Am Assoc Cancer Res 1980.
9. Pavelic ZP, Slocum H, Rustum YM, Creaven PJ, Karakousis C, Takita, H Mittelman A: Proc Am Soc Clinical Oncol, 1980.
10. Slocum HK, Rustum YM, Creaven PJ, Pavelic ZP, Siddiqui FA, Karakousis C, Takita H, Vincent RG: Proc Am Assoc Cancer Res, 1980.
11. Pluznik D, Sachs L: J Cell Comp Physiol 66:319–324, 1966.
12. Rustum YM: Cancer Res 38:543–549, 1978.

3
Protocols of Procedures and Techniques in Chromosome Analysis of Tumor Stem Cell Cultures in Soft Agar

Jeffrey M. Trent

PROTOCOL 1: PREPARATION OF TUMOR COLONIES FOR CHROMOSOME ANALYSIS

A. Procurement and Cultivation of Agar Cultures

1. Cultures should be selected for chromosome analysis during the cluster stage of colony growth to maximize the number of mitotic figures. Selection of cultures with significant colony growth is recommended for detailed cytogenetic analysis.

2. Agar cultures are initially incubated at 37°C in the presence of 2.5 ml of enriched medium CMRL-1066 containing 0.1 μM colchicine. The time of colchicine incubation varies with the proliferative capacity of the culture. Usually one hour is sufficient for rapidly proliferating colonies; however, up to 17 hours may be required for less vigorous colony samples. In early clusters, the number of mitotic figures has been shown to be directly proportional to the duration of colchicine exposure (up to 18 hours).

3. Following colchicine incubation, the entire plating layer (containing the desired colonies) is detached from the feeder layer by gently agitating the overlaying culture medium with a Pasteur pipette. Usually, the 0.3% agar plating layer quickly frees from 0.5% agar feeder layer which remains attached to the plate. However, if both layers remain attached to the plate a small rent between agar layers can be made with a Pasteur pipette. The plating layer may then be removed by gentle spurts of the overlaying culture medium by the pipette, followed by manually swirling the plate. The plating layer together with the overlaying culture medium is then gently poured into a 15-ml conical centrifuge tube and the remaining feeder layer (remaining attached to the plate) is discarded.

4. The colony suspension within the 15-ml conical centrifuge tube is then centrifuged for five minutes at 150g.

5. Following centrifugation the supernatant is carefully removed, the pelleted colonies (within the residual agar) are resuspended, and 10 ml of fresh 0.075 M KCl (prewarmed to 37°C) is added for 25 minutes.

6. Following hypotonic treatment, cultures are recentrifuged for five minutes, and the supernatant is discarded. Seven ml of fresh, cold fixative (3:1 absolute methanol to glacial acetic acid) is added and the suspension is mixed vigorously with a vortex. Cultures are then washed twice more with fixative, slides are prepared, and excess material is stored at −9°C.

B. Slide Preparation

1. Slides are first placed in a Coplin jar containing distilled water at 4°C. Seven to ten drops of the colony suspension are delivered to each slide by a Pasteur pipette. Slides are then tipped back and forth to spread the suspension further, and allowed to air-dry. A sharp burst of air may be delivered to the slide to further enhance spreading.

2. If difficulty in obtaining well-spread mitotic figures is experienced, slides can be placed into 50% glacial acetic acid prior to the addition of the colony suspension. Furthermore, in our laboratory, holding slides over boiling water while dropping the colony suspension greatly enhances chromosome spreading.

3. Flame-dried slides appear to be inappropriate in the analysis of agar cultures; the agar contracts upon flaming, causing colonies and mitotic figures within them to be obscured.

C. Biologicals and Reagents

1. **Enriched medium CMRL-1066.** Ingredients as follows: a) medium CMRL-1066 (Gibco) 100 ml, b) horse serum 15 ml, c) $CaCl_2$ 4 ml (100 mM), d) insulin 2 ml (100 μ/ml), e) vitamin C 1 ml (30 mM), f) penicillin/streptomycin 1 ml (200,000 μ/ml), and g) glutamine 2 ml (200 mM).

2. **Colchicine.** (Sigma Chemicals)

3. **Hypotonic solution (0.075 M KCl).** Prepare fresh and heat to 37° prior to use.

4. **Fixative.** Prepare fresh, 3:1 absolute methanol to glacial acetic acid.

D. References

1. Trent JM, Salmon SE: Human tumor karyology: Marked analytic improvement via short-term agar culture. Br J Cancer, in press, 1980.
2. Trent JM, Salmon SE: Potential applications of a human tumor stem cell bioassay to the cytogenetic assessment of human cancer. Cancer Genet Cytogenet 1:291–296, 1980.

PROTOCOL 2: GIEMSA-TRYPSIN BANDING MODIFICATION (G-BANDING)

A. Procedure

1. Freshly prepared slides should be placed in a drying oven at 60°C for 18 hours or allowed to air-dry a minimum of 48 hours prior to giemsa banding.

2. Prepare combined stain-enzyme solution as follows: a) phosphate buffer (0.025 M pH 6.8) 36.5 ml, b) methanol (absolute) 12.5 ml, c) giemsa stain 0.9 ml, and d) trypsin-EDTA 10X (Gibco) 0.25 ml.

3. Preincubate slides for ten minutes at 57°C in 0.025 M phosphate buffer (pH 6.8).

4. Remove slides and place horizontally on a staining rack. Gently cover the working surface of the slide with the stain-enzyme solution (approximately 3 ml). The viscosity of the stain-enzyme solution is sufficient to prevent drainage. Incubate slides for 10–12 minutes at room temperature.

5. Rinse slides gently but thoroughly with distilled water, and allow to air-dry.

6. Coverslip.

B. Biologicals and Reagents

1) Phosphate buffer 0.025 M KH_2PO_4, pH 6.8; 2) methanol (absolute); 3) giemsa stock: Giemsa, Fisher G-146 1 gm, methanol 66 ml, and glycerin 66 ml; and 4) trypsin: Trypsin-EDTA 10X (Gibco).

C. Reference

1. Sun NC, Chu EHY, Chang CC: Staining method for the banding patterns of human mitotic chromosomes. Mammal Chrom Newslett 14:26, 1973 (or their revised procedure, Sun et al: Caryologia 27:315, 1974).

PROTOCOL 3: CONSTITUTIVE HETEROCHROMATIN STAIN (C-BANDING)

A. Procedure

1. Slides should be allowed to "age" by either incubation in a drying oven at 60°C for at least 18 hours, or allowed to air-dry a minimum of 48 hours.

2. To insure satisfactory metaphase spreads, prescreen slides by phase microscopy prior to staining. Mitotic figures with long, well-spread chromosomes provide optimal results.

3. Initially, each slide should be dipped twice in 95% ethanol. Slides are then immediately dipped five to eight times in 0.9% NaCl, and placed in a saturated solution of $Ba(OH)_2$ for 8–11 minutes at room temperature. (Note: Overexposure to $Ba(OH)_2$ may produce G-banding rather than C-banding.)

4. Following exposure to Ba(OH)$_2$ slides are dipped three times in 70% ethanol, three times in a second solution of 0.9% NaCl, and incubated in 2× SSC (60°C) for four hours.

5. Following incubation slides are rinsed gently in tap water and stained for five minutes in 4% Giemsa (Gurr's R-66).

6. Coverslip.

B. Reagents

1. Barium hydroxide Ba(OH)$_2$ — prepare a saturated solution and pour the resulting supernatant into a clean coplin jar prior to staining (in order to reduce residue on slides).

2. 2× SSC — sodium chloride 1.75 gm (0.3 M), sodium citrate 0.88 gm (0.03 M), and distilled water 100 ml.

C. Reference

1. Miller DA, Tantravahi R, Dev VG, Miller OG: Q- and C-band chromosome markers in inbred strains of mus musculus. Genetics 84:67–75, 1976.

PROTOCOL 4: SILVER STAINING OF NUCLEOLUS ORGANIZING REGIONS (NOR-BANDING)

A. Procedure

1. Prepare a 50% solution of silver nitrate (1 gm AgNO$_3$ in 2 ml distilled deionized water) and let stand at room temperature 15 minutes before use. (Note: Utilize the same water source for each step in this procedure. The maintenance of a consistent water source is essential to decrease the occurrence of nonspecific silver staining.)

2. Incubate slides in distilled deionized water for 15 minutes at room temperature.

3. Prepare a moist chamber for Ag incubation. Use of a square petri dish or plastic slide mailer containing moistened filter paper in the base is satisfactory.

4. Remove slides from distilled water and allow to air-dry.

5. Add three to four drops of the 50% silver nitrate solution onto each slide using a filter syringe (0.45 μ-millipore) and cover with a 22- × 40-mm coverglass.

6. Cover moist chamber and incubate at 56°C for approximately 8–18 hours.

7. At the end of eight hours, examine each slide for the presence of silver staining. You should observe a golden brown tint to the nuclear area and dark brown-black coloration of the nucleoli. Dots of stained material should appear on the short arms of the acrocentric chromosomes at the conclusion of the incubation.

8. Following incubation, rinse slides with distilled water and counterstain in 1% Giemsa for seven seconds. Rinse slides with distilled deionized water and coverslip.

9. An alternate method to greatly decrease the time of incubation is to simply add three drops of 3% buffered formalin to the 50% $AgNO_3$ solution prior to staining. Slides may then be incubated for only 1–4 hours at 65°C.

B. Reagents

1) Fifty percent solution silver nitrate, 1 gm $AgNO_3$ in 2 ml distilled deionized water; 2) distilled deionized water; and 3) 0.45 μ filter syringe (millipore).

C. References

1. Goodpasture C, Bloom S: Visualization of nucleolar organizer regions in mammalian chromosomes using silver staining. Chromosoma 53:37–50, 1975.
2. Satya-Prakash K, Hsu TC: Behavior of the chromosome care in mitosis and meiosis. Chromosoma, in press, 1980.

4
Tabular Summary of Pharmacokinetic Parameters Relevant to In Vitro Drug Assay

David S. Alberts and H.-S. George Chen

TABLE I.

Drugs	Dosage schedule and route of administration		Peak concentration µg/ml	CXT µg · hr/ml	Range of CXT µg · hr/ml	Terminal half-life hour	Major routes of elimination	Urinary excretion % dose	References
Actinomycin D	I.V.	15 µg/kg	0.075	0.182	0.173–0.190	28.5–36.0	Urinary and fecal (14.3%)	20.3	[1]
Adriamycin	I.V.	30 mg/M²	0.50	1.38		34.0	Liver metabolism	0	[2]
		15 mg/M²	0.24	1.56		43.6			
		30 mg/M²	0.36	2.34		43.6		3.40	[3, 4]
		45 mg/M²	0.48	3.54		30.1			
		60 mg/M²	0.60	3.84		33.2			
		60 mg/M²	0.36	1.56		28.1			
Daunomycin (Daunorubicin) (Rudinomycin)	I.V.	100 mg/M²	0.41	7.32		14.5	Biliary excretion		[5] [3]
		120 mg/M²	0.647 ± 0.32	16.26		54.5 ± 2.0		13.73	
		80 mg/M²	0.398 ± 0.60	10.03		55.4 ± 5.2		12.45	[6]
Ara-C	I.V.	10 mg/kg	250.0	15.23		0.54	Deamination	5.0	[7]
		100 mg/M²	10.0	0.275	0.34–0.43	0.21			[8]
5-Azacytidine		150 mg/M²	8.45	1.56		0.13	Metabolism and degradation	73%–98% of C¹⁴	[9]
BCNU (Carmustine)	I.V.	95 mg/M² (2.25 mg/kg) Infusion for 30 min	1.97	1.02		1.13	Hydrolysis degradation		[10]
CCNU (Lomustine)	P.O.	30–100 mg/M²	No intact drug was detectable in any plasma sample as early as 10 minutes after an oral dose of CCNU or MeCCNU.						
MeCCNU (Semustine)	P.O.	120–290 mg/M²							[11]
Chlorozotocin	I.V.	120 mg/M²	47.12	3.90		0.12	Metabolism 50%–60% C¹⁴ in urine	<10.0%	[12] [13]
			30 ~ 50	6.85		0.3–0.5			

TABLE II.

Drugs	Dosage schedule and route of administration	Peak concentration μg/ml	CXT g·hr/ml	Range of CXT μg·hr/ml	Terminal half-life hour	Major routes of elimination	Urinary excretion % dose	References
Bleomycin	I.V. 15 U/M²	2.0–4.0	4.99	3.77– 6.11	4.03 ± 0.62	Renal	44.8 ± 12.6	[14]
	I.P. 60 U/M²	0.4–5.0	15.07	8.12–39.92	5.5 ± 2.6	Renal	23.5 ± 9.5	[15]
	IPL 60 U/M²		9.13	5.37–17.25	3.6 ± 0.8	Renal	17.3 ± 4.7	[15]
	Continuous I.V. infusion 30 U QD × 4D	0.28 ± 0.19	26.6	26.6 ± 7.9	3.7 ± 1.6	Renal	41.6 ± 2.5	[16]
	Subcutaneous 10 U Q 8 h × 4D	0.75 ± 0.21	26.2	26.2 ± 7.9	2.3 ± 0.6	Renal	66.7 ± 30.8	[16]
	Intramuscular 10 U Q8h × 4D	0.55 ± 0.21	26.04	26.04 ± 7.86	3.04 ± 0.54	Renal	59.2 ± 25.7	Unpublished data
Chlorambucil	Oral 0.6 mg/kg	1.10 ± 0.61	2.38	1.30–4.93	1.53 ± 0.32	Metabolism	0.54 ± 0.16	[17]
Cyclophosphamide	I.V. 720 mg	34.58	118.62		5.87		21.87	[18]
	I.V. 10 mg/kg	29.40 ± 13.00	109.29	32.19–206.57	6.46 ± 2.97	Liver metabolism	14.2 ± 5.2	[19]
	I.V. 6–80 mg/kg				6.45 ± 1.1		10.12 ± 5.22	[20]
Isophosphamide	I.V. 130 mg/kg	221.96 ± 44.67	3,175.40	2,109.55–3,867.72	15.16 ± 3.61		53.08 ± 9.65	[21]
Deazauridine	I.V. 11.7 mg	0.8	3.41		7.6 ± 3.4		90.7 ± 9.5 C¹⁴	[22]
	I.V. infusion for 1 hr 680–1,500 mg				4.4			
	Rapid infusion 4.8 g/M²	204.5	2,205.4		4.4 (2.8–5.8)		7.8	[23]
	Continuous infusion 1.5 g/M²	2.39			21.3 (8.0–44.8)		7.2	

TABLE III.

Drugs		Dosage schedule and route of administration	Peak concentration μg/ml	CXT μg·hr/ml	Range of CXT μg·hr/ml	Terminal half-life hour	Major routes of elimination	Urinary excretion % dose	References
DTIC	I.V.	4.5 mg/kg	8.8 ± 1.8	7.15		1.25	Renal	32.0 ± 4.0	[24]
	I.V.	250 mg/M²	15.28	30.72		5.02		40.0	[25]
5-Fu	I.V.	15 mg/kg	60.	16.33	13–21	0.2	Metabolism		[26]
	I.V.	10.9 mg/kg	26–123	21.42	8.0–35.5	0.19 ± 0.03			[27]
Ftorafur	I.V.	50 mg/M				1.8	Metabolism	9.0	[28]
	I.V.	5 g/M	184.0	2,335.8		8.8		10.1	[29]
Hydroxyurea	I.V.	1,000 mg/M	74.77 ± 10.6	215.55	136.48–294.62	2.0	Renal	36–82	[30]
	P.O.	1,000 mg/M	48.29 ± 3.01	431.57	343.27–519.87	6.2		28–76	
Melphalan	I.V.	0.6 mg/kg	3.38 ± 1.92	2.47	0.87– 5.50	1.44 ± 0.81	Hydrolysis	13.0 ± 5.4	[31]
	P.O.	0.6 mg/kg	0.28 ± 0.15	0.883	0.15– 2.50	1.50 ± 0.95	Hydrolysis	10.9 ± 4.9	[32]
6-Mercaptopurine	I.V.	4.2– 6.7 mg/kg	8–20	7.57		0.35		18.2	[33]
	I.V.	5.0–37.7 mg/kg	2.5–37	27.12		0.78		21.7	
(MTX) Methotrexate	I.V.	500 mg/M²	12–24	11.5 ± 1.17		0.17 ± 0.07	Renal	87.0	[34]
	I.V.	30 gm/M²	2.75	5.34		7.59			[35]
Mitomycin C	I.V.	10 mg	0.52	0.11		0.15	Metabolism	5.8	[36]
		20 mg	1.5	0.36		0.17			
		30 mg	2.7	1.10		0.18			

TABLE IV.

Drugs	Dosage schedule and route of administration	Peak concentration μg/ml	CXT μg·hr/ml	Range of CXT μg·hr/ml	Terminal half-life hour	Major routes of elimination	Urinary excretion % dose	References
PALA	I.V. 100–200 mg/M²	18–20	121.16	121.16 ± 41.61	4.42 ± 1.50	Renal	55.7 ± 26.8	Private communication
Cis-platinum	I.V. 100 mg/M²	2.49 ± 0.41	1.94	1.45– 2.52	0.54	Renal	46–51.3	[37]
Pyrazofurin	I.V. 100 mg/M²	3.32	1.45		3.85			[38]
	200 mg/M²	5.38 ± 0.38	1.25	1.14– 1.36		Metabolism	4.0–12.4	
	250	6.07 ± 0.61	6.42	3.52–10.83				
	300	13.34 ± 9.53	1.79	1.34– 2.02				
Rubidazone	I.V. 150–600 mg/M²				35.0	Liver metabolism and renal	28.0	[39]
6-Thioguanine	I.V. 135 mg/M²	12.5	71.85		1.33	Metabolism	6.67	[40]
	P.O. 135 mg/M²	0.16	5.71				2.00	
Thymidine	I.V. Infusion 75 g/M²/D/5d	242.2–726.71	581 × 10⁴	2.91 × 10⁴–8.72 × 10⁴	1.53 ± 0.13			[41]
	I.V. Infusion 90 g/M²/D	484.5	1.16 × 10⁴		1.33			[42]
Vinblastine	I.V. 0.2 mg/kg	0.783 (0.444–1.374)	0.174	0.116–0.254	19.55 ± 1.10	Fecal (99%)	12.6	[43]
Vincristine	I.V. 0.025 mg/kg	0.218	0.231		28.8			[44]
		0.372	0.064	0.025–0.103	2.58 ± 0.30	Bile	9.6 ± 5.1	[45]
		0.068	0.472		144.0			[44]

TABLE V.

Drugs	Dosage schedule and route of administration		Peak concentration $\mu g/ml$	CXT $\mu g \circ hr/ml$	Range of CXT $\mu g \circ hr/ml$	Terminal half-life hour	Major routes of elimination	Urinary excretion % dose	References
Vindesine	I.V.	1.5 mg/M²	0.603	0.150	0.066 – 0.241	20.22 ± 8.22	Bile	19.0 ± 10.8	[45]
	I.V.	2 mg/M² (0.049 mg/kg)	0.883	0.193		24.3		6.0	[44]
VM26	I.V.	67 mg/M²	23.84 ± 5.02 (20.96–32.78) 27.45	113.89 132.38	96.22 – 132.38	22.43 ± 8.73	Metabolism (86%)	12.82	[46]
VP16	I.V.	90 mg/M² 130 mg/M² 170 mg/M² 220 mg/M² 290 mg/M²	29.95 ± 4.40 30.52 ± 3.19 30.75 ± 3.02 34.18 ± 5.27	91.40 115.18 121.01 153.94	86.16 – 96.63 93.76 –132.41 111.47 –132.60 129.46 –195.54	8.19 ± 3.61	Metabolism (66%)	29.07	[46]
Pentamethyl-melamine	1 hr infusion	80 mg/M²	0.804			2.22	Liver metabolism	<0.1	[47]
Hexamethyl-melamine	P.O.	120–200 mg/M²	0.2–20.8		0.92 – 60.10	3–10			[48]
AMSA	1 hr infusion	160 mg/M²	6.7			2.5	Liver metabolism and renal	42% C¹⁴	[49]
	I.V.	120 mg/M² Normal hepatic and renal function		14.20		7.4	Liver metabolism and renal	12.0	[50]
		Mild hepatic dysfunction		13.90		9.9		12.6	
		Severe hepatic dysfunction		18.48		17.2		20.0	
		Mild renal dysfunction		14.20		7.4		<4.0	
		Severe renal dysfunction		24.97		10.8			
	I.V.	120 mg/M²	6.90	22.22		25.2			[51]

REFERENCES

1. Tattersall MHN, Sodergren JE, Sengupta SK, Trites DH, Modest EJ, Frei E III: Pharmacokinetics of actinomycin D in patients with malignant melanoma. Clin Pharmacol Ther 17: 701–708, 1975.
2. Harris PA, Gross JF: Preliminary pharmacokinetic model for adriamycin (NSC-123127). Cancer Chemother Rep, Part I, 59:819–825, 1975.
3. Bachur NR, Riggs CE Jr, Green MR, Langone JJ, Van Vunakis H, Levine L: Plasma adriamycin and daunorubicin levels by fluorescence and radioimmunoassay. Clin Pharmacol Ther 21:70–77, 1977.
4. Benjamin RS, Riggs CE Jr, Bachur NR: Pharmacokinetics and metabolism of adriamycin in man. Clin Pharmacol Ther 14:592–600, 1974.
5. Benjamin RS, Riggs CE, Bachur NR: Plasma pharmacokinetics of adriamycin and its metabolites in human with normal hepatic and renal function. Cancer Res 37:1416–1420, 1977.
6. Alberts DS, Bachur NR, Holtzman JL: The pharmacokinetics of daunomycin in man. Clin Pharmacol Ther 12:96–104, 1971.
7. Dedrick RI, Forrester DD, Ho DHW: In vitro–in vivo correlation of drug metabolism-deamination of 1-β-arabinofuranosylcytosine. Biochem Pharmacol 21:1–16, 1972.
8. van Prooijen HC, Vierwinden G, van Egmond J, Wessels JMC, Haaren C: A sensitive bioassay for pharmacokinetics studies of cytosine arabinoside in man. Eur J Cancer 12:899–905, 1976.
9. Israili ZH, Wogler WR, Mingioli ES, Pirkle JL, Smithwick RW, Goldstein JH: The disposition and pharmacokinetics in humans of 5-azacytidine administered intravenously as a bolus or by continuous infusion. Cancer Res 36:1453–1461, 1976.
10. Levin VA, Hoffman W, Weinkam RJ: Pharmacokinetics of BCNU in man: A preliminary study of 20 patients. Cancer Treat Rep 62:1305–1312, 1978.
11. Sponzo RW, DeVita VT, Oliverio VT: Physiologic disposition of 1-(2-chloroethyl)-3-cyclohexyl-1-nitrosourea (CCNU) and 1-(2-chloroethyl)-3-(4-methyl cyclohexyl)-1-nitrosourea (MeCCNU) in man. Cancer 31:1154–1159, 1973.
12. Schein PS, Bull JM, Doukas D, Hoth D: Sensitivity of human and murine hematopoietic precursor cells to 2-[3-(2-chloroethyl)-3-nitrosoureido]-D-glucopyranose and 1,3-bis (2-chlorethyl)-1-nitrosourea. Cancer Res 38:257–260, 1978.
13. Hoth D, Woolley P, Green D, Macdonald J, Schein P: Phase I studies on chlorozotocin. Clin Pharmacol Ther 23:712–722, 1978.
14. Alberts DS, Chen H-SG, Liu R, Himmelstein KJ, Mayersohn M, Perrier D, Gross J, Moon T, Broughton A, Salmon SE: Bleomycin pharmacokinetics in man: I. Intravenous administration. Cancer Chemother Pharmacol 1:177–181, 1978.
15. Alberts DS, Chen H-SG, Mayersohn M, Perrier D, Moon TE, Gross JF: Bleomycin pharmacokinetics in man: II. Intracvitary administration. Cancer Chemother Pharmacol 2:127–132, 1979.
16. Alberts DS, Peng Y-M: Effective simulation of the minimum cytotoxic concentrations of continuous infusion (CI) bleomycin (BLEO) with multiple subcutaneous injections (MSC) in cancer patients. Proc AACR and ASCO 20:C–589, 1979.
17. Alberts DS, Chang SY, Chen H-SG, Larcom BJ: Pharmacokinetics and metabolism of chlorambucil in man: A preliminary report. Cancer Treat Rev (in press).
18. Whiting B, Miller SHK, Caddy B: A procedure for monitoring cyclophosphamide and isophosphamide in biological samples. Br J Clin Pharmacol 6:373–376, 1978.
19. Cohen JL, Jao JY, Jusko WJ: Pharmacokinetics of cyclophosphamide in man. Br J Pharmacol 43:677–680, 1971.
20. Bagley CM Jr, Bostick FW, DeVita VT Jr: Clinical pharmacology of cyclophosphamide. Cancer Res 33:226–233, 1973.

21. Allen LM, Creaven PJ: Human pharmacokinetic model for isophosphemide (NSC-109724). Cancer Chemother Rep, Part I, 59:877–882, 1975.
22. Creaven PJ, Rustum YM, Slocum HK, Mittelman A: Clinical pharmacokinetics of 3-deazauridine, a new antineoplastic agent. In Siegenthaler W, Luthy R (eds): "Current Chemotherapy." Washington, DC: American Society for Microbiology, 1978, p 1208–1210.
23. Benvenuto JA, Hall SW, Farquhar D, Stewart DJ, Benjamin RS, Loo TL: Pharmacokinetics and disposition of 3-deazauridine in humans. Cancer Res 39:349–352, 1979.
24. Skibba JL, Ramirez G, Beal DD, Bryan GT: Preliminary clinical trial and the physiologic disposition of 4(5)-(3,3-dimethyl-1-triazeno)imidazole-5(4)-carboxamide in man. Cancer Res 29:1944–1951, 1969.
25. Loo TL, Housholder GE, Gerulath AH, Saunders PH, Farquhar D: Mechanism of action and pharmacology studies with DTIC. Cancer Treat Rep 60:149–152, 1976.
26. Finn C, Sadee W: Determination of 5-fluorouracil (NSC-19893) plasma levels in rats and man by isotope dilution-mass fragmentography. Cancer Chemother Rep, Part 1, 59:279–286, 1975.
27. MacMillan WE, Wolberg WH, Welling PE: Pharmacokinetics of fluorouracil in humans. Cancer Res 38:3479–3482, 1978.
28. Lu K, Loo TL, Benvenuto JA, Benjamin RS, Valdivieso M, Freireich EJ: Pharmacologic disposition and metabolism of ftorafur. Pharmacologist 17:202, 1975.
29. Benvenuto JA, Lu K, Hall SW, Benjamin RS, Loo TL: Disposition and metabolism of 1-(tetrahydro-2-furanyl)-5-fluorouracil (ftorafur) in human. Cancer Res 38:3867–3870, 1978.
30. Davidson JD, Winter TS: A method of analyzing for hydroxyurea in biological fluids. Cancer Chemother Rep 27:97–110, 1963.
31. Alberts DS, Chang SY, Chen H-SG, Moon TE, Evans TL, Furner RL, Himmelstein K, Gross JF: Pharmacokinetics of melphalan in man: Intravenous administration. Clin Pharmacol Ther (in press).
32. Alberts DS, Chang SY, Chen H-SG, Evans TL, Moon TE: Bio-availability of oral melphalan in man. Clin Pharmacol Ther (in press).
33. Loo TL, Phil D, Luce JK, Sullivan MP, Frie E III: Clinical pharmacologic observations on 6-mercaptopurine and 6-methylthiopurine ribonucleoside. Clin Pharmacol Therap 9:180–194, 1968.
34. Coffey JJ, White CA, Lesk AB, Rogers WI, Serpick AA: Effect of allopurinol on the pharmacokinetics of 6-mercaptopurine (NSC 755) in cancer patients. Cancer Res 32:1283–1289, 1972.
35. Bischoff KB, Dedrick RL, Zaharko DS, Longstreth JA: Methotrexate pharmacokinetics. J Pharmacol Sci 60:1128–1133, 1971.
36. Crooke ST, Bradner WT: Mitomycin C: A review. Cancer Treat Rev 3:121–139, 1976.
37. Patton TF, Himmelstein KJ, Belt R, Bannister SJ, Sternson LA, Repta AJ: Plasma levels and urinary excretion of filterable platinum species following bolus injection and intravenous infusion of cis-dichlorodiammineplatinum (II) in man. Cancer Treat Rep 62:1359–1362, 1978.
38. Ohnuma T, Roboz J, Shapiro ML, Holland JF: Pharmacological and biochemical effects of pyrazofurin in humans. Cancer Res 37:2043–2049, 1977.
39. Benjamin RS, Keating MJ, McCredie KB, Luna MA, Loo TL, Freireich EJ: Clinical and pharmacologic studies with rubidazone (R) in adults with acute leukemia (AL). Proc AACR and ASCO 17:285, 1976.
40. Lepage GA, Whitecar JP Jr: Pharmacology of 6-thioguanine in man. Cancer Res 31:1627–1631, 1971.
41. Chiuten DF, Wiernik PH, Zaharko DS, Edwards L: Phase I–II trial and clinical pharmacology of continuous infusion high dose thymidine. Proc AACR and ASCO 20:305, 1979.

42. Zaharko DS, Bolten BJ, Giovanella BC, Stehlin JS: Thymidine and thymine measurements in biological fluids: Mouse, monkey and man. Proc AACR and ASCO 20:250, 1979.
43. Owellen RJ, Hartke CA, Hains FO: Pharmacokinetics and metabolism of vinblastine in humans. Cancer Res 37:2597–2602, 1977.
44. Dyke RW, Nelson RL: Phase I anti-agent vindesine (desacetyl vinblastine amide sulfate). Cancer Treat Rev 4:135–142, 1977.
45. Owellen RJ, Root MA, Hains FO: Pharmacokinetics of vindesine and vincristine in humans. Cancer Res 37:2603–2607, 1977.
46. Allen LM, Creaven PJ: Comparison of the human pharmacokinetics of VM-26 and VP-16, two antineoplastic epipodophyllotoxin glucopyranoside derivatives. Eur J Cancer 11:697–707, 1975.
47. Dutcher JS, Jones RB, Boyd MR: A sensitive and specific assay for pentamethylmelamine in plasma: Application to preliminary pharmacokinetic studies in cancer patients. Proc AACR and ASCO 20:948, 1979.
48. D'Incalci M, Sessa C, Belloni C, Morasca L, Garattini S: Hexamethylmelamine (HMM) and pentamethylmelamine (PMM) levels in plasma and ascites after oral administration to ovarian cancer patients. Proc AACR and ASCO 20:185, 1979.
49. Von Hoff DD, Howser D, Gormley P, Bender RA, Glaubiger D, Levine AS, Young RC: Phase I study of methanesulfonamide, N-[4-(9-acridiylamino)-3-methoxyphenyl]-(m-AMSA) using a single dose schedule. Cancer Treat Rep 62:1421–1426, 1978.
50. Hall SW, Benjamin RS, Legha SS, Gutterman JU, Loo TL: Clinical pharmacokinetics of the new antitumor agent AMSA. Proc AACR and ASCO 20:707, 1979.
51. Malspeis L, Khan MN, Blat HB: HPLC Determination of 4'-(9 acridinyl-amino methanesulfon)m-ansidide (m-AMSA; NSC 249992) in the plasma of a patient with hepatic dysfunction. Proc ASCO and ASCO 20:C–383, 1979.

Index

Adaptation, 273
Additive effects, 240
Adherent cells, non-, 26
Adriamycin, 55, 121, 217, 224, 249, 280
Alpha-fetoprotein, (AFP), 327
Amphotericin-B, 248
AMSA, 56, 122, 236, 295, 303
Aneuploidy, 174
Anticancer drug development, 291
Area-under-survival concentration curve, 202, 211
Ascites, 63, 117
Assay data, 228
Assay procedure, flow chart of, 138
Automated
 counting, 297
 instrumentation, 320
Automation, 180
Autoradiography, 141, 268

BCNU, 253, 260, 291
BSA gradients, 127, 129
Basis of culture clonogenicity, 127
Bioactivation, 227, 302
Biochemical determinants of drug action, 339
Biological response modifiers, 256, 295
Biology of tumor growth, 321
Biophysical separation, 27
Biotransformed active product, 198

Bladder
 barbotage technique, 79
 carcinoma, 75, 146
Blast colony formation (AML), 53
Bleomycin, 217, 226
Bone marrow, 107, 113
 involvement, 45
Brain tumors, 259
Breast carcinoma, 228, 294

C-banding, 170
CCNU, 260
Calculation of sensitivity index, 211
Carcinoembryonic antigen (CEA), 141, 324
Carcinoma in situ, 81
Carmustine (BCNU), 24, 260
Catecholamines, 103, 118
Cell
 freezing, 299
 interactions, 127
 kinetics, 153, 158
 loss factor, 272
 methods of kinetics analysis, 154
 separation procedures, 267
Chemical and drugs (suppliers), 331
Chex-UP (protocol), 253
Chromosome
 banding, 166
 non-random changes, 165
Cis-platinum, 95, 121, 217, 225, 280, 291

Clinical
  correlations, 223
  staging, 224
Clonal
  evolution, 55, 172, 173
  heterogeneity, 232, 294
  progression, 239
Clonality of human tumors, 172
Cloning efficiency, 7, 130, 228, 286
Clonogenicity, 15, 17, 127
Colon carcinoma, 149
Colony
  counting, 12, 179
  size, 18
Complete response, 226
Computerized tomography, 260
Concentration-time product
  (CXT), 197, 235
Conditioned medium, 5, 30, 44, 53,
  69, 227
  preparation, 333
Conditioning factors, 321
Contact
  one hour vs continuous, 205
Continuous contact, 55, 227, 302
Coomassie blue G-250, 136
Correlative clinical trials, 234
Counting equipment, 180
Criteria of response, 226
Cross-resistance, 304
  table, 240, 241
Cryopreservation, 86, 214, 294, 298
Cumulative experience at two
  institutions, 237
Cyclic nucleotides, 319
Cyclophosphamide, 224, 321
Cystoscopy, 79
Cytogenetic
  analysis, 68, 77, 103, 165
  protocols, 345
Cytosine arabinoside, 55, 291

DCT panel of in vitro screens, 295
DCT pre-screen, 295
Dacarbazine (DTIC), 95, 225

Daunomycin, 307
Decision network, 291
Decreasing sensitivity, 239
Density gradient fractionation, 129
Depletion
  of T-lymphocytes, 54
  of phagocytic cells, 70
  of macrophages, 286
Design
  of drug screening, 218
  of preclinical studies, 209
Diagnostic applications of in vitro
  assay, 80, 107, 113, 315
Differentiation, 95
Diffuse histiocytic lymphoma
  (DHL), 234
Dihydroxyanthracenedione, 295, 307
Discriminant analysis, 214
Division of cancer treatment
  (DCT), 291
DNAse, 64
DOPA reaction, 138
Double minutes, 174
Drug
  anticancer, 351
  assay, 197
  assay kits, 320
  clearance, 255
  delivery, 271
  Developmental program, 291
  dose survival curves, 54
  effects on self renewal and
    primary clonogenicity, 58
  exposure time, 200
  flow of through DCT screens, 295
  increasing resistance, 239
  resistance (acquisition of), 238
  resistance, 12, 225, 247, 248
  sample size for identification, 218
  sensitivity, 53, 54, 95, 120, 158,
    223, 247, 259
  sensitivity assay, 335
  sensitivity boundaries, 217
  sensitivity studies, 80
  steps in drug sensitivity assay, 225

Dye exclusion, 340

Electron microscopy, 77, 142
Embryonic antigens, 321
Endometrial
  adenacarcinoma, 169
  cancer, 66
Enzymatic
  cocktail, 267, 320
  disaggregation of human solid tumors, 339
Epidermal growth factor (EGF), 57, 91, 319, 334
Epithelial vells, 63
Equivalent circular diameter (ECD), 185
Esterage
  non-specific, 44
Estrogen receptor, 243, 294
Exposure of tumor cells to drug, 226

FAS II Image Analyzer, 183
Feeder layer, 6, 16, 24, 72
Fermentation broth, 297
Fetal antigen, 322
Fibroblasts, 8
  growth factor (FGF), 91
5-fluoroutracil, 123, 226, 248, 291, 294
Fixation of agar culture, 138
Flow cytofluorometric (FCM) analysis, 268
Flow microfluorometry, 153
Follicle stimulating hormone (FSH), 91
Fractional survival curves, 54
Fractionation, 128
Future directions, 315

G-banding, 170
Gene amplification, 174
Glutaraldehyde, 138, 192
Granulocyte
  colonies, 28
  colonystimulating factor, 40

Granulopoietic progenitor (CFU-C), 53
Graphic analysis, 228
Growth
  factors, 90
  patterns of solid and ascitic forms of ovarian carcinoma, 285
  rates, 161

Heliotrine, 303
Heterogeneity, 55
Hexa-CAF (protocol), 253
Hexamethylmelamine, 291
Histograms of the size of distributions, 191
Histology, 118
Historical development of DCT
  mouse tumor screening systems, 293
Hodgkin lymphoma, non-, 43
Homogeneously staining regions (HSR), 174
Host cells, 133
Human
  chorionic gonadotropin (HCG), 324
  leukocyte interferon, 288, 308
  myeloma cell line, 228
  tumor xenografts, 292
Hydroxyurea suicide, 156
Hypernephroma, 234
Hyperthermia, 96

Image
  analysis, 12, 227, 286
  analysis system, 179
  analyzer, 161
Immunofluorescence, 28, 141
Immunoglobulin, 26
Immunological tolerance, 322
In vitro
  anticancer drug pharmacology, 200
  assay, 236
  criteria for sensitivity for new Phase I-II agents, 296
  Phase II clinical trials, 294
  problems with assay, 113, 270
  quantification of assay, 209

response production of, 284
resistance, 233
studies of combination
chemotherapy, 240, 242
In-dwelling catheter, 257
Interferon, 277, 295, 308
time course response, 284
Intermediate zone, 228
Intraperitoneal chemotherapy, 248
Intratumoral drug concentrations, 236
Investigator fatigue, 179

Karyology, 11
Karyotypic homogeneity, 17

Laboratory procedures, 332
Lactic dehydrogenase, 137
Leukemia
    acute myeloblastic (AML), 53
    L1210, 293
    P385, 292
    hemopoiesis, 53
Light microscopy, 135
"Linear Array," 291
Linear survival-concentration
    curves, 231
Liver microsomal
    cocktail, 302
    homogenate, 227
Logistic regression model, 215
Logistical problems, 297
Lomustine, 260
Lymphocyte-tumor cell interactions, 287
Lymphocytes, 283
Lymphoid cells, 130
Lymphoma, 43

Macrophages, 6, 71, 72, 130, 268, 282, 321
    hypothesis, 123
    tumor cell interactions, 70
Malignancies, adult and pediatric, 116
Malignant
    effusions, 226, 247

gliomas, 261
Marker chromosomes, 169
Maternal tolerance, 322
Mathematical model, 209
Maximum horizontal chord, 185
Mechanical disaggregation, 64
Melanin, 89, 138
    staining, 140
Melanocyte stimulating hormone
    (MSH), 91
Melanogens, 118
Melanoma, 85
    amelanotic, 85
Melatonin, 92
Melphalan, 4, 203, 217, 224, 254
Membrane immunoglobulin, 44
Mercaptoethanol, 32, 64
Mesothelial cells, 282
Metastatic melanoma, 210
Methotrexate, 226, 253, 291
Methyl ester (DOPA), 96, 97
Methylcellulose, 54, 76
Microsome fractions, 271
Minimum cytotoxic concentration, 239
Mitotic index, 167
Mixed response, 226
Monolayer, 270
    clonal assay, 271
Morphologic
    studies, 135
    variants, 86
Morphological and histochemical
    stains, 68
Mouse tumor systems, 291
Mucinous adenocarcinoma, 65, 69
Multifactorial analysis, 59
Multilog kill, 236
Multiple meyloma, 4, 23, 156, 224
Multipotent stem cell, 53
Multivariate data, 240
Myelogenous leukemia, 53
Myeloma
    associated antigens, 40
    colony-forming cell, 23, 28

stem cells, 45

National Cancer Institute (NCI), 291
National Cancer Program, 291
Natural products, 292
Negative exponential response, 54
Nerve growth factor (NGF), 91, 319
Neuroblastoma, 101
Neuroendocrine hormones, 91
Nitrosoureas, 229
Nitrosourea therapy, 299
NOR-banding, 184

Oat cell carcinoma, 234
Omnicon FAS II, 181, 286
One-hour suicide, 153
Optimizing sensitivity of
   screening, 299
Ovarian carcinoma, 63, 224, 247, 277
   response of solid/ascitic forms of
      to interferon, 284

PALA, 281
Partial response, 226
Peak plasma concentration, 198
Periodic acid schiff (PAS), 44
Peritoneal washings, 247
Peritoneoscopy, 250
Permanent slides, 138
Peroxidase staining, 28
PHA-LCM (conditioned medium), 53
Phagocytic
   cells, 64, 89, 128
   depletion, 89
   non-phagnocytic, 26, 65
Pharmaceutical industry, 291
Pharmacokinetic
   data, 277, 296
   tabular summary of parameters,
      351
Pharmacokinetics, 271, 351
Pharacologic
   applications, 247
   considerations, 197
Phase I-II
   agents, 295

   clinical trial, 303
Phytohemagglutinin (PHA), 53
Plasma cells, 23
Plateau, 217
Plating
   efficiency, 7, 58, 106–110,
      119, 251, 252, 270
   layers, 331, 332
Pleural fluid, 117
Population dynamics, 15
Predictive cancer chemotherapy, 224
Preparation of cell suspensions, 76
Primary screening for new agents, 296
Prognostic factors, 58, 216
Proliferation, 95
Prospective
   correlative trial, 316, 217
   decision-aiding trial, 315
Prostaglandins, 130, 319, 322
Pulse
   exposure, 56
   labeled mitosis curves, 153

Quality control, 298

Radiation therapy, 96
Radiosensitivity, 318
Random screening, 291
Reagent development, 320
Reconstitution studies, 128
Remission induction, 54
Renewal hierarchy, 17
Resectoscope, 75
Retinoic acid
   aromatic ethyl ester, 92
   B-trans, 92
   13-cis, 92
Retinoids, 92
Retinol, 92
Retrospective correlative trial,
   316, 317
Rhabdomyosarcoma, 150
Ribonucleoside triphosphate pools
   339

Sarcoma, 235

growth factor, 319
Scanning electron microscopy, 147
Screening, 291
 secondary of new agents, 294, 304
Second-look peritoneoscopy, 255
Secondary colony formation, 54
Sedimentation velocity, 304
Selection, 273
Self-newal
 capacity, 240
 properties, 8, 15, 54, 95, 154
Semi-solid support, 127
Sensitivity
 guidelines, 303
 index, 201, 211, 213, 228, 251
 limit, 236
 to interferon, 278
Separation procedure, 47
Serendipitious discovery, 291
Serial
 in vitro studies, 238
 passage, 273
Serous adenocarcinoma, 65, 69
Serum (suppliers), 338
Single cell suspension, 17
Soft agar methylcellulose assay, 75
Somatomedin, 319
Specimen
 source of, 117
 shipment, 298
Stages
 in new drug development, 292
 of development in clinical trials, 317
Standard solvents, 319
Statistical analysis, 209
Stem cell, 15
Strategies to optimize detection of active compounds, 301
Stroma, 272
Stromal interactions, 128
Suicide index, 56
Supplies and suppliers, 338
Survival drug concentration curve, 228

Synergistic effects, 242
Synthetic compounds, 291

T cell growth factor, 319
TCFUs
 doubling time of, 240
 from malignant washings, 247, 250
 recovery from cryopreservation, 299
Tamoxifen, 293
Technological development, 320
Tenckhoff dialysis catheter, 248
Terminal phase plasma half-life ($t1/2$), 198
Tetrazolium salts, 136
Therapeutic trials, 316
^{3}H-thymidine, 26, 40, 64, 156, 202, 230
Thyroid carcinoma, 147
Tissue culture media, 338
Toxicology testing, 303
Training sets, 214, 228
Transition
 electron microscopy, 142
 state inhibitor, 306
Transitional cell carcinoma (TCC), 75
Transport media, 298, 320
Trapezoidal method, 211
Tritiated thymidine
 labeling index, 153, 155
 suicide, 26, 40, 64, 156, 202, 230
True rate
 negative, 237
 positive, 237
Tumor
 antigens, 323
 burden, 272
 cell disaggregation, 259, 267
 cell kinetics, 153
 heterogeneity, 268
 karyology, 166
 lymphoreticular cell interactions, 28
 markers, 118
 matrix, 271

procurement, 297
progression, 172
promoter, 325
  stem cell, 15
  types cultured, 116

Underlayers, 333
Undifferentiated carcinoma, 69
Upper boundary concentration, 229
Urothelim, 75

Velocity sedimentation, 27
Video colony counters, 192
Vinblastine, 202, 206, 226, 291
Vincristine, 217
Visual counting, 179
Visual counts vs FAS II, 189
VMA, 103

Wet agar stains, 138